Applied Geochemistry in the 1980s

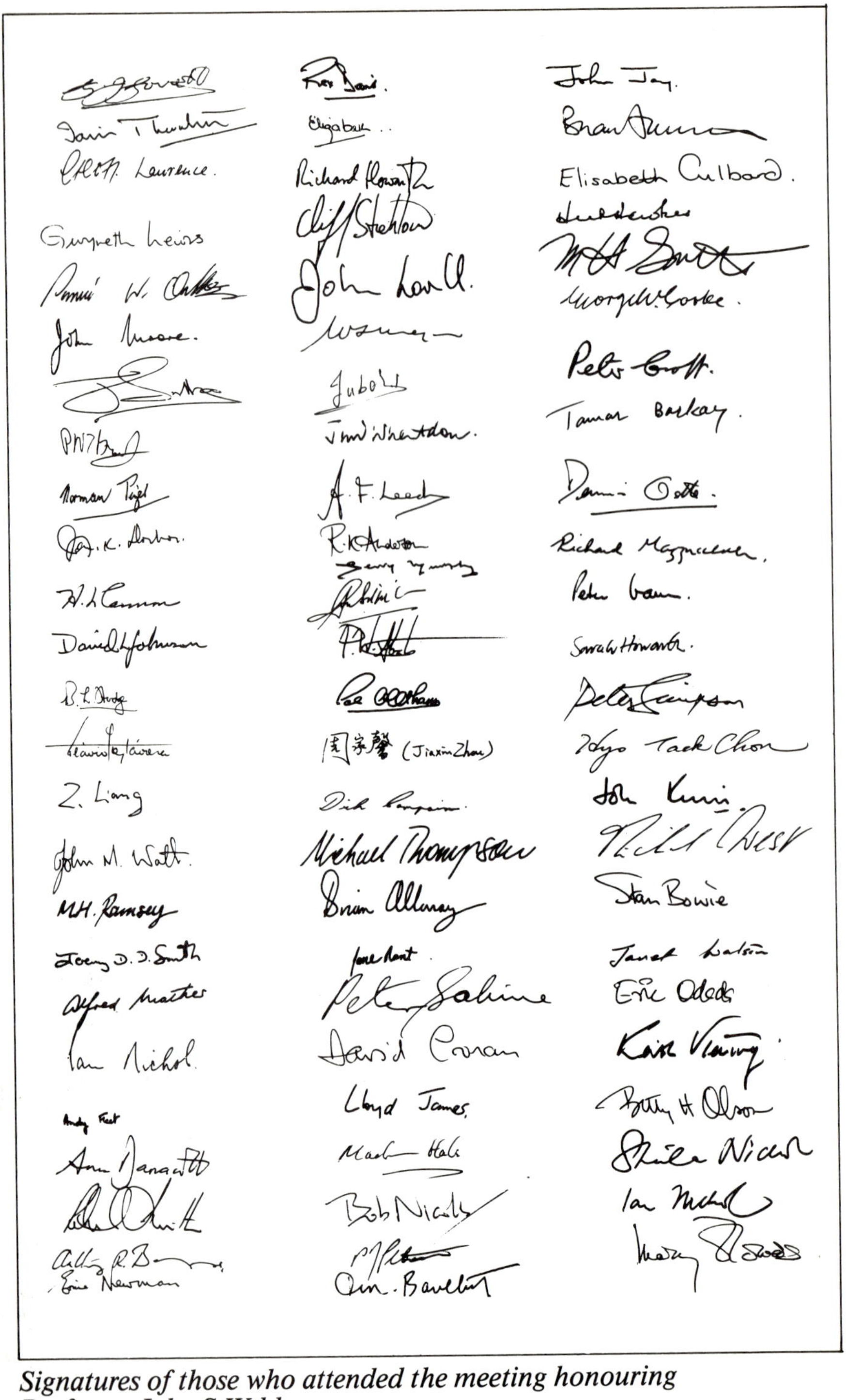

Signatures of those who attended the meeting honouring Professor John S Webb.

Professor John S Webb

Applied Geochemistry in the 1980s

Proceedings of a meeting to honour
the contribution of Professor John S Webb
to Applied Geochemistry, held on
29 April 1983 at Imperial College, London

Edited by

Iain Thornton, Reader in Environmental Geochemistry and
Head of the Applied Geochemistry Research Group
Imperial College, London

and

Richard J Howarth, Reader in Mathematical Geology
Imperial College, London

With assistance from

David S Cronan
Martin Hale
Michael Thompson
Applied Geochemistry Research Group
Department of Geology
Imperial College, London, UK.

Applied Geochemistry in the 1980s

edited by

Iain Thornton and Richard J Howarth

Graham & Trotman

First published in 1986 by
Graham & Trotman Limited
Sterling House
66 Wilton Road
London SW1V 1DE
UK

ISBN 0 86010 796 5

Typeset, printed and bound in Great Britain.

Contents

List of Contributors

P.W. ABRAHAMS
School of Environmental Sciences, University of East Anglia, Norwich NR4 7TJ, UK

P.E. CROFT
Robertson Research International Limited, 'Ty'n-y-Coed', Llanrhos, Llandudno, Gwynedd, LL30 1SA, UK

D.S. CRONAN
Applied Geochemistry Research Group, Imperial College, London SW7 2BP, UK

R.G. CROUNSE
Trace Element Centre, Department of Surgery, School of Medicine, East Carolina University, Grenville, North Carolina, USA

E. CULBARD
Applied Geochemistry Research Group, Imperial College, London SW7 2BP, UK

A. DARRACOTT
8 Guard's Club Road, Maidenhead, Berkshire SL6 8DN, UK

G.R. DAVIS
"Bredon", Deepdene, Park Road, Dorking, Surrey, UK

M.D. FORREST
British Geological Survey, Metalliferous Minerals and Applied Geochemistry Unit, 154 Clerkenwell Road, London EC1R 5DU, UK

R.G. GARRETT
Resource Geophysics and Geochemistry Division, Geological Survey of Canada, Ottawa, Ontario, Canada

G.J.S. GOVETT
School of Applied Geology, The University of New South Wales, P.O. Box 1, Kensington, New South Wales 2033, Australia

M. HALE
Applied Geochemistry Research Group, Imperial College, London SW7 2BP, UK

R.S. HATHERLEY
Institute of Mining Research, University of Zimbabwe

J.F. HODGSON
British Geological Survey, 154 Clerkenwell Road, London EC1R 5DU, UK

R.J. HOWARTH
Applied Geochemistry Research Group, Imperial College, London SW7 2BP, UK

G. LEWIS
91 Surbiton Hill Park, Surbiton, Surrey, UK

J.S. LOVELL
Barringer Resources Inc., 1626 Cole Blvde., Building No. 7, Golden Co. 80401, USA

R.H. MAZZUCCHELLI
Western Mining Corporation Limited, P.O. Box 91, Belmont, Western Australia 6104, Australia

W.T. MEYER
Barringer Resources Inc., 1626 Cole Blvde., Building No. 7, Golden Co. 80401, USA

I. NICHOL
Queen's University, Kingston, Canada

B.H. OLSON
Programme in Social Ecology, University of California, Irvine CA 92717, USA

J.A. PLANT
British Geological Survey, Metalliferous Minerals and Applied Geochemistry Unit, 154 Clerkenwell Road, London EC1R 5DU, UK

J.A. ROTHER
Science Laboratory, University of Durham, Department of Botany, South Road, Durham DH1 3LE, UK

P.A. SMITH
Superior Oil (UK) Limited, 65 Buckingham Gate, London SW1E 6AS, UK

R.T. SMITH
British Geological Survey, Metalliferous Minerals and Applied Geochemistry Unit, 154 Clerkenwell Road, London EC1R 5DU, UK

A.G. STEVENSON
British Geological Survey, Metalliferous Minerals and Applied Geochemistry Unit, 154 Clerkenwell Road, London EC1R 5DU, UK

J. SUTTON
Environmental Technology, Imperial College, Princes Gate, London SW7 2BP, UK

M. THOMPSON
Applied Geochemistry Research Group, Imperial College, Princes Gate, London SW7 2BP, UK

I. THORNTON
Applied Geochemistry Research Group, Imperial College, Princes Gate, London SW7 2BP, UK

N.J. TOPPING
Institute of Mining Research, University of Zimbabwe

K.A. VIEWING
Institute of Mining Research, University of Zimbabwe

Foreword

J. SUTTON

The papers that follow provide a permanent record of a remarkable occasion when former students and colleagues gathered at Imperial College to pay tribute to John Webb as a friend and a scientist of vision.

John Webb and I first met 44 years ago as undergraduates. One of my earliest recollections was working under the kindly eye of David Williams in the Mining Geology laboratories where John and I were endeavouring to identify minerals with the charcoal block, blow pipe and borax bead that represented the sum of our analytical equipment at that time. When John and I returned to the Department after the war, John was completing his thesis on the tin lodes of Cornwall. I asked him what he intended to do. He replied that he was going to work in applied geochemistry and I had to ask him what he meant by that expression. This is a measure of how things have moved since those far off days. One of the early operations was an investigation by John Webb and Tony Miliman carried out in West Africa analysing natural waters and plants in an area of lead zinc mineralization. I still recall the cautious way in which this pioneering effort was received at a meeting of the IMM. A long period followed when industry had to be convinced that geochemical prospecting had something to offer. John succeeded in convincing industry by his insistence on developing practical methods that could be used in the field. Never satisfied with research alone he emphasized the importance of developing a new technique to the point when it could be applied in the bush and made to work there with the resources available. This required extensive visits abroad to identify problems, to convince management that a cost-effective method of prospecting was available, and concurrently to recruit the necessary enthusiasts to carry out the work and to build up the resources at Imperial College that the projects required. With the expansion of the college in the mid-1950s, John seized the opportunity to develop laboratories on a new floor at the top of the Royal School of Mines which have housed what is now the Applied Geochemistry Research Group ever since.

It was at this time that Professor Hawkes joined forces with John Webb to produce their text book, which drew on their combined experience in a variety of terrains and which became the bible for succeeding generations concerned with exploration geochemistry.

With the laboratories in being and an eager band of colleagues around him John showed how geochemical methods could be applied to a widening range of problems. The concept of regional geochemical mapping emerged from work on the Copper Belt and led, in due course, first to the collaboration in the UK with the IGS which produced the geochemical maps of Denbighshire, Devonshire and Derbyshire, and subsequently, thanks to a grant from the Wolfson Foundation, to the *Wolfson Geochemical Atlas of England and Wales*.

Work in the British Isles raised the possibility of support for agriculture and Iain Thornton who organized this meeting was recruited to lead this part of the work. Soon the group, established in the first instance to develop ways of prospecting for minerals, became concerned with oysters, sheep and cattle and a new field of applied geochemistry was launched, subsequently extended to consider the effects, not only on animals and plants but on man himself. This was the start of environmental geochemistry. The possible applications of geochemistry to marine geology were initiated with work on placer tin deposits off the coasts of Cornwall and south east Asia. Studies of phosphorites on the continental shelf were to follow and were to lead to today's productive effort in marine geochemistry led by Dave Cronan, which have been extended into all the major oceans.

From time to time one has the good fortune to meet scientists of vision and from time to time to meet those with an unusual capacity to organize laboratories. In John, as we all know, we see a man of vision with the practical ability to turn vision into reality. This volume is a small recognition of this fact, but above all it is a tribute from John's friends.

Geochemistry: Its Achievements and Potential in Mineral Exploration

G. J. S. GOVETT

Introduction

In theory, an account of the achievements of exploration geochemistry should be a factual statement of developments about which there can be no dispute, whereas a dissertation on the future potential of exploration geochemistry is essentially an individual view of the future which can be debated until the future becomes the present. In practice, especially within the space of a single paper, the distinction is not so sharp. For a catalogue of achievements and a summary of the state-of-the-art there are a number of sources: the most recent is the Proceedings of the 'Exploration 77' meeting in Ottawa (Hood, 1979). There is clearly no point in attempting to summarize this book, and the alternative is to present a personal view of achievements—which implies identification of various landmarks along the road of development that seem to be important to the writer. In so doing it is inevitable, despite efforts to the contrary, that personal interests will cause bias. Perhaps examples from personal experience are favoured over others (which some may argue are more appropriate) derived from the literature. Nevertheless, the literature, supplemented by discussions with other geochemists, is the dominant source of data.

A paper of this kind must also provide a brief history of exploration geochemistry. Some events are obviously important—such as the demonstration of the usefulness of stream sediment as a sampling medium, the formation of the Geochemical Prospecting Research Centre at Imperial College, the development of lake sediment geochemistry. Others are less clearly defined and are difficult to date precisely, such as the use of computers; more subtle events include the mysterious change of preferred terminology from 'geochemical prospecting' to 'exploration geochemistry'.

This paper is divided into two parts. In the first part an attempt is made to identify major achievements and to interpret them in the context of the development of the science and external events. In the second part the probable role of geochemistry in the future is discussed, and areas where development is required are identified.

Achievements of exploration geochemistry

EARLY DEVELOPMENTS

There are numerous historical reviews of the development of exploration geochemistry (Boyle, 1967; Boyle and Garrett, 1970; Bloom, 1971; Hawkes, 1976; Thomson, 1982). Only the salient features of the historical development in the English-speaking world since 1945—especially as they concern some of the main personalities in the early work—are recorded here.

The first significant programme for exploration geochemistry arose from separate proposals by H. E. Hawkes, L. Huff and H. Cannon of the US Geological Survey which resulted in a budget of US $125,000 to examine geochemical techniques in 1946–47; these workers were subsequently joined by H. Lakin and F. Ward. Continuation of the programme was justified by their successful demonstration that geochemical techniques did indeed work. In particular, the development of the rapid and sensitive dithizone test by Huff (1948, 1951), and the successful demonstration of soil geochemistry in Tennessee by Hawkes and Lakin (1949) led to at least one mining company successfully using geochemistry (this was New Jersey Zinc Company which developed the Flat Gap Mine in Tennessee following a soil survey in 1950; see Hoagland, 1962). This guaranteed a future for geochemistry in mineral exploration in the US.

At about the same time, in Canada, H. V. Warren at the University of British Columbia (who was joined there by R. Delavault) had begun his experiments with biogeochemical techniques. J. S. Webb was trying hydrogeochemical and biogeochemical techniques in the Nyeba lead-zinc district of Nigeria (Webb and Millman, 1950, 1951). V. P. Sokoloff of the US Geological Survey reported favourably on the use of geochemistry in Australia (Anon, 1948). Lovering *et al.* (1950) had demonstrated the usefulness of stream sediments in exploration in the late 1940s. In 1952–53 H. Bloom developed the cold-extractable dithizone field test for heavy metals. The Geological Survey of Canada started its long and important exploration of geochemistry research with an investigation of primary halos around gold deposits at Yellowknife in 1949 and with R. W. Boyle's drainage and soil surveys at Keno Hill in 1954–55 (Boyle, 1958, 1961). The chance selection of the Nyeba lead-zinc district for some soil experiments by H.E. Hawkes (Hawkes, 1954) led to his meeting J.S. Webb and to one of the more fruitful international collaborative ventures, culminating in the publication of the first English language book on exploration geochemistry (Hawkes and Webb, 1962).

In 1953 Hawkes, Bloom, and J. Riddell experimented with drainage geochemistry in eastern Canada and found the Nash Creek deposit in New Brunswick (Hawkes and Bloom, 1956). In 1954 the team was joined by J.S. Webb (who brought substantial financial backing from Selection Trust Ltd. of London), and the Mt. Pleasant and Restigouche deposits were discovered (Hawkes *et al.*, 1960). This pioneering exploration programme collected 4937 samples (including the follow-up work in 1955) from most of the pre-Carboniferous terrain of Gaspé and New Brunswick at a cost of US$150,000 and conclusively demonstrated the efficiency of geochemistry in mineral exploration.

Another event during this period was to have far-reaching effects not only on the fledgling technique of exploration geochemistry, but on many other applied geochemical activities. In 1952 Hawkes and Webb together toured commercial and academic geochemical exploration projects in the US and Canada (Webb, 1953). The data gathered on this tour was of major importance in the establishment at Imperial College of the Geochemical Prospecting Research Centre—later reorganized as the Applied Geochemistry Research Group—under the direction of J.S. Webb.

THE LITERATURE ON EXPLORATION GEOCHEMISTRY

Results and sampling of the literature

The monumental bibliographic effort by H.E. Hawkes over the years, culminating in the publication of all citations to early 1981 (Hawkes, 1982a,b), is a rich source of information for some future scientific historian. The Hawkes bibliography was used to construct the histogram in Figure 1 which illustrates the phenomenal growth in the exploration geochemistry literature since 1950. As any geochemist knows, the choice of class intervals can influence conclusions drawn from histograms, and the data are accordingly drawn in graphical form for each year from 1945 to 1980 (Figure 2).

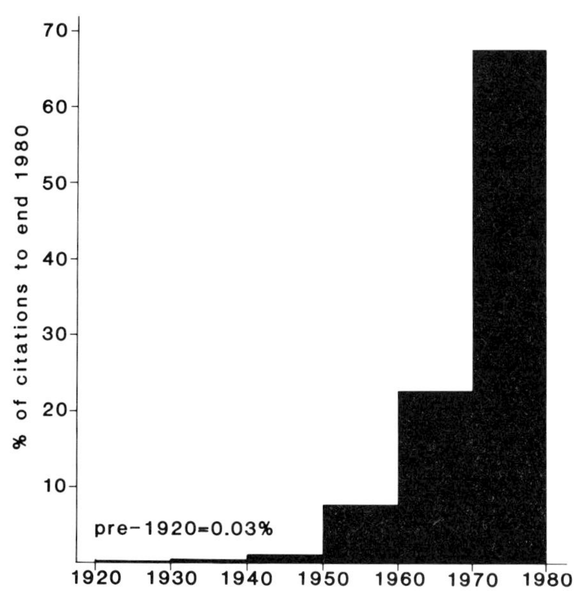

Figure 1 History of publication of exploration geochemistry to end of 1979. Total number of citations = 6430 (Hawkes, 1982a, b)

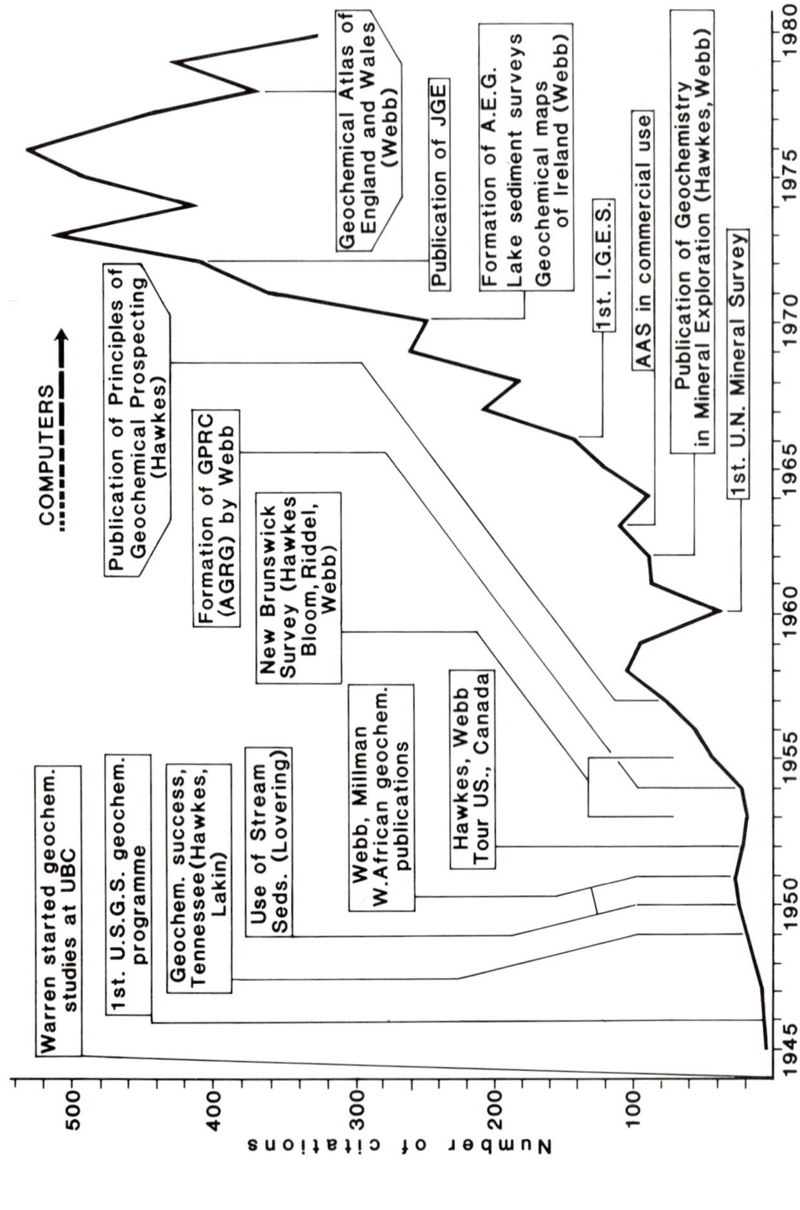

Figure 2 Number of citations in the Exploration Geochemistry Bibliography (Hawkes, 1982A, B) and some of the events in the development of exploration geochemistry.

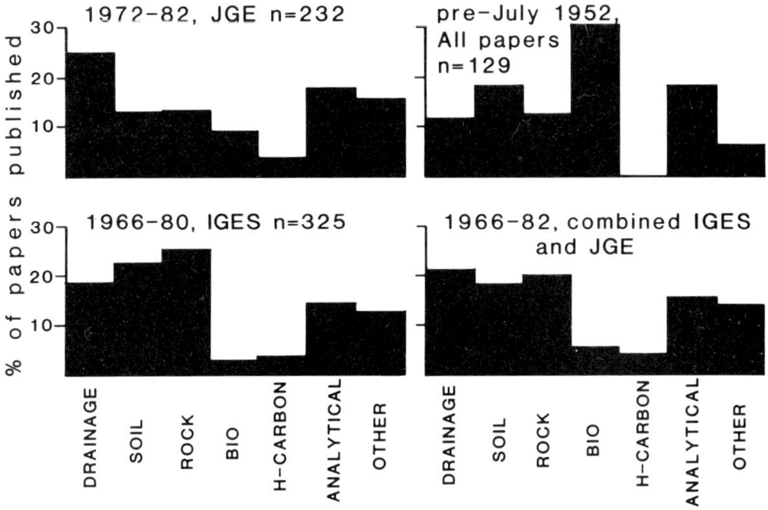

Figure 3 Distribution of exploration geochemistry publications by subject. Pre-July 1952 all sources from Harbaugh (1953). JGE—Journal of Geochemical Exploration excluding Conceptual Models and Hydrocarbon issues. IGES—Proceedings of International Geochemical Exploration Symposia

A startling development is evident in Figure 2. The total annual publication rate reached a peak in 1976 and has declined fairly steadily since that year. In 1980, the publication rate was similar to that in 1970–71. The possible significance of this pattern of publication will be examined later.

A gross comparison of *type* of exploration geochemistry papers published before and after 1952 is given in Figure 3; in constructing this diagram if a paper was concerned with three topics, one-third was assigned to each category. The category 'other' includes fundamental studies, statistics and some broad review papers. The data for the pre-July 1952 period are based on all papers cited in Harbaugh (1953); the post-1952 period is based on the proceedings of the various IGES (International Geochemistry Exploration Symposium) from 1966 to 1980, and on publications in the JGE (Journal of Geochemical Exploration) from 1972 to 1982.

The major differences between the publications in the two periods are that more than 30% of the pre-1952 papers were on biogeochemical methods compared to less than 10% in the more restricted later samples; there are no hydrocarbon papers cited for the pre-1952 period. The preponderance of biogeochemical papers in the pre-1952 period is plausibly explained by the bias toward translated Russian and Scandinavian papers in the bibliography and the influence of Scandinavian work on early geochemical experiments in the English speaking world. The fairly high proportion of rock geochemistry papers published before 1952 is a little misleading insofar as many of them are only marginally (or not at all) concerned directly with exploration.

There are also marked differences between the relative proportions of the

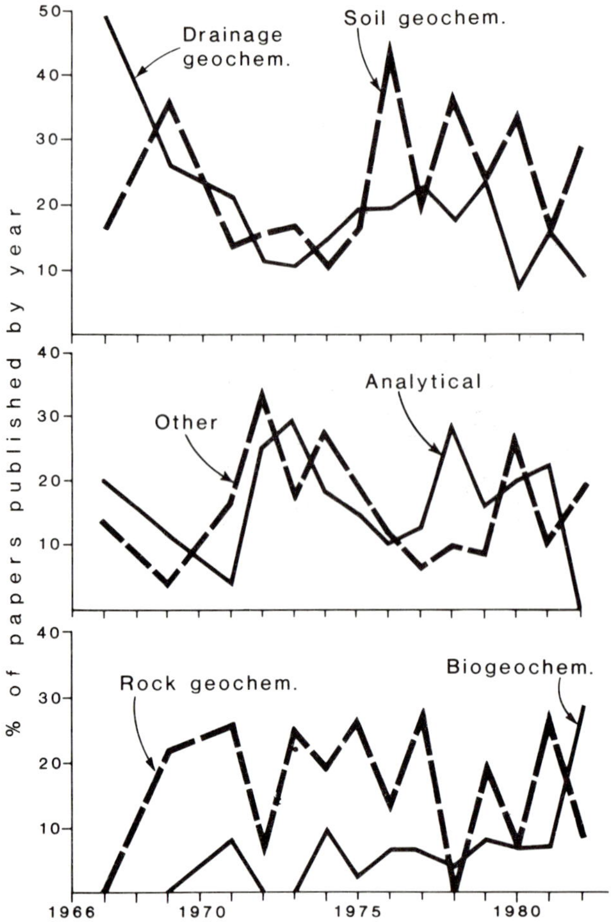

Figure 4 Variation in proportion of papers in various categories as a percentage of total papers published in a year in JGE and IGES

three main techniques between the JGE and IGES; rock geochemistry is most popular in IGES, whereas drainage surveys are most popular in JGE. The combined JGE and IGES papers, however, show that drainage, soil and rock have had about equal attention in the entire period since 1966.

The data in Figure 4 show few well-defined, consistent long-term trends. Drainage papers show a general decline over the entire period, and papers on soils have been erratically declining since 1976. Publications on biogeochemistry generally have increased since the early 1970s—but the exceptionally high proportion in 1982 is regarded as aberrant.

There is no noticeable long-term variation in the 1972–82 period in the proportion of contributors to the JGE employed by industry, governments, or universities (Figure 5). The papers by those affiliated with universities generally account for an increasing share relative to government

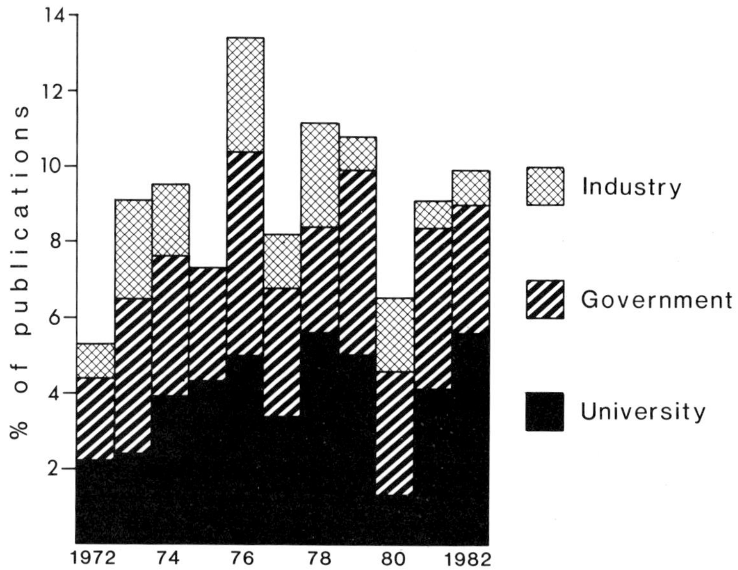

Figure 5 Affiliations of authors of papers published in JGE 1972–1982 as a proportion of total papers in this time period (excludes Conceptual Models and Hydrocarbon issues)

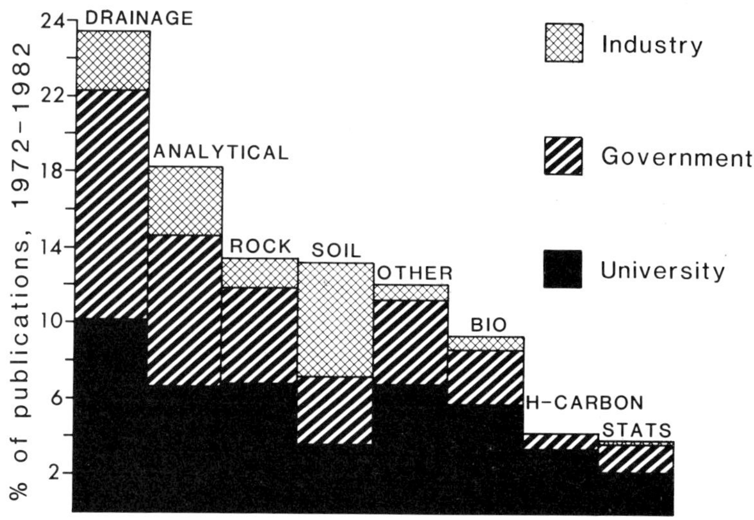

Figure 6 Papers in various categories published in JGE 1972–1982 according to affiliation of author as a proportion of the total number of papers in this time period (excludes Conceptual Models and Hydrocarbon issues)

contributions until 1979, but there is an abnormal slump in 1980, followed by a subsequent recovery. In the 1972–82 period as a whole the cumulative averages are: industry, 16.8%; government, 40.5%; and universities, 42.8%.

The data on the share of industry, government and universities in various categories of papers published in the JGE in the period 1972–82 (Figure 6) indicate that industry dominates only in soil geochemistry and is weakest in statistical methods. Universities are the major contributors in rock geochemistry, biogeochemistry, statistical methods and the category 'other'. Government are the major contributors in drainage and analytical categories.

Evaluation of the data

Although exploration geochemistry papers are published in journals other than the JGE and the proceedings of the IGES, most professional exploration geochemists belong to the AEG (Association of Exploration Geochemists) and appear to publish at least a proportion of their work in the JGE. Although the record of publication can be assumed to be reasonably representative of interests, it may possibly understate the contribution from industry (through a possible preference for publishing in the Transactions of the various Institutes of Mining and Metallurgy). Similarly, the relative importance of papers on statistical methods may be understated because of a possible preference by mathematical geochemists to publish in the *Journal of Mathematical Geology*.

The interpretation of the data is obviously open to debate, but several conclusions are suggested:

1. The relatively low contribution from industry in such a practical field is disappointing, but it is to be expected given company policies toward confidentiality and the lack of time available to write up results when such data can be published.
2. The seeming dominance of university research in rock geochemistry, statistical methods, biogeochemistry and 'other' is also expected. Rock geochemistry is the newest of the main techniques used by industry, and it is to be expected that universities should provide much of the basic research (similar comments can be made about statistical methods). Insofar as the 'other' category includes fundamental studies, it is again appropriate for university researchers to take the lead. Biogeochemistry is the least used, least successful and perhaps the most difficult technique for routine application; it is therefore not surprising that it has become a university-dominated topic.
3. The time distribution of the various categories of publications is rather puzzling, and the time period studied is too short for definitive trends to become apparent. It is perhaps significant that the period of low publication rates for drainage and soil papers in 1971–75—the two best-established techniques—coincides with a peak in analytical and 'other' papers. It might be surmised that this represented a period of reflection and an effort to provide some fundamental data. If this

interpretation is correct, then the renewed surge of case history data that began in 1976 is now declining and another period of more fundamental studies is occurring.

TECHNICAL AND SCIENTIFIC ACHIEVEMENTS OF EXPLORATION GEOCHEMISTRY

Technical advances

In the mid-1950s most of the analytical techniques were colorimetric; there was great excitement at the freedom and speed of analysis that the cold-extractable copper test allowed. Increases in productivity were achieved by such ingenious devices as multi-tube holders for primus stoves to allow more than one fusion digestion to be done at a time. Wooden racks to hold four separating funnels to increase the productivity of the tedious Pb determination were all the rage, and large hands to enable 10 test tubes to be held and shaken at a time were probably the greatest productivity aid of all. A mere decade later AAS was in commercial use—giving rapid, accurate and precise determinations with extraordinary sensitivity for a greatly expanded range of elements.

The improvements in X-ray fluorescence with computer-controlled operation and computer-controlled matrix correction has made it a powerful tool in many exploration geochemistry programmes. Similarly, plasma spectrographs are on the threshold of causing another order of magnitude change in exploration geochemical analytical data generation. These, and other analytical techniques such as the specific ion electrode, are well reviewed by Ward and Bondar (1979), and treated in detail by Fletcher (1981).

A similar dramatic change has occurred in the capacity to handle data. Again, the comparison must be made with the situation in the mid-1950s (when the height of sophistication in data analysis was to have the use of a hand-cranked calculator to compute standard deviations or analysis of variance on small data sets). Today, high-speed computers are available as super-fast calculating machines allowing statistical data manipulation that was unthinkable a few decades ago. Equally important is the use of the computer as a data storage device and as a tireless draughtsman to plot graphs and maps (see Howarth, 1983 and Howarth and Garrett, this volume).

Developments in concepts and applications

Nothing as traumatic as the demonstration, more than 30 years ago, that properly analysed soil and drainage samples could reveal the presence of hidden mineralization has occurred recently. The basic principles of exploration geochemistry taught at Imperial College in the mid-1950s are *essentially* unchanged. What *has* changed is the scope of application and the understanding of these principles—for example, the use of conceptual models introduced by Bradshaw (1975) for Canadian conditions and subsequently applied to Scandinavia (Kauranne, 1976), the Basin and Range Province in North America (Lovering and McCarthy, 1978), and Australia (Butt and Smith, 1980). The remainder of this section briefly reviews developments in specific techniques.

Soil geochemistry

Soil geochemistry has been an effective technique since the earliest days of exploration geochemistry because of the almost universal presence of soil (or overburden of some kind) as an irritating blanket obscuring bedrock. A simple technical solution to the problem of thick transported glacial drift in northern latitudes, or the equally thick leached cover in places such as Australia, has been to use overburden drilling. Great progress has been made, however, in understanding the principles of dispersion and modifying procedures to meet local conditions.

In glaciated regions a study of the glacial history has allowed intelligent sampling to capitalize upon the mechanisms of glaciation (rather than to be frustrated by an exotic cover). For example, analysis of the coarse size fraction of samples of upper till (just below the podsol soil layer) allow till to be used as a reconnaissance 'drainage' medium to trace mineralization as demonstrated around Mt. Pleasant in New Brunswick (Figure 7). Post-glacial dispersion patterns have been identified using the humus layer rather than the conventional B horizon, and by sampling of peat bogs (Bölviken and Gleeson, 1979). A number of workers have demonstrated that subtle anomalies can be

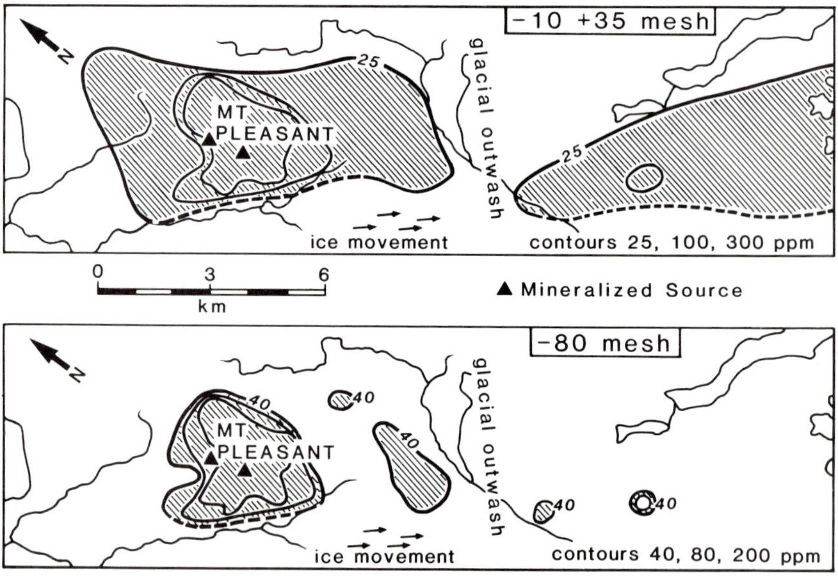

Figure 7 Distribution of Zn in minus 10 plus 35-mesh and in minus 80-mesh size fraction of till around Mt. Pleasant, New Brunswick, Canada. (Redrawn with permission from Szabo *et al.*, 1975)

detected in surface soils even overlying glacio-lacustrine deposits by various techniques (see Bölviken and Gleeson, 1979). The measurement of H-ion and conductance of aqueous water slurries has been shown to be effective in a number of instances (Figure 8) based on an electrochemical model of

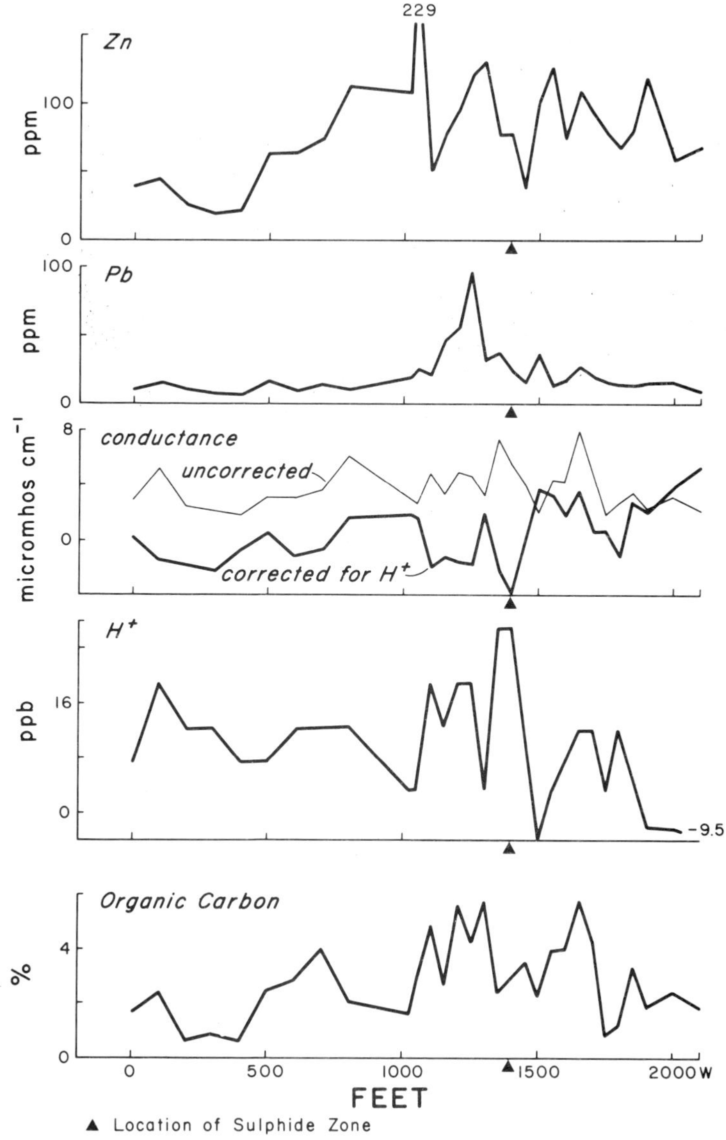

Figure 8 Distribution of organic carbon, H+, conductance Pb and Zn in B-horizon soils, Key Anacon, New Brunswick, Canada. Element and organic carbon concentration are in sample; H+ and conductance are in aqueous slurry and should be multiplied by 100 to show as same basis as elements. Reproduced with permission from Govett and Chork, 1977, *Prospecting in Areas of Glaciated Terrain,* 1977, Institution of Mining & Metallurgy, London, 1977, Figure 1, p. 49

dispersion (Govett, 1973, 1976; Govett and Chork, 1977; Bölviken and Lögn, 1975; Nuutilainen and Peuraniemi, 1977); recent work by Smee (1983) has demonstrated that whereas the dispersion *pattern* is electrochemically controlled, the dispersion *mechanism* is by diffusion.

In non-glaciated areas (where there are problems due to severe soil leaching, transported alluvium, windblown sand, loess and the development of calcrete or duricrust) useful results are being obtained by the use of metal ratios, profile sampling and, in some cases, by the use of immobile elements such as Au (see Learned and Boissen, 1973, for work on copper porphyry deposits in Puerto Rico). In Western Australia Smith *et al.* (1980) showed how a transported and consolidated pisolitic blanket around the Golden Grove Cu-rich sulphide deposit could be used to advantage by demonstrating a Bi–Sn halo 1 km × 2 km in the pisolites around the deposit. Theobald and Allcott (1973) found that in Saudi Arabia the dilution effect of windblown sand could be removed by discarding the fine fraction, and anomalies related to Mo–W mineralization could be enhanced by analysing the non-magnetic heavy mineral fraction of the minus 10 plus 30-mesh size fraction. None of these techniques are conceptually new, nor are any new principles involved. What is new is a greater awareness of the problems posed by the environment, a greater understanding of surface processes, and a determination to devise techniques to overcome the problems.

Drainage geochemistry

The potential usefulness of widely spaced sampling was first demonstrated by Webb *et al.* (1964a,b) in Zambia. Later Garrett and Nichol (1967) used samples with catchment areas of up to 39 km^2 in Sierra Leone, and Armour-Brown and Nichol (1970) covered 207 000 km^2 in central Zambia with a sample density of one stream sediment per 200 km^2. The practical outcome of these early studies is the mammoth NURE (National Uranium Resource Evaluation) Programme in the US—which has as its objective the systematic evaluation of all potential uranium areas by samples of drainage sediment, stream water and wells at a density of about 1 sample per 10 km^2. A similar programme in Canada (Federal-Provincial Uranium Reconnaissance Programme), in the mid to late 1970s, was based on about 1 sample per 13 km^2 in British Columbia and Yukon. Insofar as multi-element data are being generated, these programmes have already shown their value in mineral exploration (McConnell, 1983).

A later development of the early low density sampling programme by the Applied Geochemistry Research Group at Imperial College was the concept of the Geochemical Atlas. The first programme in Northern Ireland in 1967 (Webb *et al.*, 1973) was followed by sampling of all England and Wales at a density of 1 sample per 2.6 km^2 in 1969 which culminated in the publication of the *Wolfson Geochemical Atlas of England and Wales* (Webb *et al.*, 1978).

A major new development in drainage geochemistry—which illustrates the design of sampling and analytical procedures to suit local conditions—has been the use of lake sediments. Although many geochemists had experimented with lake sediments, the main development of the technique took place in Canada, largely for use in the Shield area. Following on the work of M.B. Mehrtens and A.G. Troup for Rio Tinto Canadian Exploration Ltd. (see Coker *et al.*, 1979), Hoffman and Fletcher (1972) in the Cordilleran region and Allan (1971) in the northern Shield area, considerable fundamental studies have shown lake sediments to be an effective reconnaissance technique at a sample density of 1 per 5 km^2 to 1 per 25 km^2 (see Coker *et al.*, 1979 for a summary).

In Scandinavia the frequent absence of inorganic sediment has led to the development of multi-element analysis of organic sediment and multiple regression to eliminate the effect of organic carbon, Fe and Mn (Larsson, 1976). The effect of Mn and Fe coatings on sediment in scavenging metals had been demonstrated by Horsnail *et al.* (1969). Since then it has been shown by many workers (e.g. Carpenter *et al.*, 1975) that deliberate selective analysis of Fe and Mn hydroxides can enhance anomalies.

Although the flat, semi-arid Tertiary pediplain area of central New South Wales is scarcely an attractive area to practise stream sediment surveys, Dunlop *et al.* (1983) have demonstrated—by sampling linear drainage depressions identified from colour aerial photographs, sieving out the coarse fraction (minus 10-mesh plus 16-mesh) as a means of concentrating pisolites, and correcting the Pb content for Fe—drainage trains of 5 km can be obtained compared to virtually no response in the minus 230-mesh size fraction (Figure 9). Thus, the understanding that Pb accumulated in Fe pisolites allowed a sampling and analytical procedure to be designed to meet the circumstances.

Rock geochemistry

Publications on rock geochemistry date back to the last century, but the development of rock geochemistry as an exploration tool essentially dates from the late 1960s. Of the more than 360 references of all types cited in Govett (1983) on rock geochemistry, 13.7% are earlier than 1960, 21% are from 1960 to 1969, and 65.3% are from 1970 onwards.

The prerequisites for the use of rock geochemistry are a geochemical signature in the rock that has a genetic association (however indirect) with mineralization, and adequate rock exposure. A better understanding of ore genesis in recent years has been a major factor in the successful interpretation of rock geochemical data and an incentive to use the technique.

The major achievement has been the demonstration that, whereas intense ore element halos may be restricted to the immediate wall rock of a deposit, more subtle multi-element halos and major element halos can be identified in whole rock analysis on a regional scale. Mineral separates have not generally been shown to be useful, but sulphide selective leaches (e.g. Cameron *et al.*, 1971; Olade and Fletcher, 1974) and even water leaches (Goodfellow and Wahl, 1976) are potentially important.

Dispersion around massive sulphides has been more extensively studied than around any other deposit type. This is probably because of the close

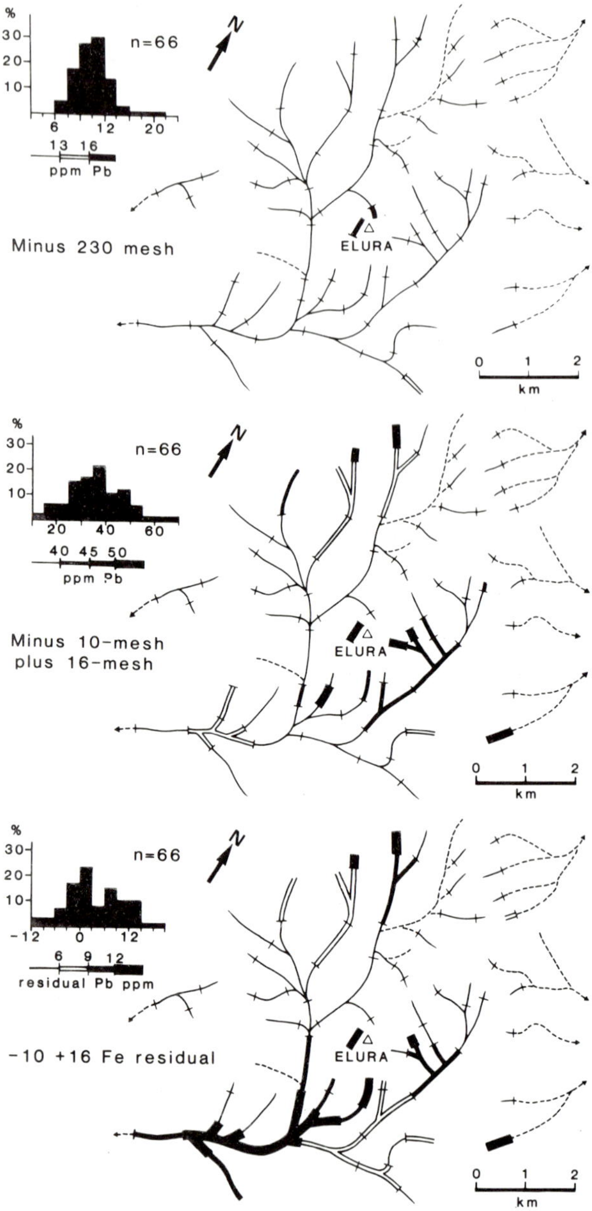

Figure 9 Distribution of Pb in minus 230-mesh, minus 10-mesh plus 16-mesh size fraction, and residual Pb in minus 10-mesh plus 16-mesh size fraction after correction for Fe content of drainage sediments, Elura zinc-lead-silver deposit, New South Wales, Australia. (Redrawn with permission from Dunlop *et al.*, 1983)

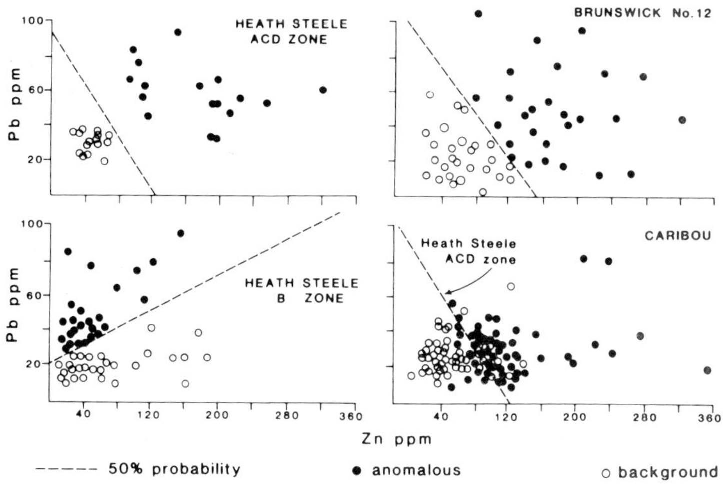

Figure 10 Variation in Pb and Zn contents in acidic volcanic rocks in relation to 50% probability line between anomalous and background conditions (calculated by linear discriminant analysis with Pb and Zn as variables) for New Brunswick, Canada massive sulphide deposits. Reproduced with permission from Govett, 1983, *Rock Geochemistry in Mineral Exploration*, Elsevier, Amsterdam, 1983, Figure 11-10, p. 283

genetic association between mineralization and the host rocks—although the writer has never quite recovered from the surprise of discovering extensive halos in post-mineralization rocks, firstly in Cyprus and later in New Brunswick. Massive sulphides have a broad similarity of response regardless of age and location. Regional responses are commonly given by ore elements (especially Zn) and in many cases by Fe and Mn. Local scale discrete halos from 100s to 1000s of metres in extent are defined by some or all of Fe, Mn, Na, K, Ca, Mg and the ore elements. The significance of understanding—and allowing for—details of ore genesis is illustrated in Figure 10. The relative proportions of Pb and Zn that define mineralization around the proximal Brunswick No. 12, Heath Steele ACD Zones, and Cariboo deposits are quite different from their mutual relation around the distal Heath Steele B Zone.

Productive intrusions can be recognized by analysis on relatively few samples for non-ore elements (K, Rb, Sr, Ba, Li) as well as ore elements. Even regions of vein and replacement deposits can normally be identified by sampling vein and fracture material at a density of 1 to 10 km^2 (i.e. the plumbing system) and analysing for As, Sb and Bi, and the expected ore elements such as Cu, Pb, Zn, Au and Ag.

Porphyry deposits have distinctive halos of K, Ca, Rb, Sr, Cu, Zn, Mo and S that extend many 100s of metres beyond mineralization. The demonstration of distinctive zoning patterns of elements by Chaffee (1976) around the Kalamazoo copper porphyry deposit is especially significant in terms of the potential to discover blind deposits. Halos around individual vein deposits are mostly for ore elements, are restricted to a few 10s of metres and show clear evidence of being diffusion-controlled by their logarithmic decay pattern.

Problems in development of rock geochemical techniques—and resistance to their wider adoption—arise from the realization that even minor variations in lithological composition and degree of fractionation necessarily cause variations in chemical composition, particularly in the level of trace and minor elements. Multi-element analysis is mandatory. Correction for lithological variation is important and, in many cases, essential.

Other techniques

The use of volatile elements (e.g. Hg, He, SO_2, radon gas) has had a rather chequered history in mineral exploration. The potential of Hg has been recognized for many years and actively used since the early 1960s (e.g. Hawkes and Williston, 1962; James, 1962), but its role and the problems of interpretation are still the subject of a large number of papers. The problems with Hg (and other volatiles) remain an incomplete understanding of their chemistry and behaviour in the secondary environment.

The development of AIRTRACE™ (Barringer, 1979) and SURTRACE™ (Barringer and Bradshaw, 1979) is intriguing. The techniques rely upon multi-element analysis of inorganic and organic particulate matter suspended in the atmosphere close to the earth's surface or lying on the surface.

Geobotanical methods have been reviewed by Cannon (1971, 1979) and biogeochemical methods by Brooks (1979). While considerable scientific advances have been made, neither method has found much favour in mineral exploration. The greatest amount of work has been in the USSR, but Boyle (1979) states that not a single deposit has been found there by these techniques. On the other hand, Helen Cannon (Cannon, pers. comm., 1983) has pointed out that Se-indicator plants have been used as guides to uranium, and that this association was directly responsible for the discovery and mining of five uranium bodies in the Yellow Cat district of Utah (Cannon, 1964) and two open-pit uranium mines in Poison Canyon, New Mexico (Cannon, 1960).

PRACTICAL ACHIEVEMENTS AND THE STATUS OF GEOCHEMISTRY IN EXPLORATION

Discoveries

Derry (1970) and Derry and Booth (1978) have attempted to classify mineral discoveries in Canada according to their chief technique of discovery (Table 1). The data make rather dismal reading for the geochemist seeking to reassure himself of the usefulness of his trade. The data do *not* show, however, how much geochemistry was used to screen areas before a discovery was made, nor how much geochemistry was used to delineate target areas before other techniques were used. The effective integration of all techniques in exploration programmes make it difficult to ascribe a discovery to a particular technique.

Table 1 **Mineral discoveries in Canada from Derry (1970) and Derry and Booth (1978)**

	No. of discoveries			
	Conventional Prospecting	Geological	Geophysical*	Geochemical
To 1950	64	11	—	—
1951–55	16	14	5	—
1956–60	6	4	14	—
1961–65	5	3	7	1
1966–70	2	4	13	1
1971–75**	1	5	15	3

*including radioactivity methods
**plus 2 discoveries by unknown methods

Boyle (1979) gives a fairly lengthy list of geochemical discoveries. They include:

1. The Cortez gold deposit in the US (surface rock and soil surveys).
2. The Bougainville copper-gold porphyry deposit in PNG (steam and soil surveys).
3. The Woodlawn copper-lead-zinc deposit in NSW, Australia (stream and soil surveys).
4. The Casino copper-molybdenum porphyry deposit in Canada (drainage, soil, and rock geochemistry).
5. Tynagh and other lead-zinc deposits in Ireland (discovered mainly by soil geochemistry).

Most of the disseminated gold deposits that are in production or under development in Nevada and other western US states were found with the assistance of bedrock geochemistry (A.W. Rose, pers. comm., 1983). Most of the lead-zinc discoveries in the Selwyn Basin of northeastern Yukon result from stream or soil geochemical surveys (D. Derry, pers. comm., 1983).

Boyle (1979) estimates that about 100 000 surface anomalies in North and South America, Australia, Asia, Africa and Europe delineated in the 20 years to 1977 led to 150 orebodies; he similarly gives figures of about 80'000 anomalies in USSR leading to 220 orebodies. The magnitude of the geochemical effort can be further gauged by the fact that in the first 10 years of United Nation mineral surveys (1960–70) 89% of ground surveys used geochemistry; 430 000 samples were collected and 1 900 000 determinations were made (Lepeltier, 1971).

Expenditure on exploration geochemistry

Whatever the actual discovery rate, exploration geochemistry is clearly

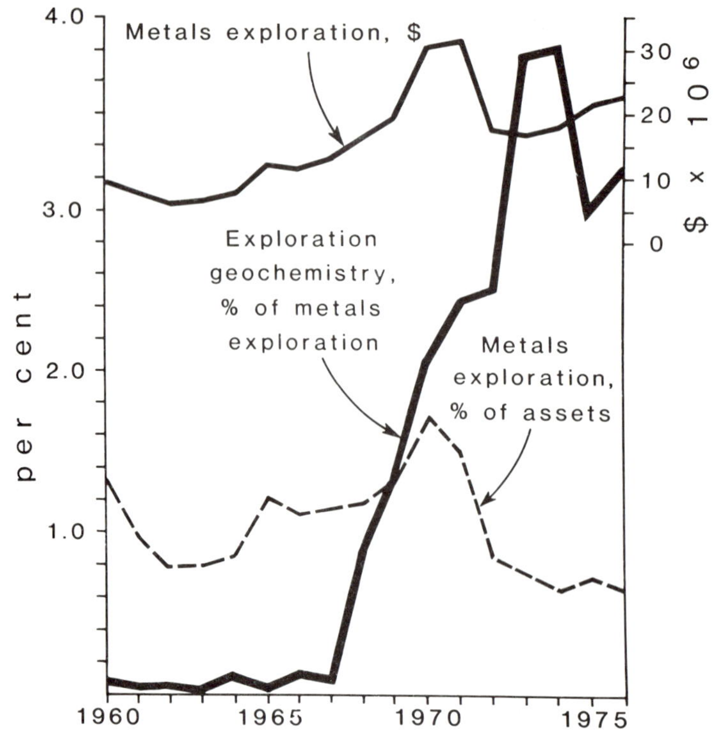

Figure 11 Exploration expenditure by a major producing Canadian mining company. Note that 40–50% of metals exploration expenditure was on mine development, and therefore the proportion spent on geochemistry should be approximately doubled to obtain a reasonable figure

regarded by the mining industry as worthwhile. The historical expenditure on geochemistry from 1960 to 1976 by a very large Canadian company is illustrative of the pattern of development (Figure 11). A large Australian mining company with strong international connections spent less than A$4,000 on geochemistry in 1975 and about A$223,000 in 1982; in 1983 this company's expenditure on geochemistry in Australia and off-shore areas is likely to exceed A$300,000. Both the Australian and the Canadian company include development drilling in their total exploration expenditures which makes it difficult to estimate the real contribution of geochemistry to exploration.

One very large multi-national company with major mining interests in Australia has stated that 'pre-Webb' their expenditure on geochemical exploration was nil, rose to about 40% of total exploration expenditure in the late 1960s and early 1970s, and has settled down to about 15% of total exploration expenditure since the mid-1970s. A large Australian mining company with a cumulative exploration budget of A$92 million since mid-1950s similarly now spends 15 ± 3% of its budget of geochemistry. It is likely that in the US geochemistry accounts for about 50% of exploration expenditures excluding drilling—i.e. about 15% including drilling.

The advent of geochemistry has had a significant impact on the effectiveness of the small company in exploration. It is a technique that can be effectively practised by intelligent geologists without the need for the high capital cost equipment and the highly specialized personnel used in geophysical programmes. Data supplied by a small company in Australia indicate that geochemistry accounts for approximately 15% to 20% of salaries and 10% to 15% of total expenditure (including drilling).

All companies contacted by the writer in Australia stress that the diminution in expenditure from the mid-1960s to early 1970s does not represent a decline in importance of geochemistry, but is rather a reflection of the changing needs of the exploration industry. In well-prospected parts of the world every stream has been sampled at least once (in many cases many times) on a reconnaissance basis. Large regional drainage surveys are a diminishing necessity, and only fill-in sampling is normally required. Also, the accumulated case history experience allows a far more efficient planning of exploration surveys. It should be pointed out that whereas this may be desirable from the point of view of the company involved, it could have a serious impact on the development of exploration geochemistry to the extent that there is less willingness to experiment.

Australian and Canadian non-fuel exploration expenditures for 1967–80 are shown in Figure 12. The peak for geochemistry publications was about 5 years after the peak in exploration in the late 1960s and early 1970s—a predictable time lag if the surge in publications was fueled by accumulation and publication of some exploration data and considerably increased industry-sponsored research in universities (a seemingly natural outcome of exploration booms). If this interpretation is correct—and geochemistry was a significant factor in exploration expenditures—an increase in publications is

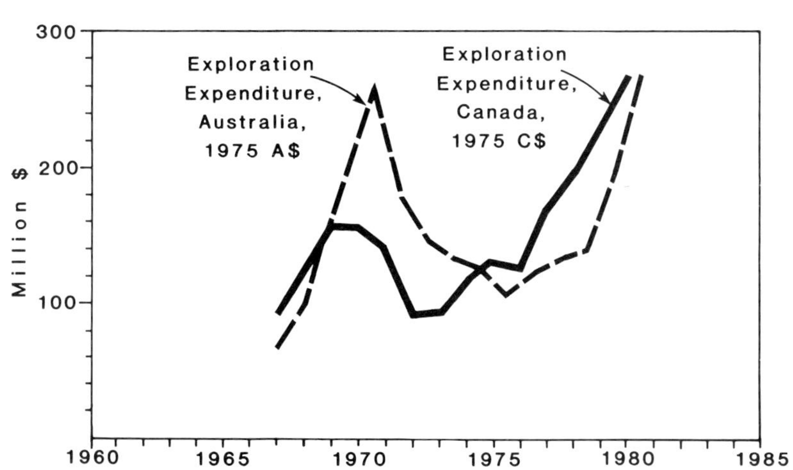

Figure 12 Exploration expenditures for Australia and Canada (data from various sources)

due and should reach a peak in 1985–86, corresponding with the peak in exploration around 1980–82.

The number of exploration geochemists

There are no data on the number of people who regard themselves as exploration geochemists in the western world. If the results of a recent survey undertaken by the Australian Geoscience Council (Sant, pers. comm., 1983) are indicative of the situation in other countries, the total number of geochemists employed by industry and government is small. The survey indicates that in Australia geochemists account for about 3.5% of the total

Table 2 **Distribution of geoscientists employed by industry, as consultants, and by government in Australia**

	A Mining/ Petroleum	B Consultants	A + B	C Government	Total A + B + C
Geochemists	1.1	0.6	1.7	1.8	3.5
Geophysicists	6.7	3.5	10.3	4.0	14.2
Geologists	46.1	8.4	54.6	14.7	69.3
Other geoscientists	4.6	3.5	8.1	4.9	13.0
Total	58.5	16.1	74.7	25.3	100.0

Number in sample—3413 (preliminary data from Australian Geoscience Council)

Table 3 **Percentage of geochemists in Australia employed in industry, government or as consultants**

	A Mining/ Petroleum	B Consultant	A + B	Government
Research	5.1	4.2	9.3	32.2
Exploration	9.3	1.7	11.0	3.4
Lab. analyses	8.5	5.9	14.4	13.6
Prospect evaluation	2.5	2.5	5.1	0.8
Mine/field development	0.8	0.0	0.8	0.0
Admin/management	2.5	2.5	5.1	0.8
Other	2.5	0.8	3.4	0.0
Total	31.2	17.6	49.1	50.8

Number in sample—118 (preliminary data from Australian Geoscience Council)

population of geoscientists outside of tertiary institutions and about 2.3% of the total number of geoscientists employed in industry; most of the geochemists who work in industry are engaged in exploration or analytical work, whereas most of the government geochemists are engaged in research. The preliminary results of the survey are summarized in Tables 2 and 3.

Interestingly, only 46% of the members of AEG list geochemistry as their main interest (Table 4). Of the latter group 87% state their main geochemical interest to be exploration, although none list biogeochemistry as a main interest (Table 5). The age distribution of geochemists who are members of AEG is shown in Figure 13. 70% of them obtained their first degree 1961–75 (with a peak of 1965–70), and only 8% obtained their first degree in 1976–80.

The decline in the number of geochemists is in part a reflection of the decline in the total number of geology graduates in Australia (which is itself a direct

Table 4 **Main interests of Australian members of the Association of Exploration Geochemistry**

Field of main interest	%
Geochemistry	46
Economic geology-metals	36
Geology—other	8
Geophysics	8
Other	2

Number in sample—50 (preliminary data from Australian Geoscience Council Survey)

Table 5 **Main geochemical interests of those Australian members of the Association of Exploration Geochemists who stated that their main interest is geochemistry (Table 4)**

Field of geochemistry	Principal interest %	Subsidiary interest, number*
Exploration	87.0	9
Inorganic	4.3	5
Isotope/geochronology	8.7	4
Physical	1.0	3
Organic	0.0	1
Biogeochemistry	0.0	2
Other	0.0	6

Number in sample—23 (preliminary data from Australian Geoscience Council Survey)
*Respondents were asked to list up to 4 subsidiary interests

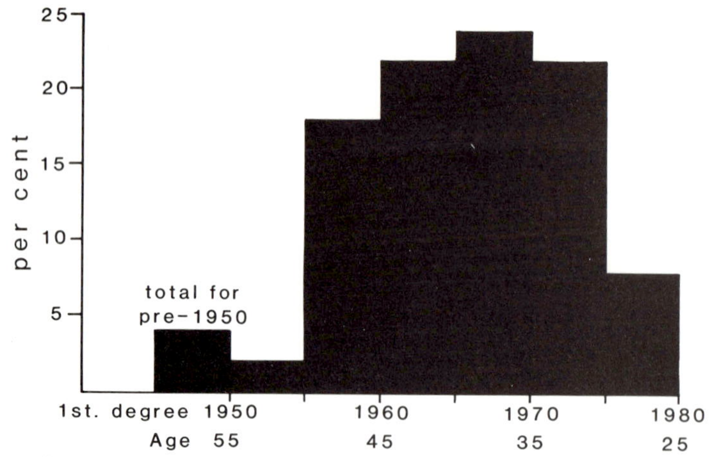

Figure 13 Distribution of year of obtaining first degree and age of Australian members of Association of Exploration Geochemists (n = 50; preliminary data from Australian Geoscience Council)

result of the severe decline in exploration in the 1970s). A contributory cause is undoubtedly that relatively few of the geochemists who graduated in early years have formal training in geochemistry, but gravitated to it by experience during their career.

Potential of geochemistry in mineral exploration

INTRODUCTION

A fairly consistent theme of Presidential Addresses to the Association of Exploration Geochemists from 1974 to 1980 has been that there is a pressing demand for minerals and the geochemist should help find them (e.g. Nichol, 1974; Govett, 1977a; Horsnail, 1978; Theobald, 1981). Then in 1981 Rose (1982) suggested that, for copper at least, supplies were adequate. This conclusion is consistent with growing evidence that the demand per capita for metals in many industrialized countries has not only reached a plateau, but in some cases is declining. Furthermore, the anticipated vastly increased demand by newly industrialized countries is not occurring for a variety of reasons, including the fact that their route to industrialization is different from that of the old industrial countries.

In 1982 base metal prices in real terms were lower than they had been since the end of World War II (Figure 14). Copper prices, which showed an upward trend from 1945 to the mid-1970s and must have fueled much exploration, declined, albeit erratically, after 1974. Lead and zinc prices, apart from a brief increase in 1974–79, have been similarly depressed. The only bright prospects for metals exploration have been gold and silver (Figure 14). A.W. Rose (pers. comm., 1983) states that in the US mining companies generally are interested

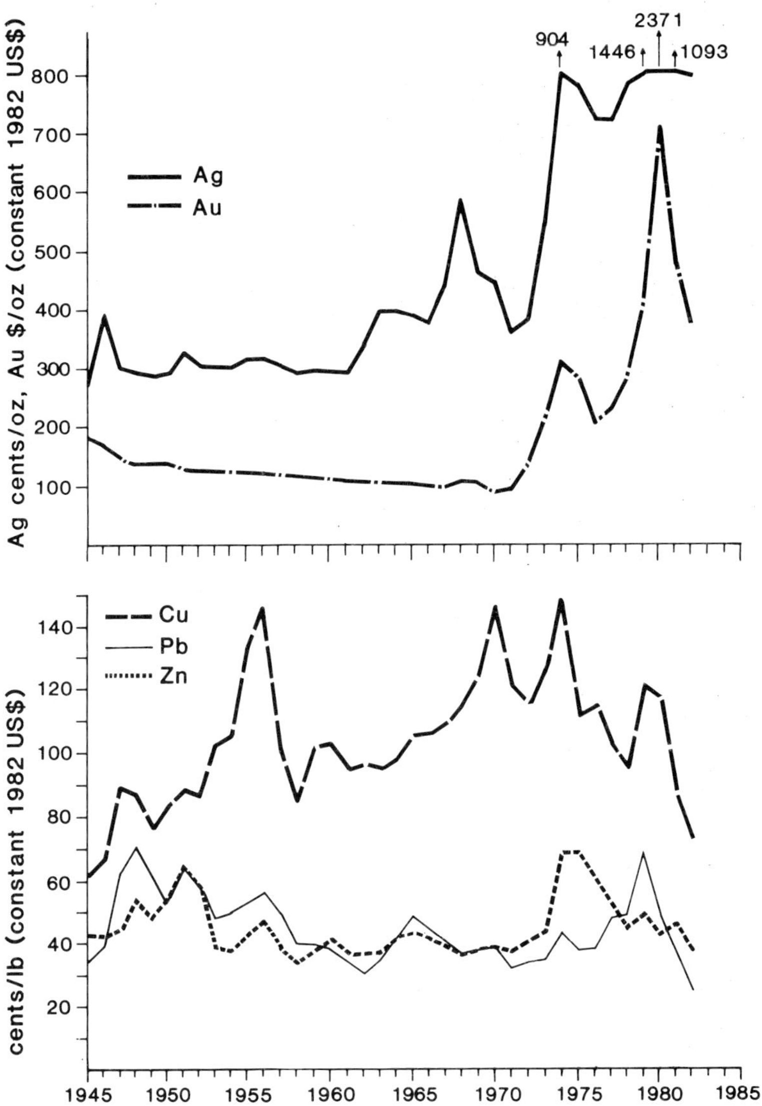

Figure 14 Metal prices in constant 1982 US dollars, 1945–1982

in new discoveries only if their production costs are likely to be in the lower half of the present range of production costs. These factors, together with the fact that exploration costs are rising faster than unit metal prices (Bradshaw, 1979) suggest that a high level of exploration activity for base metals is not to be expected for some years; such exploration as does occur will be highly specific for high-value targets. Thus, the amount of *routine* exploration geochemistry is likely to be at a relatively low level.

Factual support for these conclusions appears to be provided by data from

Australia where the pattern of distribution of exploration expenditures is changing. For example, the proportion of the total non-petroleum exploration expenditure on metals has declined from about 91% in 1973–74 to 70% in 1980–81. The major increase in the expenditures for non-metallic minerals is accounted for by coal (expenditures increased nine-fold in real terms 1973–74 to 1980–81)—which is an expected consequence of the oil price increases in the mid-1970s—although diamond exploration is also a significant factor. Detailed long term data on the proportion of exploration expenditure accounted for by specific metals are not available; however, in 1979–80, exploration for copper, lead, zinc, silver, nickel and cobalt accounted for 46% of the total metals expenditure compared to 40% in 1980–81. In the same time period, expenditures for gold exploration increased from 15% to 22% of total metals exploration; it is expected that gold exploration has been accounting for an increased proportion of metals exploration in a number of countries since the price of gold started to increase significantly in the late 1970s.

Given the investment climate, the future potential for exploration geochemistry lies in the realm of development of specialized techniques for specific purposes. The great challenges for exploration geochemistry in the 1980s are to develop techniques capable of detecting concealed mineralization and to improve the capability of discriminating between true and false anomalies. To achieve these goals it is clear that a major research effort is necessary in all aspects of geochemistry, and real progress will only occur if geochemical research is integrated with developments in all aspects of earth science.

EMPIRICISM AND FUNDAMENTALISM

From its beginning exploration geochemistry has been criticized for being too empirical—for being a technique rather than a science. Those within the profession have been its harshest critics: a consistent theme of Presidents of the Association of Exploration Geochemists for the past 10 years—whether their affiliation is with industry, government or universities—has been that there is a desperate need to undertake more research on processes of element dispersion, and to apply the results of fundamental studies by other earth science disciplines (e.g. Erickson, 1973; Govett, 1977a, b; Bradshaw, 1979; Theobald, 1981). Interestingly, this plea is no different in intent than that of Hawkes in 1950 who stated ' . . . a better understanding is needed of the chemical factors that affect the ore metals in the cycle of weathering' (Hawkes, 1950, p. 550).

Obviously advances in fundamental knowledge have been made in the past 30 years, but as knowledge accumulates, the gaps in our understanding are brought more sharply into focus. It is inevitable that in the early stages of a new science work is empirical: it is a data-gathering stage upon which hypotheses and laws are later based. It is also undeniable that it is easier to conduct an orientation programme in a new area or to analyse samples from a well-known area for a new set of elements and call it 'research' than to deliberately set out to find the chemical and physical controls on the movement of an element across the bedrock-soil interface.

The concern for more fundamental research is inspired by very practical considerations. The need to develop techniques to detect buried and blind deposits is frequently stated; this formidable task can only successfully be tackled on the basis of a thorough understanding of the processes of element dispersion. Moreover, the costs of locating near-surface deposits would be sharply reduced by more reliable interpretation of routine geochemical data—it is becoming prohibitively expensive to drill so-called anomalies just to find out that they are due to weathering processes or barren pyrite. Thus, data processing and interpretative procedures need to be improved; particular attention should be given to pattern-recognition techniques and to identifying subtle anomalies within the background range.

Concern for fundamental research is sufficiently widespread to ensure that attempts will be made to meet the demand. It is possible that the decline in publications in exploration geochemistry noted above is at least partly a reflection of editors' and reviewers' weariness with 'just another empirical case history' and consequent rejection of the manuscript. There is a danger in this too: it is essential that well executed empirical studies continue because they provide much of the data upon which more theoretical studies are based. More importantly, they are the means of communication between practising exploration geochemists and of bringing new techniques—or useful minor modifications to old techniques—to the attention of exploration managers.

DEVELOPMENT OF VARIOUS TECHNIQUES

Computer data processing

The development that will ultimately lead to far greater use being made of all geochemical data is the availability of inexpensive but powerful desk-top computers. Within the span of the writer's career data handling has progressed from the situation of having to borrow a hand-cranked mechanical calculator, through laboriously punching data cards and dubiously submitting them to a batch station of a giant mainframe computer (and returning the next day to find the program aborted), to sitting at home in front of a personal computer, processing the data and printing and plotting the results. The example in Figure 15 was computed and plotted on a 48K microcomputer. The raw data—pH and conductance of soil slurries—is filed on mini-disk. The pH values were converted to H+ion and corrected for the H+ion content of deionized water on the computer; conductivities were corrected for the conductivity of deionized water and for the contribution due to H+ion and OH-ion. The data were finally plotted out on a computer-driven printer-plotter as a three-point moving average profile. In this particular example no complex statistics were performed, but the calculations would have been very tedious by hand. Moreover, the choice of vertical and hori-

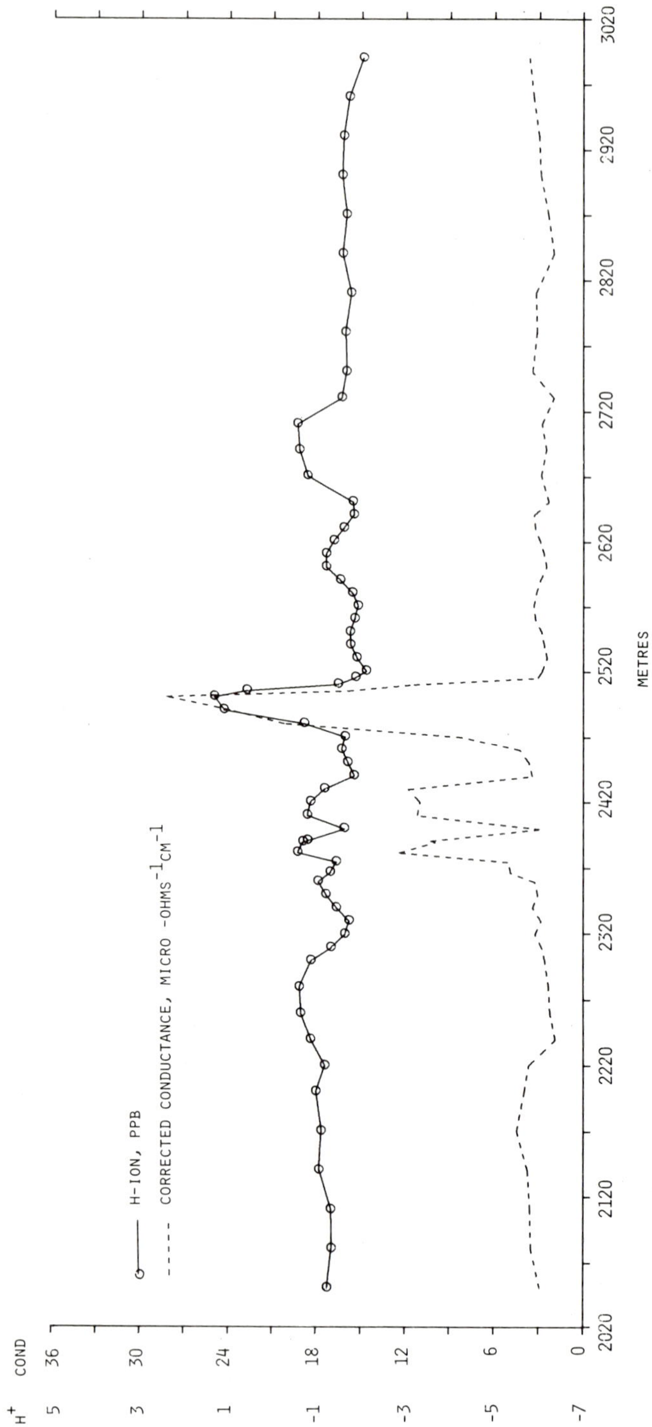
H-ION, PPB
CORRECTED CONDUCTANCE, MICRO -OHMS^{-1}CM^{-1}
METRES
H^+ COND
2020
2120
2220
2320
2420
2520
2620
2720
2820
2920
3020

zontal scale for maximum effect was determined by trial runs on the video screen before finally plotting hard copy.

The advantages of the wide availability of desk-top computers are many. The geologist who is responsible for collecting data can do his own data processing quickly. The interactive capability of the computer and the speed of computation allows many interpretative techniques (from simple ratios to discriminant analysis) to be tried as a matter of routine rather than as a rare event. Moreover, sample outliers can readily be identified and data sets can be quickly modified by elimination of particular samples.

The use of small personal computers is currently limited only by the paucity of interactive user-oriented software. The development of the Q-GAS system (Lavin and Nichol, 1981) for a larger office desk-top Digital Equipment Corporation system is a step in the right direction, but more flexible packages (in terms of scope of data processing and compatability with smaller and other computers) must also be developed.

Analysis and sampling

The trend of improved sensitivity and precision that has characterized the development of instrumental techniques will undoubtedly continue. There is no virtue, however, in seeking to improve analytical techniques to the point that their variance is far lower than the sampling variance; it is probable that routine analytical precision for many techniques is better than many samples deserve (Govett, 1983). This conclusion is particularly relevant to the constituents present in very small concentrations (e.g. Bi, Se, Te, Tl, Au) that are increasingly being used as pathfinder elements.

The analytical improvements that have occurred and will continue to occur will only be useful in exploration geochemistry if greater consideration is given to ensuring that the physical size of the initial sample is adequate to provide a representative sample of the constituent of interest. Similarly, subsequent sample treatment in terms of size reduction, concentration and specific components, and ultimate aliquot for analysis must be designed to meet the capabilities of the analytical technique and nature of occurrence of the component of interest (Fletcher, 1981; Govett, 1983).

The usefulness of multi-element analysis has been demonstrated many times since the advent of AAS; the imminent routine availability of inductively coupled plasma atomic emission spectrometry (ICPAES) to determine reliably and simultaneously a wide range of elements in geological materials will give impetus to the trend towards multi-element analysis. The multi-element approach should not, however, be uncritically accepted. Not least is the problem of interpretation of multi-element data: many *practising* geochemists have difficulty in interpreting the information provided by half a dozen elements. In the present stage of data evaluation it is doubtful whether analyses for 20 or 30 elements would materially improve the efficiency of exploration geochemistry; indeed, it is possible that the sheer mass of data would have an adverse effect on interpretation by overwhelming the exploration geologist.

There is another point to be considered. In a recent review of rock geochemistry in mineral exploration, Govett (1983) showed that whereas many elements give measurable geochemical responses, a relatively small number of elements are adequate in any particular case—and most of these are common to many situations. It is suggested, therefore, that the usefulness of ICPAES will be to make a wide range of elements *available,* but that the next stage in improved anomaly recognition must be dependent on the geochemist intelligently selecting a small number of elements appropriate to the target and scale of exploration from the large number of elements offered.

Drainage surveys

The role of geochemistry in mineral exploration has been subtly changing for some time. Large-scale conventional ore or pathfinder element regional surveys are becoming much less important, whereas large-scale multi-purpose 'geochemical map' type surveys by non-industry organizations are increasing in importance. This trend will continue unless it can be demonstrated that regional scale multi-element drainage surveys can detect buried or blind target zones or reliably identify prospective intrusions or horizons. Local scale drainage surveys still have a role for specific targets, but their application in well-prospected regions—where the only remaining mineral deposits are deeply buried—will decline.

Soil surveys

Local scale soil and overburden surveys will continue to have a major role in mineral exploration. Even in well prospected areas of relatively simple residual soil cover, new surveys and analyses for different elements and other parameters with more advanced interpretative techniques will prove to be useful. In areas of glacial overburden continued research on glacial processes and chemical dispersion will improve the discrimination between syngenetic and epigenetic dispersion, and hence increase the reliability of interpretation. Considerably more work needs to be done in alluvium covered areas, in areas of thick, deeply leached soil, and in lateritic terrains.

In all areas of very thick overburden—whether residual or transported—there is scope for use of gaseous and volatile elements and compounds. Where the hydro-geological conditions are suitable direct hydrogeochemical techniques may prove to be useful (see below). Under some circumstances there is also scope for detecting subtle hydromorphic halos through selective analysis of the appropriate mineralogical fraction of the soil. In areas of deeply leached soil the use of immobile elements is an obvious approach. A neglected, but potentially useful, area of research in these circumstances is the use of quantitative X-ray diffraction examination of the clay-sized fraction of the overburden to detect either residual alteration products directly associated with mineralization, or secondary alteration products (such as alunite) that might be expected to form from an oxidizing sulphide.

Rock geochemistry

There is tremendous potential for development of rock geochemistry where there is adequate outcrop. Applications include identification of the appropriate favourable environment of deposition for specific types of mineral deposits and the direct use of rock geochemistry in locating individual deposits. An extension of the type of work done by Garrett (1971, 1973) on identification of prospective granitoids in the Canadian Cordillera, by Davenport and Nichol (1973) on discrimination between barren and productive volcanic cycles in the Canadian Shield, or of the low density (1 sample/5 km^2) rock survey of Govett and Pwa (1981) to define prospective mineralized zones in the Palaeozoic volcanic rocks of New Brunswick is desirable in this context. Developments of these techniques will inevitably involve the application of rather more sophisticated geochemical principles. For example, A.W. Rose (pers. comm., 1983) points out that magmas with low O fugacity which were formed by assimilation of sedimentary rock appear to be favourable for Sn and W; these may be identified by mineral assemblages (ilmenite, high content Fe-mafic minerals, aluminous phases) and by Sr and O isotope ratios.

Identification—and understanding—of element zoning and discrimination between syngenetic and epigenetic primary halos should considerably improve the capability for detection of blind deposits. It should also help to differentiate between halos resulting from potential economic mineralization and those due to mineralization that has been truncated or completely removed by erosion. A determined effort should be made to investigate the claims in Soviet literature that size and grade of deposits can be estimated from geochemical relations.

One of the most useful potential applications of rock geochemistry is in drilling. It is also the one application where practical difficulties militate against its use—the time required in sampling core, crushing, grinding and analysing (to say nothing of interpretation) is so great that in most situations the data are available far too late to be of any value in guiding a drilling programme. It can be speculated that this particular problem will be overcome by development of analytical instrumentation (e.g. portable on-site laser plasma spectrographs linked directly to a mini-computer with plotting and printing peripherals).

Hydrogeochemistry and vapour geochemistry

In the past hydrogeochemistry has not played a significant role in exploration in English-speaking countries; it is now receiving increased attention for a variety of reasons. The availability of very low detection limits on small samples provided by flameless AAS and plasma spectrometers has removed the major practical difficulty. The research on understanding the chemistry of natural aqueous systems (e.g. Mann and Deutscher, 1977) and the recognition that reliable interpretation generally can be made only by consideration of mineral equilibria in aqueous systems (e.g. Langmuir and Chatham, 1980) is restoring faith in the possibilities of overcoming some of the apparent interpretive problems of hydrogeochemical data. Most importantly, the pressure to develop techniques to detect concealed mineralization has focused attention on ground-water as a potential sample medium. Hydrogeochemistry will probably play an increasing role in exploration for specific targets in appropriate hydrogeological circumstances.

Vapour phase and volatile element geochemistry will obviously attract increasing attention—because of the possibility of detecting blind deposits—as analytical capabilities improve. As with more conventional geochemical techniques, multicomponent analysis may be expected to increase reliability of interpretation. An example of the type of approach needed is the work by Hale and Moon (1983) who measured desorbed Hg, oxygen deficiency, carbon dioxide enrichment and carbonyl sulphide in soil to detect buried mineralization at Keele, Eire.

Geobotanical and biogeochemical techniques

Although geobotanical and biogeochemical techniques were some of the earliest investigated, they have not prospered in modern exploration geochemistry (although there is a large amount of literature on the subjects). There remains considerable scope for the development of applications of both techniques for particular situations. The inhibiting factor will continue to be the specialized botanical knowledge required for successful application of the principles. The main areas of potential development in the near future are likely to be indirect—spectral analysis of remote sensing imagery (geobotany) and analysis of plant exudates (biogeochemistry).

Conclusions

It is difficult to appreciate the significance of the statistic that more than 95% of the literature on exploration geochemistry has been published in the past 30 years. At a time when geochemistry is used routinely by most of the mining industry all over the world as one of the main techniques of exploration it is hard to remember that 30 years ago there was probably no company with a single member of staff who would call himself a geochemist. It is also fairly safe now to dignify exploration geochemistry as a science on its own right, judging by the number of universities that now admit its existence in both undergraduate courses and postgraduate research.

The history of exploration geochemistry may conveniently be summarized in relation to the graph of exploration geochemistry publications shown in Figure 2 above. The broad, low peak in publications between 1945 and 1953 (with the peak central around 1950) was a period of discovery of the techniques of geochemical exploration in the western world. During the latter part of this period university and government experimental programmes began; industry, tentatively in many cases, began using simple techniques for exploration. A surge of case histories and other papers emerged between 1955 and 1960 as a result of these activities. Publications remained fairly steady in the period 1960–65 as the results of the developmental and testing work became available. At the same time the constraints of *relatively* unproductive and restrictive colorimetric tests were being removed by a gradual introduction of AAS on a commercial basis following its development by A. Walsh at the CSIRO in Australia. A parallel technical development was the increasing availability of computers to assist in handling the torrent of data about to be created by the advent of AAS.

By 1966 there was a sizeable body of geochemists in government, industry and universities who had experience with geochemistry and a conviction of its usefulness. Communication between these people was established at the inaugural biennial IGES in Ottawa in 1966, the formation of AEG in 1970, and the beginning of publication of JGE in 1972. This was also a period of intense exploration activity and expanding exploration budgets. All these factors contributed to the growth curve in publication of exploration geochemical data that continued to the mid-1970s. This unprecedented activity was largely a period of data gathering, refining techniques and developing new techniques (although studies on processes were not entirely neglected).

The sharp decline in publication activity in the late 1970s is perhaps a reflection of saturation with case histories, but it is also certainly a function of the decrease in base metal exploration that began in 1971. Since there has been a significant increase in exploration for precious metals since 1976, it could be expected that the publication rate will again increase in the next few years—although the level of increase will depend on the proportion of increased exploration expenditure that is assigned to geochemistry.

The *apparent* simplicity of exploration geochemistry is both its great merit and its disadvantage. The basic principles are indeed simple and lead the unwary into the erroneous belief that uncritical collection of samples, analysis and subsequent examination of results (or, worse, uncritical submission of data to computer analysis) will lead to instant recognition of anomalies. When the result is either a multitude of anomalies or none, or a false anomaly is drilled, discontent sets in, and geochemistry is abandoned as an exploration tool. The solution to this particular problem is, of course, training of professional exploration geochemists. The need for this is increasing as more and more subtle anomalies are sought to detect concealed and more deeply buried mineralization.

To achieve these ends significant funding is necessary and theoretical geochemists must be attracted to exploration geochemistry research. Reliable global statistics on expenditures on exploration geochemistry should be compiled, and at the national level a determined effort should be made to compile geochemical case history data—the raw material for generalized conceptual models.

Industry must be prepared to make appropriate contributions. Mining companies tend to be relatively generous in good times, but become lamentably unforthcoming during exploration down-turns. This is just the time that they should be giving serious consideration to their exploration problems. It would be much more intelligent—and productive—to send two bright young geologists to university on half pay to do research on an exploration geochemistry problem than to dismiss one of them in hard times. The science of exploration geochemistry would benefit, exploration managers would solve some problems, and companies would have more useful geologists ready to rejoin them for the next up-turn in exploration.

If the Australian statistics on numbers of geochemists in industry are representative, then only about 2% of the geoscience industry workforce are geochemists. Given the significance of geochemistry in exploration this could be taken as a flattering statistic. It is also, perhaps, an indication that industry does not even now have a sufficiently high regard for geochemists, nor appreciate how much more effective their exploration might be if they employed more professional geochemists.

There have been a number of pioneers in exploration geochemistry, but most workers agree that the role of Professor John S. Webb in setting up—and maintaining—a research and training centre at Imperial College was vital to establishing geochemistry as one of the main techniques of exploration. The contributions from the development of techniques by the Research Centre have been considerable, but it has been the training of well in excess of 100 postgraduate exploration geochemists for which Professor Webb must be especially honoured. These graduates have literally permeated the entire western world mining establishment, and many are now in top management positions. Some of the Imperial College graduates have set up their own university groups on the Imperial College model, and some of *their* graduates are now doing the same.

The achievements of exploration geochemistry in the past 30 years have been staggering—a global multi-million dollar industry has come into being, and a whole new branch of science has been established. The potential for the future is limited only by the imagination of geochemists. To ensure that the next 30 years are as productive as the last 30 years, there must be an accelerated research effort into geochemical processes and a determined attempt to quantify geochemical responses. Most of all, however, geochemists must continue to be fearless innovators and willing to try everything—even if pure science dictates it can't work. In summary, brilliant fundamental research must be matched by imaginative empirical studies.

ACKNOWLEDGEMENTS

The writer gratefully acknowledges the data and information provided by friends and colleagues in many parts of the world, in particular: R.J. Allan, L.G. Closs, R.G. Garrett, J.F. Gilfillan, D.R. Derry, A.W. Rose, P.K. Theobald, F.N. Ward and M.H. Govett. Many people in the mining industry cannot be named to maintain the anonymity of the company concerned; their help was invaluable. The Australian Geoscience Council is thanked for permission to publish preliminary data from their Geoscience

Human Resources Survey. The assistance of Mr P.R. Atherden in library research, Mrs M. Horvath for draughting, Mr G. Small for photography, and Mrs M. Valentine for producing the manuscript is sincerely appreciated.

References

Allan, R.J. (1971). Lake Sediment: a medium for regional geochemical exploration of the Canadian Shield, *Can. Min. Metall. Bull.*, **64(715)**, 45–99.

Anon (1948). Geochemical prospecting—investigations in Australia, *Chemical Engineering and Mining Review*, 10 Nov. **41(2)**, 43–44.

Armour-Brown, A. and Nichol, I. (1970). Regional geochemical reconnaissance and the location of metallogenic provinces, *Econ. Geol.*, **65**, 312–330.

Barringer, A.R. (1979). The application of atmospheric particulate geochemistry in mineral exploration (abs.). In: *Geophysics and Geochemistry in the Search for Metallic Ores* (Ed. P.J. Hood), pp. 363–364; Geol. Surv. Can., Econ. Geol. Rep., **31**.

Barringer, A.R. and Bradshaw, P.M.D. (1979). Surtrace—an airborne and ground exploration technique based on surface micro-layer geochemistry (abs.). In: *Geochemical Exploration 1978* (Eds. J.R. Watterson and P.K. Theobald), pp. 243–249; Association of Exploration Geochemists, Toronto, Ontario.

Bloom, H. (1971). Exploration geochemistry—past, present and future, *The Mines Magazine*, **May**, 12–14.

Bölviken, B. and Gleeson, C.F. (1979). Focus on the use of soils for geochemical exploration in glaciated terrain. In: *Geophysics and Geochemistry in the Search for Metallic Ores* (Ed. P.J. Hood), pp. 295–326; Geol. Surv. Can., Econ. Geol. Rep., **31**.

Bölviken, B. and Lögn, O. (1975). An electrochemical model for element distribution around sulphide bodies. In: *Geochemical Exploration 1974* (Eds. I.L. Elliott and W.K. Fletcher), pp. 631–648; Elsevier, Amsterdam.

Boyle, R.W. (1958). Geochemical prospecting in permafrost regions of Yukon, Canada, *Int. Geol. Cong. 20th, Mexico*, **1**, 175–188.

Boyle, R.W. (1961). The geology, geochemistry and origin of the gold deposits of the Yellowknife district, *Geol. Surv. Can. Mem.*, **310**.

Boyle, R.W. (1967). Geochemical prospecting—retrospect and prospect, *Can. Inst. Min. Metall. Bull.*, **60(657)**, 44–49.

Boyle, R.W. (1979). Geochemistry overview. In: *Geophysics and Geochemistry in the Search for Metallic Ores* (Ed P.J. Hood), pp. 25–31; Geol. Surv. Can., Econ. Geol. Rep., **31**.

Boyle, R.W. and Garrett, R.G. (1970). Geochemical prospecting—a review of its status and future, *Earth-Science Reviews*, **6**, 51–75.

Bradshaw, P.M.D. (Ed.) (1975). Conceptual models in exploration geochemistry—The Canadian Cordillera and Canadian Shield, *J. Geochem. Explor.*, **4**, 1–213.

Bradshaw, P.M.D. (1979). Presidential Address to the Association of Exploration Geochemists, *J. Geochem. Explor.*, **12**, 1–7.

Brooks, R.R. (1979). Advances in botanical methods of prospecting for minerals Part II—advances in biogeochemical methods of prospecting. In: *Geophysics and Geochemistry in the Search for Metallic Ores* (Ed. P.J. Hood), pp. 397–410; Geol. Surv. Can., Econ. Geol. Rep., **31**.

Butt, C.R.M. and Smith, R.E. (Eds.) (1980). Conceptual models in exploration geochemistry—Australia, *J. Geochem. Explor.*, **12**, 89–365.

Cameron, E.M., Siddeley, G. and Durham, C.C. (1971). Distribution of ore elements in rocks for evaluating ore potential: nickel, copper, cobalt, and sulphur in ultramafic rocks of the Canadian Shield. In: *Geochemical Exploration* (Eds. R.W. Boyle and J.I. McGerrigle), pp. 298–313; Can. Inst. Min. Metall. Spec. Vol., **11**.

Cannon, H.L. (1961). The development of botanical methods of prospecting for uranium on the Colerado Plateau, *US Geol. Surv. Bull.*, **1085-A**.

Cannon, H.L. (1964). Geochemistry of rocks and related soil and vegetation in the Yellow Cat area, Grand County, Utah, *US Geol. Surv. Bull.*, **1176**.

Cannon, H.L. (1971). The use of plant indicators in ground water surveys geological mapping, and mineral prospecting, *Taxon*, **20 (2/3)**, 227–256.

Cannon, H.L. (1979). Advances in botanical methods of prospecting for minerals Part I—advances in geobotanical methods. In: *Geophysics and Geochemistry in the Search for Metallic Ores* (Ed. P.J. Hood), pp. 385–395; Geol. Surv. Can., Econ. Geol. Rep., **31**.

Carpenter, R.H., Pope, T.A. and Smith, R.L. (1975). Fe–Mn oxide coatings in stream sediment geochemical surveys, *J. Geochem. Explor.*, **4**, 349–363.

Chaffee, M.A. (1976). The zonal distribution of selected elements above the Kalamazoo porphyry copper deposit, San Manuel District, Pinal County, Arizona, *J. Geochem. Explor.*, **5**, 145–165.

Coker, W.B., Hornbrook, E.H.W. and Cameron, E.M. (1979). Lake sediment geochemistry applied to mineral exploration. In: *Geophysics and Geochemistry in the Search for Metallic Ores* (Ed. P.J. Hood), pp. 435–478; Geol. Surv. Can., Econ. Geol. Rep., **31**.

Davenport, P.H. and Nichol, I. (1973). Bedrock geochemistry as a guide to areas of base-metal potential in volcano-sedimentary belts of the Canadian Shield. In: *Geochemical Exploration 1972* (Ed. M.J. Jones), pp. 45–57; Institution of Mining and Metallurgy, London.

Derry, D.R. (1970). Exploration expenditure, discovery rate and methods, *Can. Inst. Min. Metall. Bull.*, **63(695)**, 362–366.

Derry, D.R. and Booth, J.K.B. (1978). Mineral discoveries and exploration expenditure—a revised review 1966–1976, *Mining Mag.*, **May**, 430–433.

Dunlop, A.C., Atherden, P.R. and Govett, G.J.S. (1983). Lead distribution in drainage channels about the Elura zinc-lead-silver deposit, Cobar, New South Wales, Australia, *J. Geochem. Explor.*, **18**, in press.

Erickson, R. (1973). Presidential Address to the Association of Exploration Geochemists, *J. Geochem. Explor.*, **2**, 175–191.

Fletcher, W.K. (1981). *Analytical Methods in Geochemical Prospecting;* Elsevier, Amsterdam.

Garrett, R.G. (1971). Molybdenum, tungsten and uranium in acid plutonic rocks as a guide to regional exploration, S.E. Yukon, *Can. Min. J.*, **92**, 37–40.

Garrett, R.G. (1973). Regional geochemical study of Cretaceous acidic rocks in the Northern Canadian Cordillera as a tool for broad mineral exploration. In: *Geochemical Exploration 1972* (Ed. M.J. Jones), pp. 203–219; Institution of Mining and Metallury, London.

Garrett, R.G. and Nichol, I. (1967). Regional geochemical reconnaissance in eastern Sierra Leone, *Inst. Min. Metall., Trans. Sect.* B, **76**, 97–112.

Goodfellow, W.D. and Wahl, J.L. (1976). Water extracts in volcanic rocks—detection of anomalous halos at Brunswick No. 12 and Heath Steele B-zone massive sulphide deposits, *J. Geochem. Explor.*, **6**, 35–59.

Govett, G.J.S. (1973). Differential secondary dispersion in transported soils and post-mineralization rocks: an electrochemical interpretation. In: *Geochemical Exploration 1972* (Ed. M.J. Jones), pp. 81–91; Institution of Mining and Metallurgy, London.

Govett, G.J.S. (1976). Detection of deeply buried and blind sulphide deposits by measurement of H+ and conductivity of closely shaped surface soil samples, *J. Geochem. Explor.*, **6**, 359–382.

Govett, G.J.S. (1977a). World mineral supplies—the role of exploration geochemistry, *J. Geochem. Explor.*, **8**, 3–16.

Govett, G.J.S. (1977b). Presidential Address to the Association of Exploration Geochemists, *J. Geochem. Explor.*, **8**, 591–599.

Govett, G.J.S. (1983). *Rock Geochemistry in Mineral Exploration;* Elsevier, Amsterdam.

Govett, G.J.S. and Chork, C.Y. (1977). Detection of deeply-buried sulphide deposits by measurement of organic carbon, hydrogen ion and conductance of surface soils. In: *Prospecting in Areas of Glaciated Terrain* (Ed. G.R. Davis), pp. 49–55; Institution of Mining and Metallurgy, London.

Govett, G.J.S. and Pwa, A. (1981). Regional reconnaissance exploration geochemistry for massive sulphides, New Brunswick, Canada, *J. Geochem. Explor.*, **15**, 139–158.

Hale, M. and Moon, C.J. (1983). Geochemical expressions at surface of mineralization concealed

beneath glacial till at Keele, Eire. In: *Prospecting in Areas of Glaciated Terrain—1982* (Ed. P.H. Davenport), pp. 228–239, Can. Inst. Min. and Metall.
Harbaugh, J.W. (1953). Geochemical Prospecting Abstracts through June 1952, *US Geol. Surv. Bull.*, **1000-A**.
Hawkes, H.E. (1950). Geochemical prospecting for ores. In: *Applied Sedimentation, a Symposium* (Ed. P.D. Trask), pp. 537–555; John Wiley, New York.
Hawkes, H.E. (1954). Geochemical prospecting investigations in the Nyeba lead-zinc district, Nigeria, *US Geol. Surv. Bull.*, **1000-B**, 51–103.
Hawkes, H.E. (1957). Principles of Geochemical Prospecting, *US Geol. Surv. Bull.*, **1000-F**, 225–355.
Hawkes, H.E. (1976). The early days of exploration geochemistry, *J. Geochem. Explor.*, **6**, 1–11.
Hawkes, H.E. (1982a). *Exploration Geochemistry Bibliography, AEG Special Volume 11*; Association of Exploration Geochemists, Toronto.
Hawkes, H.E. (1982b). Exploration Geochemistry Bibliography—a 'Review', *Assoc. Explor. Geochem. Newsletter*, **40**, 9.
Hawkes, H.E. and Bloom, H. (1956). Heavy metals in stream sediment used as exploration guides, *Min. Eng.*, **8**, 1121–1126.
Hawkes, H.E. and Lakin, H.W. (1949). Vestigal zinc in surface residuum associated with primary zinc ore in east Tennessee, *Econ. Geol.*, **44**, 286–295.
Hawkes, H.E. and Williston, S.H. (1962). Mercury vapour as a guide to lead-zinc-silver deposits, *Mining Cong. J.*, **48**, 30–32.
Hawkes, H.E. and Webb, J.S. (1962). *Geochemistry in Mineral Exploration;* Harper and Row, New York.
Hawkes, H.E., Bloom, H., Riddell, J.E. and Webb, J.S. (1960). Geochemical reconnaissance in eastern Canada, *Int. Geol. Cong. 20th, Mexico*, **3**, 607–621.
Hoagland, A.D. (1962). Distribution of zinc in soil overlying the Flat Gap Mine, *Min. Eng.*, **14**, 56–58.
Hoffman, S.J. and Fletcher, W.K. (1972). Distribution of copper at the Dansey-Rayfield River property, south-central British Columbia, Canada, *J. Geochem. Explor.*, **1**, 163–180.
Hood, P.J. (Ed.) (1979). Geophysics and Geochemistry in the Search for Metallic Ores, *Geol. Surv. Can., Econ., Geol. Rep.*, **31**.
Horsnail, R.F. (1978). Mineral development in the 1980s, *J. Geochem. Explor.*, **10**, 195–202.
Horsnail, R.F., Nichol, I. and Webb, J.S. (1969). Influence of variations in secondary environment on the metal content of drainage sediments. In: *Proceedings International Geochemical Exploration Symposium* (Ed. F.C. Canney), pp. 307–322; Golden, Colorado, 1968, Q. Colo. Sch. Mines. V. 64.
Howarth, R.J. (1983). *Statistics and Data Analysis in Geochemical Prospecting;* Elsevier, Amsterdam.
Huff, L.C. (1948). A sensitive field test for heavy metals in water, *Econ. Geol.*, **43**, 675–684.
Huff, L.C. (1951). A sensitive field test for detecting heavy metals in soil or sediment, *Econ. Geol.*, **46**, 524–540.
James, C.H. (1962). A review of the geochemistry of mercury and its application to geochemical prospecting. Geochemical Prospecting Research Centre, Imperial College, (London), Tech. Commun. 41.
Kauranne, L.K. (Ed.) (1976). Conceptual Models in Exploration Geochemistry—Norden, *J. Geochem. Explor.*, **5**, 175–420.
Langmuir, D. and Chatham, J.R. (1980). Groundwater prospecting for sandstone-type uranium deposits: a preliminary comparison of the merits of mineral-solution equilibria, and single-element tracer methods, *J. Geochem. Explor.*, **13**, 210–219.
Larsson, J.O. (1976). Organic stream sediments in regional geochemical prospecting, Precambrian Pajala District, Sweden, *J. Geochem. Explor.*, **6**, 233–249.
Lavin, O.P. and Nichol, I. (1981). Q'GAS: A minicomputer-based system to aid in the interpretation of exploration-oriented geochemical data, *J. Geochem. Explor.*, **15**, 521–537.
Learned, R.E. and Boissen, R. (1973). Gold—a useful pathfinder element in the search for porphyry copper deposits in Puerto Rico. In: *Geochemical Exploration 1972* (Ed. M.J. Jones),

pp. 93–103; Institution of Mining and Metallurgy.

Lepeltier, C. (1971). Geochemical Exploration in the United Nations Development Programme. In: *Geochemical Exploration* (Eds. R.W. Boyle and J.I. McGerrigle), pp. 24–27; Can. Inst. Min. Metall. Spec. Vol., 11.

Lovering, T.G. and McCarthy, J.D. (Eds.) (1978). Conceptual Models in Exploration Geochemistry—the Basin and Range Province of the Western United States and Northern Mexico, *J. Geochem. Explor.*, **9**, 113–276.

Lovering, T.S., Huff, L.C. and Almond, H. (1950). Dispersion of copper from the San Manuel copper deposit, Pinal County, Arizona, *Econ. Geol.*, **45**, 493–514.

Mann, A.W. and Deutscher, R.L. (1977). Solution geochemistry of copper in water containing carbonate, sulphate and chloride ions, *Chem. Geol.*, **19(4)**, 253–266.

McConnell, J.W. (1983). Regional lake sediment and water geochemistry as a guide to bedrock composition and mineralization in Labrador. In: *Prospecting in Areas of Glaciated Terrain—1982* (Ed. P.H. Davenport), pp. 132–159; Can. Inst. Min. Metall.

Nichol, I. (1974). The challenge to exploration geochemists in a changing climate of resource development, *J. Geochem. Explor.*, **3**, 389–397.

Nuutilainen, J. and Peuraniemi, V. (1977). Application of humus analysis to geochemical prospecting: some case histories. In: *Prospecting in Areas of Glaciated Terrain* (Ed. G.R. Davis), pp. 1–15; Institution of Mining and Metallurgy, London.

Olade, M.A. and Fletcher, W.K. (1974). Potassium chlorate-hydrochloric acid: a sulphide-selective leach for bedrock geochemistry, *J. Geochem. Explor.*, **3**, 337–344.

Rose, A.W. (1982). Mineral adequacy, exploration success, and mineral policy in the United States, *J. Geochem. Explor.*, **16**, 163–182.

Smee, B.W. (1983). Laboratory and field evidence in support of electrogeochemically enhanced ionic diffusion through glaciolacustrine sediment, *PhD Thesis, University of New Brunswick, Canada* (unpubl.).

Smith, R.E., Frater, K.M. and Moeskops, P.G. (1980). Golden Grove Cu–Zn deposit, Yilgarn Block, W.A., *J. Geochem. Explor.*, **12**, 195–198.

Szabo, N.L., Govett, G.J.S. and Lajtai, E.Z. (1975). Dispersion trends of elements and indicator pebbles in glacial till around Mt. Pleasant, New Brunswick, *Can. J. Earth Science*, **12**, 1534–1556.

Theobald, P.K. (1981). One man's view of the research gap, *J. Geochem. Explor.*, **15**, 1–20.

Theobald, P.K. and Allcott, G.M. (1973). Tungsten anomalies in the Uyaijah Ring Structure, Kushaymiyah igneous complex, Kingdom of Saudi Arabia; Ministry of Petroleum and Natural Resources, Jeddah, Saudi Arabia.

Thomson, I. (1982). Accomplishments in exploration geochemistry: a Canadian perspective, *Canadian Mining Journal*, **April**, 78–81.

Ward, F.N. and Bondar, W.F. (1979). Analytical methodology in the search for metallic ores. In: *Geophysics and Geochemistry in the Search for Metallic Ores* (Ed. P.J. Hood), pp. 365–383; Geol. Surv. Can., Econ. Geol. Rep., **31**.

Webb, J.S. (1953). A review of American progress in geochemical prospecting, *Inst. Min. Metall. Trans.*, **62**, 321–348.

Webb, J.S. and Millman, A.P. (1950). Heavy metals in natural waters as a guide to ore; a preliminary investigation in West Africa, *Inst. Min. Metall. Trans.*, **59**, 323–336.

Webb, J.S. and Millman, A.P. (1951). Heavy metals in vegetation as a guide to ore, *Inst. Min. Metall. Trans.*, **60**, 473–504.

Webb, J.S., Fortesque, J.A.C., Nichol, I. and Tooms, J.S. (1964a). Regional geochemical reconnaissance in the Namwala Concession Area, Zambia, *Tech. Comm. Geochem. Prospect. Res. Centre*, **47**.

Webb, J.S., Fortesque, J.A.C., Nichol, I. and Tooms, J.S. (1964b). Regional geochemical maps of the Namwala Concession Area, Zambia, Nos. 1–8, *Lusaka Geological Survey*, Lusaka.

Webb, J.S., Lowenstein, P.L., Howarth, R.J., Nichol, I. and Foster, R. (1973). Provisional geochemical atlas of Northern Ireland, *Applied Geochemical Research Group*, Imperial College, Tech. Comm., 60.

Webb, J.S., Thornton, I., Thompson, M., Howarth, R.J. and Lowenstein, P.L. (1978). *The Wolfson Geochemical Atlas of England and Wales;* Oxford University Press, Oxford.

A Perspective on the Application of Geochemistry in Mineral Exploration South of the Equator

R. H. MAZZUCCHELLI

Introduction

Mineral exploration environments in the Southern Hemisphere have provided a number of scenarios which were influential in the early development of geochemical exploration technology. Geochemical methods have played an important and successful role in subsequent exploration for mineral resources. This review attempts to describe what might be regarded as typical applications of geochemistry in mineral exploration and some which may have features unique to the Southern Hemisphere. A brief assessment is made of the past achievements and current status of geochemical exploration in the Southern Hemisphere and some predictions are made as to future developments in this area. Examples are drawn from the literature covering all of the main environments of geochemical exploration but, in view of the writer's main area of experience, this account will inevitably be seen from an Australian perspective.

The Antipodean exploration environment

The environment of exploration geochemistry south of the Equator includes three major continental land masses—Australia, southern Africa and South America—and a multitude of oceanic islands. This review will necessarily exclude Antarctica. In terms of climatic environments, the areas to be dealt with are dominated by those with tropical and sub-tropical conditions, with moisture regimes ranging from humid through semi-arid to very arid conditions. Although subordinate in the context of total area, alpine and temperate to cool-temperate humid climates are also represented to a significant extent in the area under consideration.

Geologically, ancient terrains are of great importance, both in terms of areal extent and mineral endowment. These include the Archaean Shield areas of Australia, southern Africa and South America, and the flanking areas of Proterozoic rocks, consisting of platform and geosynclinal sediments and

Table 1 **Important styles of mineralization in the Southern Hemisphere**

Geological environment	Deposit type	Examples
Archaean shield	Vein and lode gold	Barberton (South Africa); Kalgoorlie, Norseman, (Australia); Araci (Brazil)
	Nickel sulphides	Trojan, Shangani, (Zimbabwe); Kambalda, Agnew, (Australia); Pikwe-Selebi (Botswana); Americano do Brasil (Brazil)
	Volcanogenic copper-zinc massive sulphides	Golden Grove, Teutonic Bore (Australia); Palmeiropolis (Brazil); Prieska (South Africa)
	Chromite	Selukwe (Zimbabwe)
Proterozoic platform sediments	Stratiform base metals and related deposits	Copper Belt (Zaire/Zambia); Tsumeb, Otjihase (Namibia); Mt. Isa, McArthur R., Broken Hill, Roxby Downs (Australia); Salobo, Vazante, Irecê (Brazil)
	Stratabound gold	Witwatersrand (South Africa); Jacobina, Passagem (Brazil); Telfer (Australia)
	Vein and disseminated uranium	Ranger, Jabiluka (Australia); Rossing (Namibia)
Proterozoic intrusive complexes	Magmatic segregation deposits of chromite platinum group metals and vanadium	Bushveld (South Africa); Great Dyke (Zimbabwe)
Phanerozoic fold belts	Volcanogenic copper-lead-zinc massive sulphides	Mt. Lyell, Rosebery, Que River, Woodlawn (Australia)
	Vein and placer gold deposits	Coromandel, Otago (New Zealand); Vatukoula (Fiji); Ballarat, Bendigo (Australia)
	Porphyry copper-gold--molybdenum deposits and related veins and skarns	Chuquicamata, El Teniente (Chile); Cerro de Pasco (Peru); Panguna, Ok Tedi (Papua New Guinea)
Phanerozoic platform sediments	Stratabound and diapiric base metal deposits	Camaqua (Brazil); Sorby Hills, Cobar, Elura (Australia)
	Uranium in terrestrial sediments	Yeelirrie, Beverley (Australia); Karroo Supergroup (South Africa)
Various	Tin-tungsten-tantalum veins, pegmatites, lodes and related placers	Rondonia, Amazonia (Brazil); Oruro, Potosi (Bolivia); Kibara (Namibia); Billiton, Bangka (Indonesia); Renison, King Is., Greenbushes (Australia)

related mobile belts. Younger sedimentary basins and fold belts also occupy significant areas within the three continental land masses, while Cainozoic fold belts, usually with a significant volcanic component, are important in the oceanic islands and South American Cordillera.

The geomorphology of the areas considered is, as would be expected, exceedingly diverse. Perhaps the most notable common feature of the continental areas is the presence of extensive plateaux of low relief. In the case of central Africa and Brazil, these areas are characterized by deep lateritic soil profiles, whereas in Australia and southwestern Africa, arid to semi-arid climatic conditions predominate and surface cover is more variable.

From the viewpoint of the exploration geochemist, much of the terrain lends itself to the classical geochemical search technology of stream sediment sampling for reconnaissance and soil sampling for follow-up and detailed exploration as pioneered in Zambia by Webb and his co-workers in the 1950s. However, problems are encountered in areas characterized by a deep and complex regolith, which may include the superimposed remnants of multiple past climatic cycles, areas of wind-blown sand and veneers of variable thickness of fluviatile and lacustrine sediments.

The most important types of mineralization represented in the Southern Hemisphere are listed in Table 1. These to a large extent have determined the objectives and hence the models for geochemical exploration programmes in the Southern Hemisphere.

Typical applications of geochemistry in mineral exploration in the Southern Hemisphere

Geochemical exploration techniques which might be regarded as typical applications of the technology in the Southern Hemisphere may be described by selecting examples from a list of successful case histories (Table 2). The list, which does not purport to be all-inclusive, nevertheless gives an impressive array of mineral resources, most of which have already come into production or are in the advanced stages of development. The environments in which these discoveries were made range in climate from humid to arid, tropical to temperate and in terrain from rugged to subdued. It is nevertheless notable that the principal metals in each case come from the group Ni, Cu, Pb, Zn and all were found by combinations of stream sediment and soil sampling programmes.

The Americano do Brasil deposit was found in 1973 as a result of soil sampling (30 cm depth, –80 + 120 mesh fraction) on a 50 × 50 m grid over a differentiated complex of mafic-ultramafic rocks. Four coincident Ni–Cu anomalies were delineated and tested by trenching, in each case revealing oxidized Ni–Cu mineralization, assaying up to 4.9% Ni and 9.6% Cu (Ribeiro *et al.*, 1979).

The huge Salobo 3A copper deposit is just one of a series of recent discoveries made by the Brazilian exploration company Docegeo, who have persevered since 1971 with a very ambitious geochemical programme in the

Table 2 **Some discoveries resulting from exploration programmes with significant geochemical contribution**

Deposit	Location	Orebody parameters	Geochemical technique	Source
Americano do Brasil	Goiás, Brazil	1 Mt. 1.5% Ni, 1% Cu	Soils	Ribeiro *et al.*, 1979
Salobo 3A	Pará, Brazil	1.1. Bt. 0.8% Cu (+Au, Mo)	Stream Sed., Soils	Meyer and Gonçalez, 1982
Palmeiropolis	Goiás, Brazil	10 Mt. 0.8% Cu, 0.9% Pb, 3.6% Zn	Stream Sed., Soils	de Andrade and Araujo, 1982
Fitula	Zambia	4.5 Mt. 5.3% Cu	Soils	Cornwall, 1969
Otjihase	Namibia	13 Mt. 2.5% cu (+Zn, Ag, Au)	Soils	Scott, 1975
Panguna	Bougainville, PNG	900 Mt. 0.5% Cu, 0.5 ppm Au	Stream Sed., Soils	Baumer and Fraser, 1975
Lady Loretta	Queensland, Australia	8.7 Mt. 6.7% Pb, 18.1% Zn, 110 ppm Ag	Soils	Cox and Curtis, 1977
Woodlawn	NSW, Australia	6.3 Mt. 1.7% Cu, 5.5% Pb, 14.4% Zn	Soils	Ryall and Nicholas, 1979
Que River	Tasmania, Australia	6 Mt. 13% (Pb + Zn)	Stream Sed., Soils	Webster and Skey, 1979
Foster Shoot	Western Australia	> 100 000 t contained Ni metal	Soils	Mazzucchelli, 1979
Shangani	Zimbabwe	16 Mt. 0.9% Ni	Soils	Philpott, 1975
Kalengwa	Zambia	0.6 Mt. 16% Cu	Stream Sed., Soils	Ellis and McGregor, 1967
Pikwe-Selebi	Botswana	46 Mt. 1.2% Ni, 1.2% Cu	Soils	Gordon, 1973

Amazon region of Brazil, surely one of the most daunting of exploration environments in the world. Because of access difficulties and the considerable thickness of lateritic soils, costs of exploration are high and a multi-media approach is used to maximize effectiveness (Meyer and Gonçalez, 1982). The basic programme involves sampling of stream sediments at a density of 1 sample per 5 km^2, a −80 mesh fraction and a panned concentrate sample being taken at each site. Because of the dense vegetation, surveyed lines have to be cut to provide access between drainage channels. Soil samples are collected from a depth of 50 cm at 200 m intervals along all cut lines during the reconnaissance stage of exploration. Experience has shown that near-surface soil samples reliably reflect bedrock features even though soil profiles are up to 115 m thick and average 20 m. Any outcrops encountered, usually in incized drainage channels, are sampled. The −80 mesh fraction stream sediment and soil samples, and crushed rocks, are analysed for Cu, Pb, Zn, Ni, Co, Cr, Fe, Mn, Ag and Au using a hot concentrated acid leach and determination by atomic absorption spectrophotometry (AAS). The heavy mineral concentrates have in the past been assessed visually for detrital minerals, but are increasingly being subjected to chemical analysis as well.

Follow-up sampling in areas of interest consists of stream sediment sampling at 500 m intervals along all drainage channels and soil sampling on a regular grid pattern with progressively decreasing spacings between samples. The Salobo area was initially selected as an area of interest because of its proximity to the Carajas iron ore deposits and the presence of well-defined aeromagnetic anomalies in the area. Stream sediment sampling at 500 m intervals was therefore carried out from the outset, one sample from a very short drainage tributary yielding a result of 2700 ppm Cu compared with a threshold of 15 ppm. At the same time reconnaissance soil samples and rock samples containing 1600 ppm Cu and 3500 ppm Cu respectively were encountered.

Delineation drilling has so far proved reserves of 1.1 billion tonnes of 0.83% Cu, which include 400 m at 1.3% Cu with economically significant amounts of gold and molybdenum (Lindenmeyer, pers. comm., 1982). The host stratigraphy for Salobo 3A is iron-rich sediments and volcanics. At least two other significant Cu deposits have been discovered in the Carajas region using soil and stream sediment geochemistry but delineation has not yet proceeded sufficiently far to determine whether these bodies are also economic. Over a five-year period Docegeo collected a total of 33 000 geochemical samples of which 68% consisted of soils, 13% rocks and 19% stream sediments as part of this project.

The discovery of the Palmeiropolis volcanogenic base metal deposits by the government exploration company CPRM follows the classical geochemical prospecting sequence (de Andrade and Araujo, 1982). An area of airborne magnetic response over a poorly defined volcano-sedimentary sequence of Lower Proterozoic age was singled out for reconnaissance exploration by means of stream sediment geochemistry. The results were subjected to moving average smoothing, which delineated a NNW-trending belt of volcanic rocks,

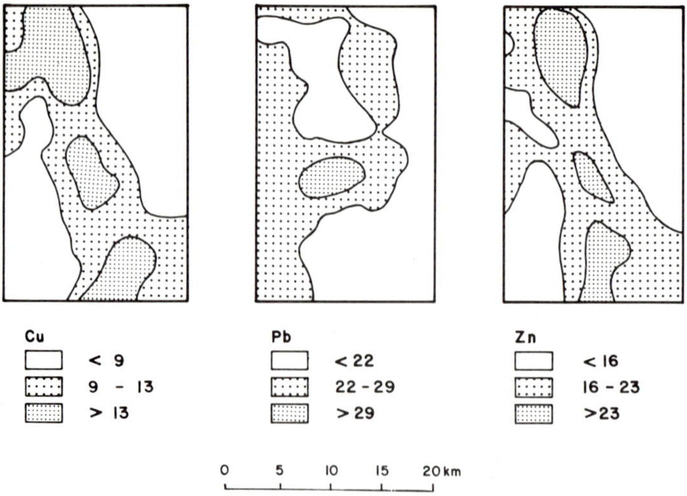

Figure 1 Moving average smoothed contours for Cu, Pb and Zn (ppm) in stream sediments, Palmeiropolis (after de Andrade and Araujo, 1982)

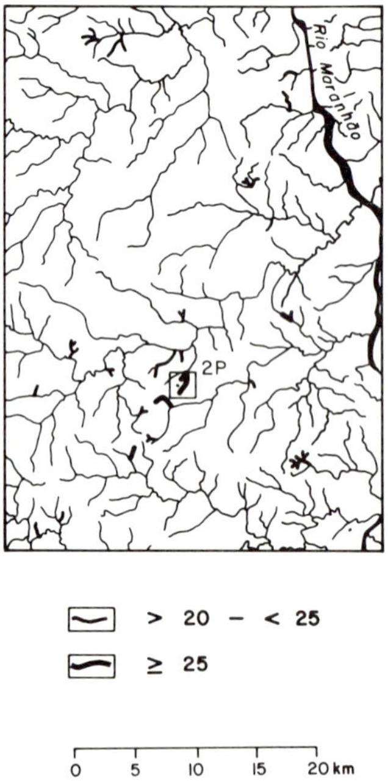

Figure 2 Distribution of Pb (ppm) in stream sediments, Palmeiropolis (after de Andrade and Araujo, 1982)

within which three nuclei stood out (Figure 1). Soil samples were collected at 60 cm depth on a 100 × 40 m grid over the catchment of a discrete anomaly within the central nucleus which showed out both in terms of absolute metal abundance and as a residual (Figure 2). The −40+80 mesh fraction of the soil samples was assayed by AAS for Cu, Pb, Zn following a hot concentrated HNO_3 acid attack. The results clearly defined a NNE-trending anomaly for all

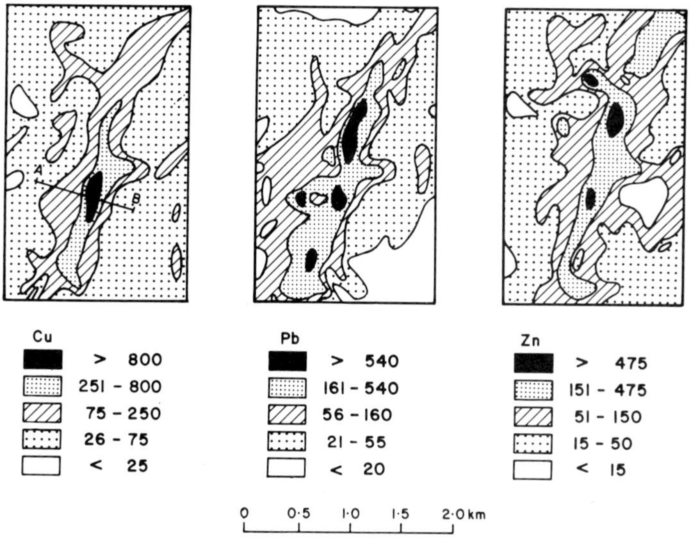

Figure 3 Distribution of Cu, Pb and Zn (ppm) in soils of Target 2P, Palmeiropolis (after de Andrade and Araujo, 1982)

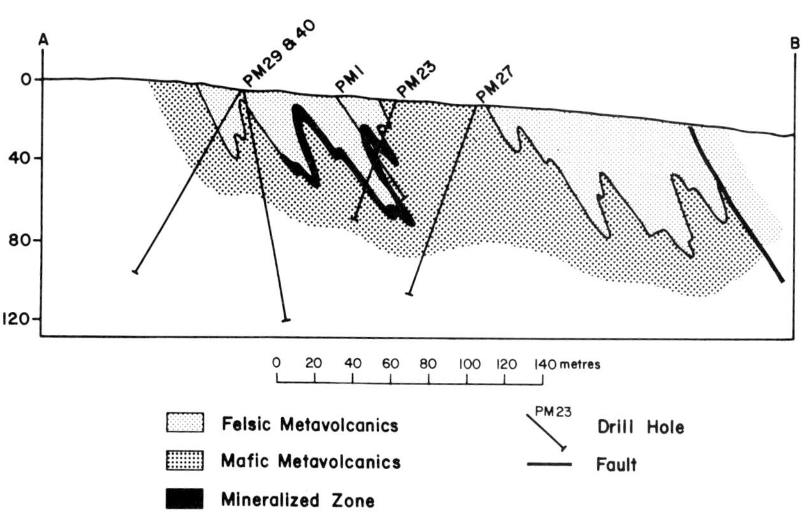

Figure 4 Schematic geological section A–B, Palmeiropolis (by courtesy, CPRM)

three elements (Figure 3). Pits over the anomaly peaks produced values up to 2.5% Cu, 42% Pb and 2.2% Zn in weathered rock. A subsequent drilling programme defined a significant tonneage of massive and disseminated base metal sulphides in a classical volcanogenic configuration (Figure 4).

The predictive potential of these techniques is well illustrated by the quantitative relationships between anomalies in soils and stream sediments established by McKenzie (1977) in the exploration carried out for porphyry copper deposits by CRA in the Western Pacific. Results quoted from exploration case histories at Bougainville and Central New Britain show that the copper concentration in stream sediments is a direct reflection of the average content of Cu in soils in the catchment (Figure 5). A similar study by Hawkes (1976) has led to a formula relating the metal content of an anomalous stream sediment sample with the grade and surface area of mineralization relative to that of the catchment in which it occurs. The understanding of such relationships has led to a more confident approach in the planning and interpretation of low density drainage reconnaissance surveys.

The case histories described above all occur in seasonal to humid tropical environments with moderate to strong relief. The same basic techniques have proved equally effective in exploration for base metals under arid conditions as demonstrated by case histories from SW Africa and Australia.

Scott (1975) reported on the role of soil geochemistry in the discovery of the Otjihase Cu–Zn deposit in magnetite-bearing quartzites of the Precambrian Khomas Series, some 15 km NE of Windhoek. The climate is semi-arid to arid and soils are relatively thin residual calcareous lithosols but generally incorporate a variable proportion of wind-blown material. Initial interest in the Otjihase area was generated by the discovery by a prospector of gossans

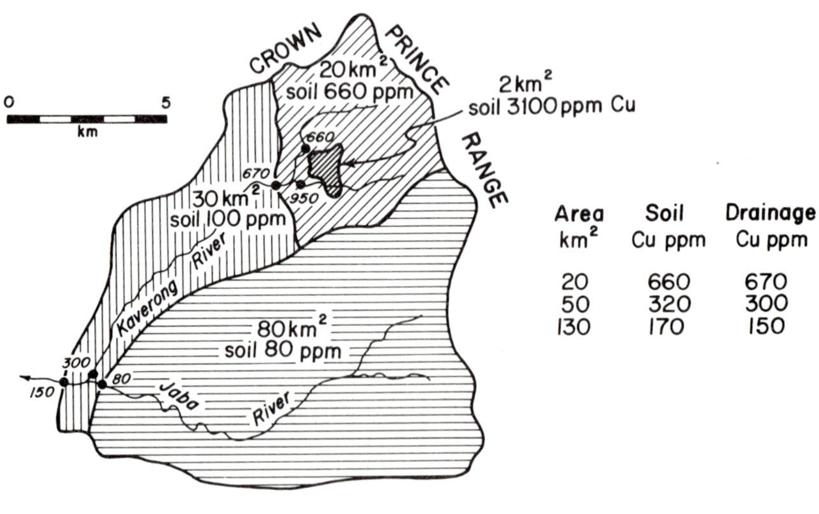

Figure 5 Catchment area soil and stream sediment relationships for copper at Panguna, PNG (after McKenzie, 1977)

with significant Cu and Zn values. Since outcrop constituted no more than 15%, detailed soil sampling on a 50 × 20 m grid was employed to delineate the subcrop of mineralization prior to drilling. Samples were collected at 15 cm depth, just below the 'A' and usually above the 'C' horizon. The −80 mesh fraction was routinely analysed for total Cu and Zn. Comparative results for Cu in three size fractions of soil samples collected on a test line show that the tenor of Cu in the −140 mesh fractions exceeds the mean Cu content of the oxide zone of the orebody which itself is severely depleted with respect to the primary ore grade. The detailed distribution of Cu in bedrock was determined by the sampling of 25 m-spaced vertical percussion drill holes at intervals of 50 cm, and deeper diamond drill holes intersecting primary mineralization. Although primary dispersion related to the mineralization is very restricted, a broad halo is developed in the oxidized zone, almost certainly due to mobilization of Cu within the zone of oxidizing sulphides and its subsequent migration and precipitation in the alkaline wallrocks. The peak of the soil anomaly appears to be closely related to the subcrop of the ore zone but the lateral extent of soil anomaly is no doubt enhanced by dispersion from the broad anomaly in oxidized bedrock. The Cu distribution in plan accurately delineated the subcrop of the ore horizon, assisting in the delineation drilling of the deposit.

The discovery of the Lady Loretta deposit in NW Queensland, Australia, is a familiar story. A programme of soil sampling on a 120 × 30 m grid was carried out on leases surrounding the Lady Annie Copper Mine, primarily in the hope of locating a rich copper deposit (Cox and Curtis, 1977). The area has a semi-arid tropical climate and consists of an eroded and dissected plateau with generally low relief 300–400 m above sea level. The soils are shallow lithosols, generally less than 15 cm thick. Orientation studies indicated a significant proportion of the soil was made up of coarse (+12 mesh) material which contained a large proportion of the Pb and Ag, usually in the form of rock and gossan fragments. Samples were taken immediately below the surface and the whole soil sample was crushed for routine analysis. All samples were analysed for Cu but the visual similarity of the rock succession to the host rocks for the Mt. Isa Pb–Zn deposits prompted the analysis of every tenth sample for Pb, Ag and Zn, i.e. effectively a 300 × 120 m grid. As a result of this work it was established that Pb and Ag provide a reliable guide to the subcrop of the ore deposit whereas the Zn anomaly shows significant downslope displacement.

The discovery of the nickel-sulphide deposits at Kambalda in 1966 set in train a very intensive phase of geochemical exploration over the ultramafic rocks occurring in the Archaean Greenstone Belts of Western Australia. Orientation studies at Kambalda showed that mineralization is indicated by very discrete coincident nickel-copper anomalies, within a broader zone of elevated Ni, related to the ultramafic host rocks (Mazzucchelli, 1972). The data acquired from the Kambalda investigation were applied in conjunction with geological and geophysical exploration methods in programmes which were successful in locating other deposits of nickel-sulphide mineralization.

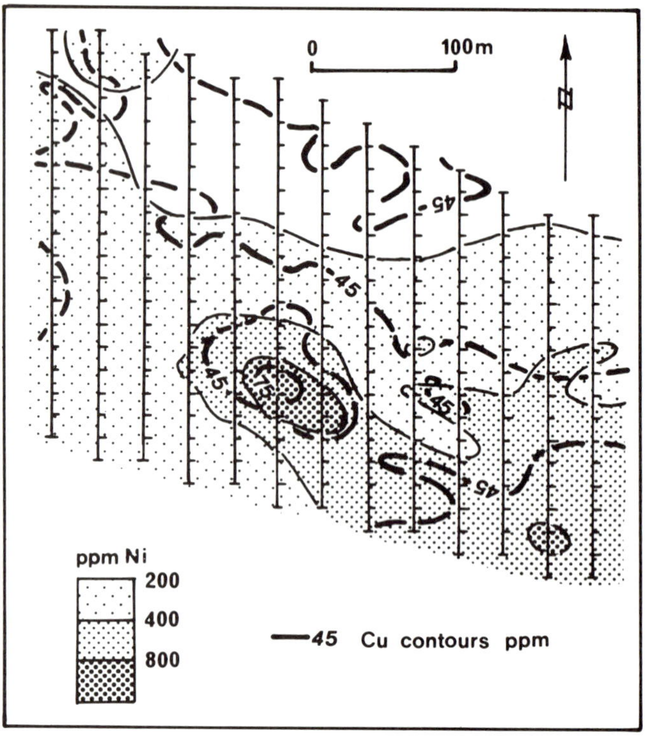

Figure 6 Soil geochemistry in Foster Shoot area, St. Ives (after Mazzucchelli, 1979)

One discovery in completely soil-covered terrain which owed more than most to geochemical techniques was the Foster Shoot at St. Ives (Mazzucchelli, 1979). The geochemistry of this occurrence is shown in plan and section in Figures 6 and 7.

Drainage reconnaissance techniques are widely employed in base metal exploration in the humid sub-tropical to temperate marine environments in South America and southeastern Australia. This is illustrated by the case history of the Que River base metal deposit in NW Tasmania, the discovery of which is attributed to follow-up of a coincident airborne EM and stream sediment geochemical anomaly (Webster and Skey, 1979).

The discovery of the Woodlawn base metal deposit in NSW is attributed to conventional soil geochemistry (Ryall and Nicholas, 1979). Samples of rock, stream sediment and 'B' horizon soils were collected along road verges in farmlands to the north of the Captains Flat lead-zinc deposit and thought to contain similar geology. Of a total of 240 soil samples collected within an area of 980 km^2 around Woodlawn, the presence of mineralization signified by a group of samples averaging 200 ppm Cu, 800 ppm Pb and 300 ppm Zn from roadside sites approximately 500 m north of the subsequently identified ore zone. Later grid sampling has documented the geochemical expression of mineralization in soils associated with the deposit which contains proven

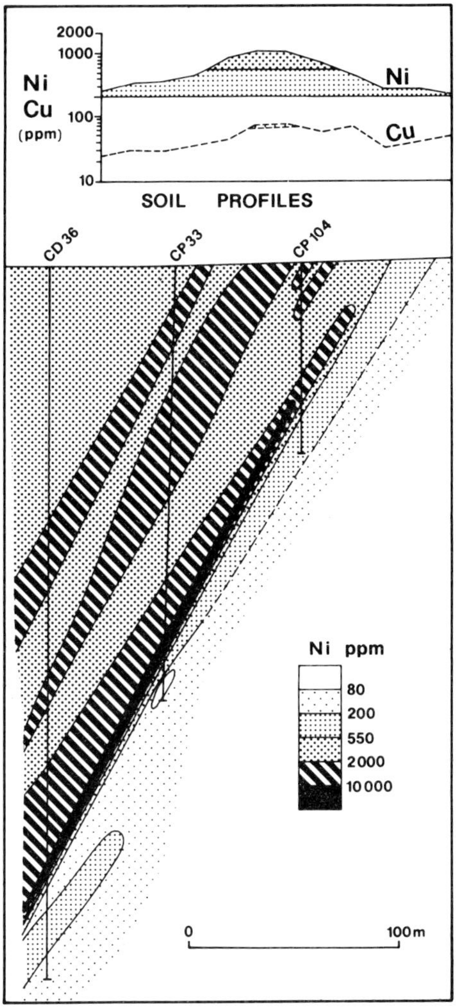

Figure 7 Geochemical cross section and soil profiles, Foster Shoot, St.Ives (after Mazzucchelli, 1979)

reserves of 6.3 million tonnes of Cu–Pb–Zn ore (1.7% Cu; 5.5% Pb; 14.4% Zn) and 3.7 million tonnes of Cu ore (1.9% Cu). The mineralization, which occurs within felsic tuffs and tuffaceous sediments of Silurian age, is most closely reflected in the results for Cu and Pb although anomalous values for all elements extend well beyond the limits of ore. Ryall and Nicholas (1979) attribute the extensive dispersion to a primary halo of mineralization in the footwall and hangingwall of the main deposit, combined with extensive hydromorphic dispersion. In the vicinity of mineralization, Sn, Ba, Sb, Bi, Se and Cd are also anomalous.

Specific Southern Hemisphere developments in geochemical exploration

It is pertinent to examine some of the directions taken by geochemical exploration in response to particular conditions which have been encountered in Southern Hemisphere environments. One such area of difficulty arose from the extension of the nickel search into areas of Western Australia mantled by remnants of deep lateritic weathering episodes dating from the Tertiary and perhaps Mesozoic. The geochemical expression of mineralization in such areas may be severely suppressed by extreme leaching or rendered indistinguishable from spurious concentrations of elements, depending on which horizon of the old lateritic weathering profile forms the parent material for contemporary soil development. The extent of this problem in Australia can be judged from Figure 8 in Butt and Smith (1980). Similar problems have been reported in parts of the Brazilian interior. The 'safe' solution to the problem of geochemical sampling in both lateritic and aeolian terrains is to use a light drilling rig to obtain geochemical samples from the weathered bedrock sometimes at depths as great as 100 m. A typical programme was that used to explore the nickel-sulphide potential of the Windarra area in a deeply weathered terrain with extensive concealment by wind-blown sand NE of Kalgoorlie. Holes were drilled at spacings of 60 × 6 m, reducing to 30 × 3 m in areas of interest (Smith *et al.*, 1980).

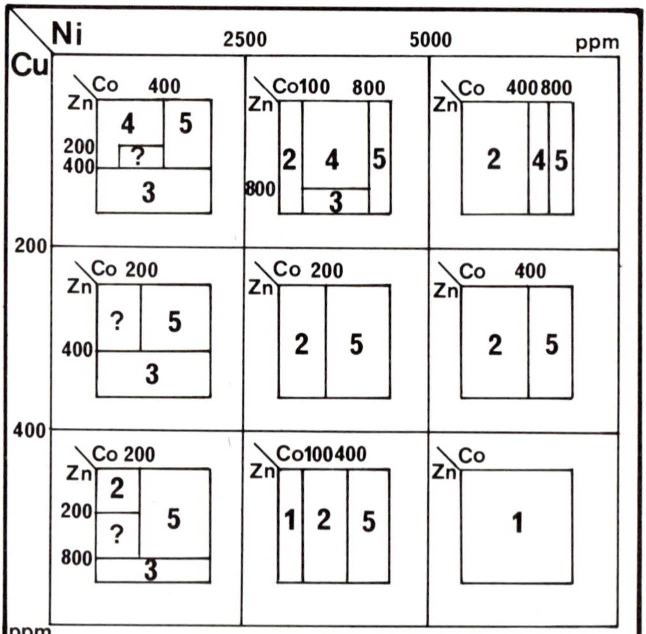

Figure 8 An idealization of the discriminant function for the evaluation of Ni gossans (after Mazzucchelli, 1979)

The results in an area where subeconomic nickel sulphides are concealed by a complete laterite profile (Smith, 1977) indicate that although weathering extends to a maximum depth of 30 m, some subdued expression of the nickel-sulphide mineralization, as indicated by the copper and nickel values, can be seen within the ferruginous zones of the laterite profile at depths less than 5 m below surface. However, once the need for a drill to obtain geochemical samples is accepted it is usual to drill holes to weathered or even fresh bedrock where possible, to obtain geological information, even though drill hole geochemical surveys are many times more expensive than those involving the collection of near-surface soils.

The existence of gossans related to Ni mineralization and the difficulties in recognizing them in a deeply weathered, lateritic environment was another challenge provided by the Ni search in WA. It will be appreciated that most previous Ni exploration had been conducted in recently glaciated terrains where fresh sulphides were frequently encountered at the surface or immediately underlying glacial deposits.

Many geochemical techniques for discrimination of gossans related to nickel sulphides were proposed, some advocating multi-variate statistical techniques based on the analysis of base metals such as Ni, Cu, Co, Cr, Zn and Mn (Clema and Stevens-Hoare, 1973; Bull and Mazzucchelli, 1975) and others proposing the use of special pathfinder elements such as Te (Bull and Mazzucchelli, 1975) and platinum group metals (Travis *et al.*, 1976). A graphical depiction of the discriminant function derived by Bull and Mazzucchelli (1975) is shown in Figure 8.

An extension of both the gossan and soil investigations has been to focus attention increasingly on the coarse fraction of soils and surface lag deposits. It had been a matter of observation that many metals associated with mineralization in arid region soils are concentrated in the coarse fractions, with secondary maxima in the silt and clay fractions, the intermediate sand sized fractions usually being dominated by quartz with low metal content (Figures 9 and 10). This is true of both lateritic and non-lateritic soils. In laterized environments it is commonly found that trace elements are preferentially concentrated in the sesquioxide phase relative to the more clay-rich and siliceous materials (Mazzucchelli and James, 1966). In non-lateritic terrain, the weathering products of sulphide-associated mineralization also tend to be dominated by fragments of iron or silica-rich gossan (Mazzucchelli, 1972). Moreover, under the prevailing arid conditions the coarse ferruginous and siliceous materials, lateritic and otherwise, tend to form a stable lag deposit at the surface. Selective sampling and analysis of such ironstones has yielded positive results in both nickel exploration, where results for Ni, Cu, Co, Cr can be evaluated by the same discriminant function used in gossan evaluation (Bull and Mazzucchelli, 1975), and in gold exploration using As as a pathfinder element. A schematic illustration of how such anomalies might form is shown in Figure 11.

One area in which ironstone sampling appears likely to play a crucial role is in exploration for volcanogenic copper-zinc mineralization in the Archaean

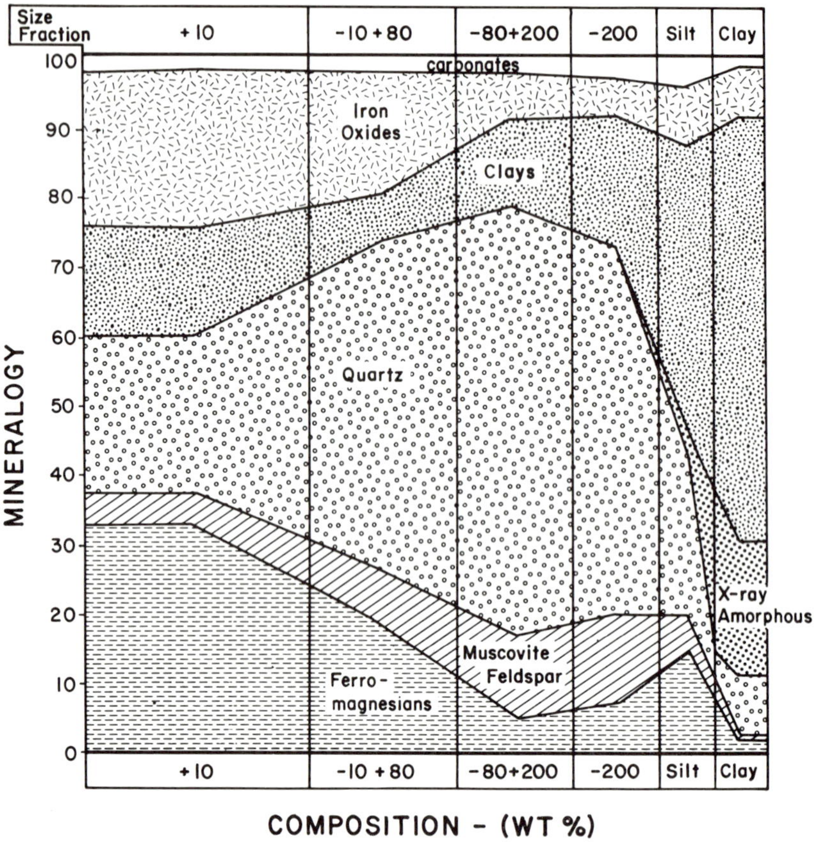

Figure 9 Mechanical and mineralogical composition of 'average' soil over 'greenstone' bedrock, Archaean Yilgarn Block (after Mazzucchelli, 1978)

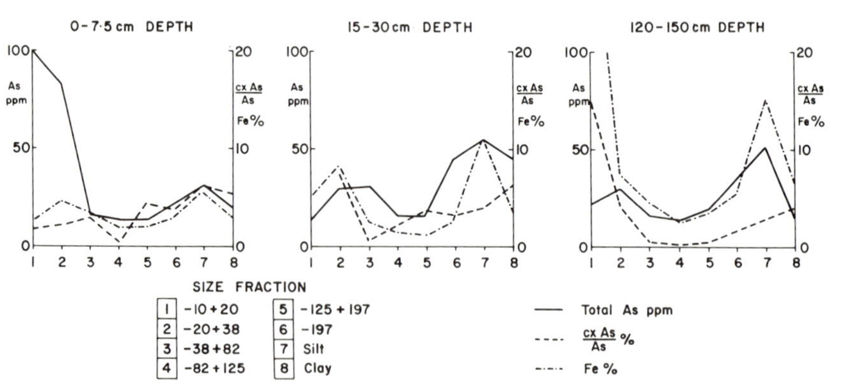

Figure 10 Distribution of As, cx As/As and Fe according to particle size at different depths in an anomalous soil profile, Norseman (after Mazzucchelli, 1965)

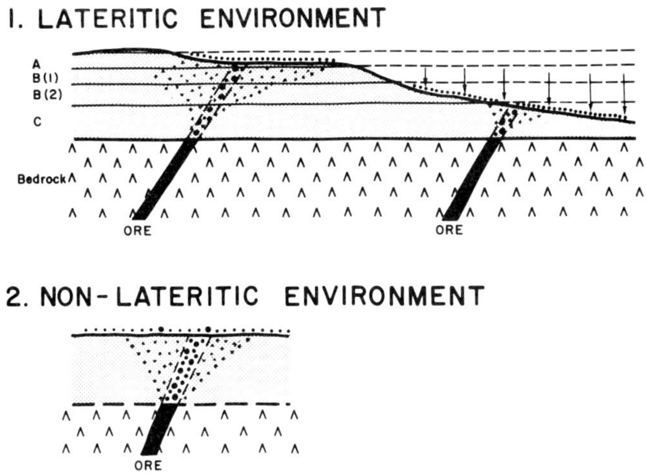

Figure 11 Schematic sketch for the development of ironstone anomalies

Yilgarn Block of Western Australia. Such ores appear to have been subjected to extreme leaching of base metals during weathering such that the geochemistry of gossans, even where strongly developed, and related soils do not always reflect the potential of the underlying sulphides. Orientation studies reported by Smith and Perdrix (1983) demonstrate that strong anomalies for pathfinder elements associated with the massive volcanogenic base metal sulphide mineralization at Golden Grove are present in pisolitic ironstone which forms a blanket 1–3 m thick throughout the area. The authors show that the geochemical expression of mineralization can be optimized by plotting a function which combines the values for base and pathfinder elements in a subjective 'chalcophile index'.

Aeolian sands present severe problems to exploration geochemistry in semi-arid to arid areas of both southern Africa and Australia. One mechanism whereby bedrock materials may be made available at the surface for sampling in such terrain is through the activities of termites. Termites are ubiquitous in the sandy desert terrains of Africa and Australia, erecting conspicuous earth mounds which may reach heights of 3 m or more above the land surface. Below the surface a network of tunnels is excavated presumably to provide access to groundwater. The spoil from the excavation of the tunnels is, in part at least, incorporated in the superstructure of the termite mound and often includes rock fragments up to 2 mm in diameter. A spectacular demonstration of the potential of termite-generated anomalies is reported by West (1970) who claims to have discovered an economic gold deposit in Zimbabwe through some 6 m of Kalahari sand by drilling a target based on the gold content of termitaries. Termite activity was subsequently observed in underground workings at depths of 19 m below surfaces giving an indication of the depths from which surface material may be derived by the agency termites. D'Orey (1975) obtained good indications of the presence of copper mineralization concealed by up to 15 m of colluvium in Mozambique, by sampling and analysis of termitaries. Conflicting results have been reported by other workers (Tooms and Webb, 1961 and Watson, 1970) whose results suggested that most of the material incorporated in termitaries is derived from within 2 m of the surface. Even so there is good evidence that termite activity effects sufficient vertical redistribution of weathered materials to be of benefit in aeolian terrains, especially if the expression of bedrock geochemistry can be enhanced by selective sampling of the constituents of termitaries.

The effectiveness of geochemical exploration techniques is not only measured in terms of successful mineral discoveries but also the cost-effective elimination of non-prospective ground during exploration programmes. An example which illustrates this comes from the exploration programme carried out for calcrete uranium deposits on the Yilgarn block. A number of deposits including the major Yeelirrie occurrence (46 000 tonnes of contained U_3O_8) were located by direct follow up of the airborne radiometric anomalies. However as the mineralization was associated with the low-lying drainage system within a region of generally very subdued relief there seemed a reasonable likelihood that some undiscovered occurrences could be concealed to radiometrics by superficial deposits. Hydrogeochemical orientation studies were carried out around Yeelirrie by collecting samples of groundwater from the unconfined aquifer in weathered rock and overburden exploited by the local pastoral industry. This demonstrated the existence of uranium anomalies in groundwater which reached maxima in the vicinity of the Yeelirrie calcrete-uranium occurrence. The survey was extended over a total area of some 600 000 km^2 at a sample density, which was determined largely by the distribution of bores, averaging one sample per 170 km^2. Drainage channels which contained known carnotite/calcrete mineralization were found to be generally characterized by groundwater containing more than 100 parts per billion uranium compared with background values of <5–15 ppb U (Cameron *et al.*, 1980). Although no new economic uranium occurrences were located by the hydrogeochemical programme, the technique afforded a systematic and effective coverage of a vast area at minimal cost.

There is increasing emphasis in the Southern Hemisphere on the use of heavy mineral concentrates as a sampling medium for drainage reconnaissance. It is notable that the use of heavy mineral concentrates was first advocated in the early 1800s for mineral exploration in Brazil (Overstreet, 1962). In a reconnaissance survey in Sumatra, Watters and Sagala (1979) made the observation that the metals of principal interest in geochemical exploration occur either as discrete heavy mineral particles or are associated with iron or manganese oxides which themselves behave as heavy detritals in dispersion processes. This appears to be true not just in the humid tropics but also in many arid environments. The collection of heavy mineral concentrates eliminates the variable dilution caused by clays and light silicate minerals, the resulting enhancement of contrast allowing the recognition of anomalies at much lower sample densities than is possible with conventional fine fraction sampling. At the same time problems of inadequate analytical sensitivity are overcome and sampling errors reduced. Reconnaissance of large areas using low density heavy mineral concentrate drainage surveys also has advantages in cost-effectiveness, the escalating costs of geochemical analysis becoming a major deterrent to surveys which involve large numbers of samples.

Status of geochemical exploration in the Southern Hemisphere

One way to measure the status achieved by geochemical exploration is to resort to statistics. However, reliable and comprehensive statistics are very difficult to come by. A survey carried out by the Association of Exploration Geochemists in the early 1970s, although incomplete, indicates that the use of geochemistry in mineral exploration in the Southern Hemisphere was at least as well entrenched as in the Northern Hemisphere (Table 3). At this time the Australian figures were boosted by exceptional soil sampling activity related to the intensive nickel exploration. A matter of some 3 years, Buhlmann *et al.* (1975) estimated a total of 1.3 million samples collected by 5 companies operating in Angola, Botswana, Mozambique, Rhodesia, South Africa and Namibia. Zambia, a country which relied on geochemical exploration methods to a possibly greater degree than any of those noted, was not included in either tabulation. Bruni (1982) estimated that 300 000 samples are currently being analysed in Brazil, for some 1 500 000 determinations, a notable increase on the figures reported in Table 3.

Perhaps more relevant are the indications of acceptance by individual companies. Again, reliable statistics are difficult to obtain. Docegeo, the largest and most successful exploration company in Brazil, claims to be spending about 90% of its exploration budget of currently about $US9m per annum, on projects based primarily on geochemistry (Lindenmeyer, pers. comm., 1982). In Australia, Western Mining Corporation, one of the largest

Table 3 **Geochemical samples collected: June 1970–May 1971**

Country	Total Samples	Type (%)			
		Rock	Soil	Stream Sed.	Other
Argentina	63 000	21	29	50	—
Australia	1 516 000	34	46	9	11
Brazil	21 000	15	80	5	—
Ecuador	17 000	20	12	68	—
New Zealand	11 000	20	40	40	—
South Africa	277 000	4	78	18	—
Total: S. Hemisphere	1 905 000	29	50	12	9
Total: N. Hemisphere	2 111 000	21	43	26	10

and most successful exploration companies, consistently spends some 15% of its budget on geochemical exploration, a proportion similar to that for geophysics.

Another measure of status, although again difficult to define, is to list the record of geochemical successes in exploration. Table 2 attempts to convey an indication of the mineral discoveries in the Southern Hemisphere to which geochemistry has made a significant contribution. There are doubtless others which could, and will in the future, be added. However, it should be emphasized that success is not only measured by discoveries made with geochemistry but also the ability to test an exploration concept effectively at low cost, and if negative results are obtained, to withdraw and re-allocate financial resources to other more promising projects.

There are naturally many areas in which improvements are required. The main problem relates to a lack of understanding of the scope and limitations of the technology, both by its proponents and by others in the organization in which it is used. This translates simply into the need for more research into the problems related to the application of geochemical methods in specific Southern Hemisphere environments and improved education at all levels, leading to better communication on matters of geochemical exploration.

Geochemistry, particularly in Australia, has been through a period of negative thinking which has resulted in excessive use of drilling to obtain geochemical samples. The high costs involved in drilling have inhibited the use of geochemistry in many projects and where it has been used, have reduced the area covered for the money spent. A more positive approach is now in evidence and current research is concentrating efforts on aspects of deep weathering profiles that can be exploited to advantage in geochemical exploration.

Future developments

A number of areas in which current work is proceeding have already been mentioned. The sampling of ironstones, the coarse fractions of soils and heavy mineral concentrates of stream sediments show considerable promise of countering some of the problems of Southern Hemisphere environments. More fundamental research on element behaviour and mobility during the weathering of different styles of mineralization is needed. Studies similar to those by Thornber (1975) at Kambalda should be extended to other deposits in order to increase our understanding of the factors which determine why, in some ore environments, base metals are leached during weathering, resulting in suppression of geochemical response, and in other cases enriched. Hydrogeochemical exploration would appear to have considerable potential in the exploration for base and precious metals in deeply weathered, peneplaned areas. In fact very little work has been done and the problems look formidable.

The investigation of primary dispersion patterns remains an area of neglect.

It seems highly probable that many leakage halos associated with viable ore deposits may have been drilled and dismissed as mineralization 'sub-economic' for want of a better understanding of primary zoning patterns.

Analytical advances have always exerted a profound influence on the course of exploration geochemistry. In recent years, the ability to obtain reliable gold analyses at ppb levels has enabled us for the first time to measure background for that element in geochemical samples. This has rejuvenated the use of geochemical exploration techniques for gold deposits and, given current activity levels, may lead to discoveries outside the narrow group of base metal successes recorded to date. One of the major challenges of the future will be to make effective use of the multi-element output of the ICP emission spectrograph. One effect of both of these examples is to force a re-examination of sampling techniques, to ensure that the quality of the sample matches the quality and cost of the analytical work to be expended on it.

Conclusion

Geochemical exploration methods have contributed to some notable mineral discoveries in the Southern Hemisphere and have achieved a satisfying level of acceptance by industry. Successes to date have been restricted largely to deposits of base metal sulphides and have generally occurred in areas amenable to the straightforward application of stream sediment and soil sampling. Although there is much room for improvement, considerable progress has been made in understanding the more difficult environments encountered south of the Equator and in developing appropriate geochemical exploration techniques. This gives reason for optimism that geochemical methods will play an increasingly important role in future mineral exploration in the Southern Hemisphere.

ACKNOWLEDGEMENTS

Western Mining Corporation Limited is thanked for making this contribution possible. Particular thanks are due to Mr T. W. Robbins for helpful comments on the manuscript, Mr. B. Dance for preparation of the illustrations and Ms S. M. Mead for typing the text. Contributions of unpublished data by Dr R. E. Smith and Mr C. N. Middleton are gratefully acknowledged.

References

Bruni, D.C. (1982). O status da exploração geoqúimica no ambito atividades minero-geológicas brasileiras, *Anais do Congr. Bras. Geol.*, **XXXII V5**, 1789–1795.

Baumer, A. and Fraser, R.B. (1975). Panguna porphyry copper deposit, Bougainville. In: *Economic Geology of Australia and Papua New Guinea* (Ed. C.L. Knight), pp. 855–865; Aust. Inst. Min. Metall., Melbourne.

Buhlmann, E., Philpott, D.E., Scott, M.J. and Sanders, R.N. (1975). The status of exploration geochemistry in southern Africa. In: *Geochemical Exploration 1974* (Eds. I.L. Elliott and W.K.

Fletcher), pp. 51–64; Elsevier, Amsterdam.

Bull, A.J. and Mazzucchelli, R.H. (1975). Application of discriminant analysis to the geochemical evaluation of gossans. In: *Geochemical Exploration 1974* (Eds. I.L. Elliott and W.K. Fletcher), pp. 219–228; Elsevier, Amsterdam.

Butt, C.R.M. and Smith, R.E. (Eds.) (1980). Conceptual models in Exploration Geochemistry—Australia, *J. Geochem. Explor.*, **12**, 89–367.

Cameron, E., Mazzucchelli, R.H. and Robbins, T.W. (1980). Yeelirrie calcrete uranium deposit, Murchison region, WA, *J. Geochem. Explor.*, **12**, 350–354.

Clema, J.M. and Stevens-Hoare, N.P. (1973). A method of distinguishing nickel gossans from other ironstones on the Yilgarn Shield, Western Australia, *J. Geochem. Explor.*, **2**, 393–402.

Cornwall, F.W.D. (1969). Discovery and exploration of the Fitula copper deposit, Nchanga area, Zambia, *9th Comm. Min. Metall. Cong. London Proc.*, **2**, 535–560.

Cox, R. and Curtis, R. (1977). The discovery of the Lady Loretta zinc-lead sulphide deposit, northwest Queensland, Australia—a geochemical exploration case history, *J. Geochem. Explor.*, **8**, 189–202.

de Andrade, R.S. and Araujo, E.S. (1982). Geoquímica de sedimentos de corrente e solos na região de Palmeirópolis—Go, *Anais Congr. Bras. Geol.*, **XXXII V5**, 1808–1820.

D'Orey, F.L.C. (1975). Contribution of termite mounds to locating hidden copper deposits, *Inst. Min. Metall. Trans.*, **84**, B150–B151.

Ellis, M.W. and McGregor, J.A. (1967). The Kalengwa copper deposit in northwestern Zambia, *Econ. Geol.*, **62**, 781–797.

Gordon, P.S.L. (1973). The Selebi-Pikwe nickel-copper deposits, Botswana, *Geol. Soc. Sth. Afr. Spec. Pub.*, **3**, 167–187.

Hawkes, H.E. (1976). The downstream dilution of stream sediment anomalies, *J. Geochem. Explor.*, **6**, 345–358.

Mazzucchelli, R.H. (1965). Geochemical dispersion related to gold mineralization, Western Australia, *PhD Thesis, University of London* (unpubl.).

Mazzucchelli, R.H. (1972). Secondary geochemical dispersion patterns associated with the nickel sulphide deposits at Kambalda, Western Australia, *J. Geochem. Explor.*, **1**, 103–116.

Mazzucchelli, R.H. (1978). Geochemical dispersion in arid regions. In: *Notes for Workshop Course 'Geochemical Exploration'*, pp. 18–38; Aust. Min. Foundation, Adelaide.

Mazzucchelli, R.H. (1979). Geochemical technology in mineral exploration. In: *Mineral Resources of Australia* (Eds. D.F. Kelsall and J.T. Woodcock), pp. 103–120; Aust. Acad. Technol. Sci., Melbourne.

Mazzucchelli, R.H. and James, C.H. (1967). Arsenic as a guide to gold mineralization in laterite-covered areas of Western Australia, *Trans. Inst. Min. Metall. (Sect. B) Applied Earth Sci.*, **75**, 285–294.

McKenzie, D.H. (1977). Empirical assessment of anomalies in tropical terrain, *Assoc. Explor. Geochem. Newsletter*, **21**, 6–10.

Meyer, D.J.K. and Gonçalez, R. (1982). A descoberta da jazida de cobre Salobo 3A através da geoquímica, *Anais do Cong. Braz. Geol.*, **XXXII V5**, 1795–1800.

Overstreet, W.C. (1962). A review of regional heavy-mineral reconnaissance and its application in the southeastern Piedmont, *Southeastern Geol.*, **3**, 133–172.

Philpott, D.E. (1975). Shangani—a geochemical discovery of a nickel-copper sulphide deposit. In: *Geochemical Exploration 1974* (Eds. I.L. Elliott and W.K. Fletcher), pp. 503–510; Elsevier, Amsterdam.

Ribeiro, M.J., dos Santos, M.M. and Ronan, S. (1979). Geochemical exploration over a mafic-ultramafic complex, Americano do Brasil, Goias, Brazil, *J. Geochem. Explor.*, **12**, 9–19.

Ryall, W.R. and Nicholas, T. (1979). Surface geochemical and biogeochemical expression of base metal mineralization at Woodlawn, New South Wales, Australia, *J. Geol. Soc. Aust.*, **26**, 187–195.

Scott, M.J. (1975). Case histories from a geochemical exploration programme—Windhoek District, South West Africa. In: *Geochemical Exploration 1974* (Eds. I.L. Elliott and W.K. Fletcher), pp. 481–492; Elsevier, Amsterdam.

Smith, B.H. (1977). Some aspects of the use of geochemistry in the search for nickel sulphides in lateritic terrain in Western Australia, *J. Geochem. Explor.*, **8**, 259–281.

Smith, B.H., Redell, C.T. and Cleghorn, J.H. (1980). Mt. Windarra Ni deposit, Yilgarn Block, W.A. *J. Geochem. Explor.*, **12**, 186–189.

Smith, R.E. and Perdrix, J.L. (1983). Pisolite laterite geochemistry in the Golden Grove massive sulphide district, Western Australia, *J. Geochem. Explor.*, **18**, 131–164.

Thornber, M.R. (1975). Supergene alteration of sulphides, I: A chemical model based on massive nickel sulphides deposits at Kambalda, Western Australia, *Chem. Geol.*, **15**, 1–14; II: A chemical study of the Kambalda nickel deposits, *Chem. Geol.*, **15**, 117–144.

Tooms, J.S. and Webb, J.S. (1961). Geochemical prospecting investigations in the Northern Rhodesian Copperbelt, *Econ. Geol.*, **56** 815–846.

Travis, G.A., Keays, R.R. and Davison, R.M. (1976). Palladium and iridium in the evaluation of nickel gossans in Western Australia, *Econ. Geol.*, **71**, 1229–1243.

Watson, J.P. (1970). Contribution of termites to the development of zinc anomaly in Kalahari sand, *Inst. Min. Metall. Trans.*, **79**, B53–B59.

Watters, R.A. and Sagala, F.P. (1979). Geochemical reconnaissance for uranium and base metals using heavy mineral separates in central and southern Sumatra. In: *Geochemical Exploration 1978* (Eds. Watterson and Theobald), pp. 317–327.

Webb, J.S. and Tooms, J.S. (1959). Geochemical drainage reconnaissance for copper in Northern Rhodesia, *Trans. Inst. Min. Met.*, **68**, 125–144.

Webster, S.S. and Skey, E.H. (1979). Geophysical and geochemical case study of the Que River Deposit, Tasmania Australia. In: Geophysics and Geochemistry in the search for metallic ores (Ed. P.J. Hood), *Geol. Surv. Can. Econ. Geol. Rep.*, **31**, 697–720.

West, W.F. (1970). Termite prospecting, *Chamber Mines J. (Salisbury, Rhodesia)*, **12**, 32–35.

Geochemical Exploration for Gold: A Special Problem

I. NICHOL

Introduction

In selecting a topic to discuss to commemorate the pioneering contribution John Webb has made to the development of applied geochemical techniques in a considerable range of fields, I chose 'Geochemical exploration for gold: a special problem' for a number of reasons. These included:

1. The topicality of the problem in view of the current concentration of exploration on searching for gold deposits.
2. Historically my first involvement at the Geochemical Prospecting Research Centre in 1961 was concerned with gold exploration.
3. At the same time, but in a different context, I was introduced to the classical work of Pierre Gy on sampling representivity, which takes on a new magnitude of importance in gold exploration.
4. A personal perception that some of the procedures currently being employed in gold exploration are inadequate and a hope that airing these ideas might serve some useful function.

The exploration for gold deposits is undoubtedly the major current focus of attention of the exploration industry. The objective of this review is to draw attention to the need for particular care to be taken in ensuring adequate sample representivity in situations where gold occurs in the particulate form. Consideration is given to theoretical aspects, followed by case history examples and a discussion of procedures available for improving the reliability of geochemical exploration for gold.

Background

The twenty-five year period 1955-80 witnessed the development of geochemical exploration methods suited to the search for a wide variety of deposit types in a range of surface environments. However, prior to the dramatic increase in the price of gold in the late seventies, gold exploration, with certain notable exceptions, had attracted only limited interest. In terms of geochemical exploration, this had resulted in relatively little emphasis being given to developing techniques suited to the search for gold deposits. Such geochemical exploration that was applied to the search for gold mostly involved the detection of anomalous zones of pathfinder elements, such as arsenic, antimony or mercury, on account of the frequent association of these elements with gold deposits, their more homogeneous distribution in mineral deposits than gold and the easier analysis of these elements. However, the weakness of this approach lies in the fact that not all gold deposits have associated pathfinder elements and by the same token not all anomalous zones of pathfinder elements are associated with gold mineralization. Therefore, the use of pathfinders as indicators of gold mineralization is of limited effectiveness. The late sixties and early seventies were characterized by the increasing use of atomic absorption analysis in geochemical exploration and routine methods for gold analysis became available with detection limits in the order of 10 ppb or less.

The dramatic increase in the price of gold in the late seventies and early eighties, coupled with the diminishing attractiveness of many other metal prices, was accompanied by a very marked increase of interest in gold exploration. This activity was characterized by a greater emphasis on the direct analysis of gold on account of the availability of highly sensitive analytical procedures, rather than the traditional use of pathfinder elements.

Although the problems of collecting and analysing a representative sample for gold have long been recognized by geologists and assayers involved in the evaluation of gold deposits, the same does not appear always to be the case with those involved in the search for gold deposits. In many instances the procedure adopted has involved the collection of a 'standard sized' sample of some 500 g, appropriate for base metal exploration, and the analysis of a 10–20 g sub-sample of the minus 177 μm (80 mesh) fraction. Frequently the adoption of such sampling and analytical procedures is highly unrepresentative of the population it is considered to represent. For a reasonable level of reliability to be placed on the results of exploration it is clearly essential that the results reflect the sought-after mineral deposit within pre-determined levels of representivity defined by the geochemical contrast between the response to the deposit and background areas.

Geology of gold deposits

The geological character of gold deposits in terms of host rocks, form, texture and control of mineralization has been classified into seven main types (Boyle, 1979):

1. Auriferous porphyry dykes, sills and stocks, coarse-grained granitic bodies, aplites and pegmatites.
2. Auriferous skarn-type deposits.
3. Gold-silver and silver-gold veins, stockworks, lodes, mineralized pipes and irregular silicified bodies in fractures, faults, shear zones, sheeted zones and breccia zones essentially in volcanic terrains.
4. Auriferous veins, lodes, sheeted zones and saddle reefs in faults, fractures, bedding plane discontinuities and shears, drag folds, crushed zones and

openings on anticlines essentially in sedimentary terrains; also replacement tabular and irregular bodies developed near faults and fractures in chemically favourable beds.

5. Gold-silver and silver-gold veins, lodes, stockworks, silicified zones, etc. in a complex geological environment, comprising sediments, volcanics and igneous or granitized rocks.
6. Disseminated and stockwork gold-silver deposits in igneous, volcanic and sedimentary rocks.
7. Gold deposit in quartz-pebble conglomerates and quartzites.

Considering areas of mineralization on a regional scale, deposits are geologically variable and are polygenetic in that they have formed at different times by different processes. For example, within the Superior Province of the Canadian Shield, on the regional scale the most important geological associations of gold deposits are mafic volcanic sequences with associated ultramafic rocks; a sedimentary volcanic sequence contact with major faults being a less characteristic feature, Hodgson *et al.* (1982). On a local or property scale the presence of felsic intrusive and extrusive rocks is important together with carbonate-dominated alteration and structural complexity. On the detailed or deposit scale, structure appears the most important characteristic.

In terms of genesis, hydrothermal gold deposits, including deep-seated and epithermal deposits, occur in a range of geological settings characterized by a variety of hydrothermal alteration and mineralization assemblages, Lewis (1982). Deep-seated deposits may be characterized by: chlorite-carbonate alteration, e.g. Yellowknife, Kalgoorlie; silicate alteration, e.g. Rossland; or iron formation, e.g. Homestake. Similarly, in epithermal deposits characteristic alteration may vary from propylitic to alunitic in passing from the fresh rock towards the centre of mineralization. Hydrothermal gold deposits are also characterized by a variety of mineral or element associations intimately associated with the mineral deposits.

In a review of the mode of occurrence of gold in different types of gold deposits, the possibly greater importance of invisible gold than previously recognized is described by Springer (1983). The marked association of gold with base metal sulphides, from submicroscopic (< 0.1 μm) to macroscopic (>100 μm) is noted, together with the potential for gold occurrences in mica schists and carbonaceous or graphite schists.

Analyses of gold grains have indicated the existence of variations in element associations with metallogenetic province, Warren and Thompson (1944), but also with deposit type within a given area, Gay (1963), Denisov *et al.* (1966), and Antweiler and Campbell (1977). In a compilation of information on gold deposits in Ontario, Hodgson *et al.* (1982) classified deposits into four types according to geological characteristics and noted considerable variations in the 'ore' minerals associated with the gold deposits, such as galena, molybdenite, tellurides, scheelite, arsenopyrite and sphalerite, even within a given deposit type. The composition of gold grains in terms of minor element contents and element ratios from gold deposits of Ontario was shown to vary according to deposit type based on mineralogy and host rock, Guindon and Nichol (1983).

In areas where a high level of geological information is available, and gold deposits are of a type characterized by minerals or elements that can be used as pathfinders, then the use of pathfinders in geochemical exploration is advantageous. On the other hand, where the base level of geological information in an area of interest is low, or there is a diversity in the character of gold deposits, associated minerals and element associations, there are obvious limitations to basing exploration on the determination of pathfinders and clear advantages in directly determining gold contents.

Geochemistry of gold

In so far as one of the keys to successful gold exploration is an understanding of the mineralogical mode of occurrence of gold, the mode of transport and accumulation in a given environment, consideration of the geochemistry of gold, or more specifically the behaviour of gold, in the surface environment is in order.

Traditionally, the view that gold is insoluble in the surface environment and is transported mechanically has dominated the design of exploration programmes aimed at discovering bedrock sources of alluvial gold. However, if one examines records of early gold mining and exploration it is evident the validity of this concept has been questioned. For example some of the early miners involved in the California gold rush postulated that some of the gold had been concentrated by solution and reprecipitation. The occurrence of larger nuggets in alluvial deposits than any found in bedrock deposits together with higher fineness of the alluvial gold relative to associated bedrock occurrences, was suggestive of the alluvial gold being formed, at least in part, by precipitation from solution.

Landsweert (1869) quoted experimental data that gold is precipitated on a variety of minerals from dilute gold chloride solutions in reducing conditions brought about by the presence of organic matter. Landsweert (1869) quotes Bischoff as citing the considerable distance in many cases between alluvial deposits and source reefs as evidence against mechanical transport, having regard to the high specific gravity of gold. Subsequently, Clarke (1908) drew attention to the dissolution of Cripple Creek ore in a solution of ferric sulphate, sodium chloride and sulphuric acid and subsequent precipitation in the presence of organic matter. Thus from the early days of geochemistry there was evidence that gold, under certain conditions, was in fact soluble and thus could be dispersed chemically.

Gold is both monovalent and trivalent but the respective aurous and auric ions cannot exist in significant concentrations in aqueous solution. In solution gold occurs as a variety of complex anions including chlorides, bromides, iodides, cyanides, thiocyanates, thiosulphates and hydroxides, Boyle (1979), Lakin *et al.* (1974). The ease of oxidation of gold to form complex anions increases in the order chloride, bromide, thiocyanate, thiosulphate, iodide and cyanide. However, in terms of relative abundance in nature of the different anions, the chloride and cyanide complexes are the most important because of the greater natural abundance of the chloride and cyanide.

The stability of different complex anions was examined by mixing portions of individual powdered minerals with solutions of individual complex anions and then determining the amount of gold removed from solution, Lakin *et al.* (1974). The results indicated the tendency for gold to be removed from solution (either by sorption or reduction to gold metal) decreases in the order: chloride, bromide, iodide, cyanide, thiocyanate and thiosulphate. In particular, the halide anion complexes are much less stable and thus less mobile than the cyanide, thiocyanate and thiosulphate complexes.

The principal modes of occurrence of gold in the surface environment are as free gold or as a minor constituent of secondary iron and manganese oxides, organic matter and clay minerals.

Geochemical exploration practice

In many instances, a common practice in gold exploration involves the collection of a sample of approximately 500 g weight, the sieving of the minus 177 μm (80 mesh) fraction and the analyses of a 10 g sub-sample, in situations where the concentration of interest is in the range 25–100 ppb. In cases where the gold has dispersed chemically and is currently associated with the iron or manganese oxides, organic matter or clays (i.e. as a minor component of a major constituent of the sample) this approach will be adequate. However, in situations where gold has dispersed mechanically or chemically and is present in the sample as free gold, the collection of a 500 g sample and the analysis of a 10 g sub-sample will, in many instances, give rise to unrepresentative gold values.

Theoretical considerations

The level of representivity of analysis required in exploration is in part dependent on the contrast between the target or anomalous population and the background population, a feature that would be expected to vary from area to area. In the case of there being a relatively small difference between the mean background ($\bar{x}$) and anomalous populations defined by the threshold (T), as in population 'A' (Figure 1), a higher degree of representivity or precision would be required reliably to identify anomalous concentrations than in the case of a situation where there is a wide difference between mean background and anomalous populations, as in case 'B'. If the contrast between anomaly and background is large, even relatively poor representivity should not obscure the anomaly. If the contrast between anomaly and background is small, then it is important that the distinction of anomalous from background populations is not obscured by poor precision.

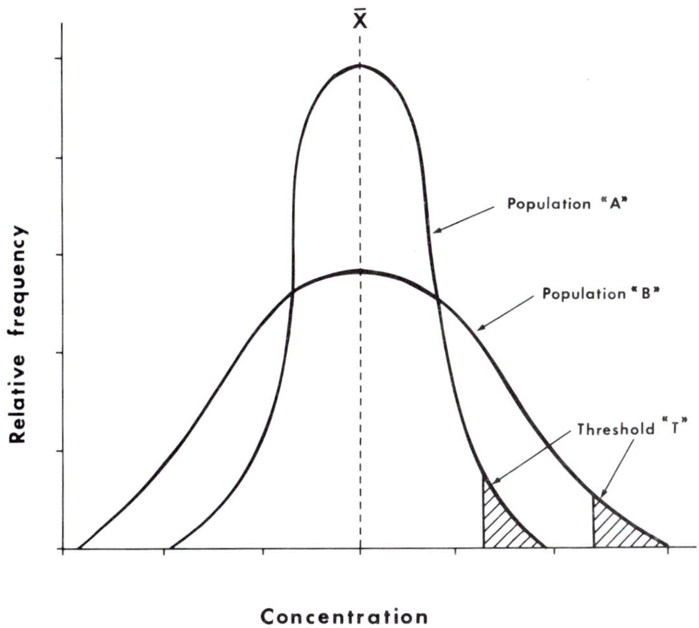

Figure 1 Hypothetical variation in range of gold contents in two populations, necessitating different precisions to distinguish anomalous concentrations

Variations in an element's content within a set of data are related to two main sources: (a) the natural or regional variation in the element content across the area and (b) variability introduced by sampling, sample processing and analyses. The overall variability in a data set can be expressed in terms of the variance S^2 where:

$$S^2 = S^2_{ntr} + S^2_{smpl} + S^2_{prc} + S^2_{anal}.$$

where:

S^2_{ntr}—natural or regional variance of the element content in the sample medium;
S^2_{smpl}—variance due to sampling;
S^2_{prc}—variance due to sample processing;
S^2_{anal}—variance due to analyses.

For significant features to be identifiable in a set of data it is important for the natural variability to exceed that introduced by sampling, processing and analyses by a factor of two. The natural variability is a feature of the area and cannot be changed. The only control we have on the relation between natural and introduced variability is the variability introduced as a result of sampling, processing and analyses. If the contrast between the response due to mineralization and background is low, then to increase the chance of recognizing mineralization the variability introduced due to sampling, processing and analyses must be lowered.

Pierre Gy introduced an equation in 1954 by which it is possible to calculate the representivity of a particular sample as follows (Ottley, 1966):

$$S^2 = \frac{Cd^3}{M}$$

where:

S^2— the variance of analyses;

S— is the standard deviation; 95% of the values of replicate analyses should fall in the range $x \pm 2S$;

C— a sampling constant for the particular material to be sampled and is related to four parameters characteristic of the sample f, g, l and m ($C = f \times g \times l \times m$). All are dimensionless except m which is $g\,cm^{-3}$ hence C is expressed in $g\,cm^{-3}$;

- f— shape factor relating to the form of mineral grains containing metal of interest—generally taken as 0.5, but 0.2 for gold (dimensionless);
- g— particle size distribution factor, generally 0.25, but 0.5 for well-sorted material (dimensionless);
- l— liberation factor, related to the ratio of the maximum particle size and the liberation size of the mineral and varies from 0 to 1 (dimensionless);
- m— mineralogical factor; related to the s.g. of both the mineral hosting the element of interest and the gangue and the content of the mineral of interest (dimensionless);

d— aperture of sieve passing 95% of sample (cm);

M— sample weight (g).

The formula can be used to determine:

1. The representivity or variability to be expected using a particular sample weight and type of sample (equation 1 above).
2. The weight of sample required to give a particular representivity:

$$M = \frac{Cd^3}{S^2} \qquad (2)$$

3. The necessary mechanical size for a given sample weight to give required representivity:

$$d = \sqrt[3]{\frac{MS^2}{C}} \qquad (3)$$

Table 1 **Relative standard deviation (%) of gold analyses**

Size	Concentration (ppb)						
	1000	500	250	100	50	25	10
420 μm (35 mesh)	230	326	460	730	1030	1450	2300
177 μm (80 mesh)	63	89	124	195	280	398	630
62 μm (250 mesh)	13	19	26	40	60	84	134

The use of the formula assumes random sampling, implies no bias in the sampling process and furthermore takes no account of any analytical errors. Considering a method for gold analysis involving a 10 g sample and assuming the gold occurs variously at 420, 177 and 62 μm, (35, 80 and 250 mesh, respectively) and at various concentrations between 1000 and 10 ppb, using Gy's equations the relative deviation ($S/\bar{x}$) shows some very marked variations (Table 1). For example, at the 1000 ppb level with gold in particles of 62 μm diameter, the variation is ±13%, which would be acceptable in most exploration programmes concerned with determining gold at this concentration level. In contrast, at the 25 ppb level and gold particles of 62 μm the variation is ±84%, which would be of questionable acceptibility in most reconnaissance level programmes.

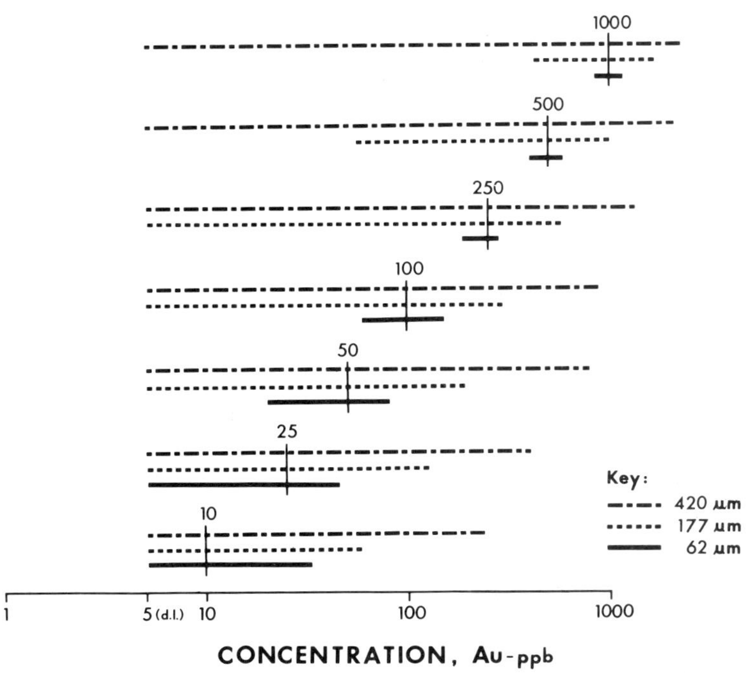

Figure 2 . Variation of gold analyses with concentration and particle size at 68% confidence levels. (d.l.—analytical detection limit)

Table 2 **Size of gold particles (μm) to provide various levels of precision**

Relative precision	Content (ppb)						
	1000	500	250	100	50	25	10
±50%	151	120	95	70	56	44	33
±100%	240	190	151	111	88	70	52

The range of gold concentration at the single standard deviation limits or 68% confidence limits is shown in Figure 2, according to gold content and size of gold particles. It is clear that the range of gold contents decreases with decreasing particle size for a given concentration. Similarly, the range of gold contents increases with decreasing gold content for a constant particle size.

As discussed previously, in geochemical exploration one is faced with the task of achieving a particular level of precision to distinguish with confidence between anomalous and background levels of concentration. In this connection, for levels of relative precision of ±50 and 100% the necessary size of the gold particles is shown in Table 2, which indicates the need for gold to be finer grained, the lower the gold content, to achieve a particular level of precision.

A similar type of approach has been adopted by Clifton *et al.* (1969), but the information is set out graphically and is somewhat easier to follow. The precision is determined by the number of gold particles within a sample, assuming:

1. Gold particles are of uniform size.
2. Gold particles comprise less than 0.1% of all particles.
3. The sample contains a total of over 1000 particles.
4. Analytical errors are disregarded.
5. Gold particles are randomly distributed within the sample.

Accepting that a relative deviation of ±50% at the 95% confidence limits might be required for most purposes, Clifton *et al.* (1969) have shown that to attain this precision, a sample must contain twenty gold particles of a given size. However, a sample with fewer gold particles may be tolerable in reconnaissance programmes where less stringent representivity requirements are acceptable. As noted previously, the necessary precision is related to the contrast between anomalous and background concentrations. It is possible to determine the weight of the sample that will give the required level of representivity for different combinations of gold content and particle size. Using the approach of Clifton *et al.*, if the size of the gold particles is known, together with the general range of gold content of the sample, then the sample weight necessary to include twenty gold particles and thus the prescribed level of precision can be determined from Figure 3. For example, assuming gold occurs in spheres of diameter 62 μm (250 mesh) a sample containing 250 ppb

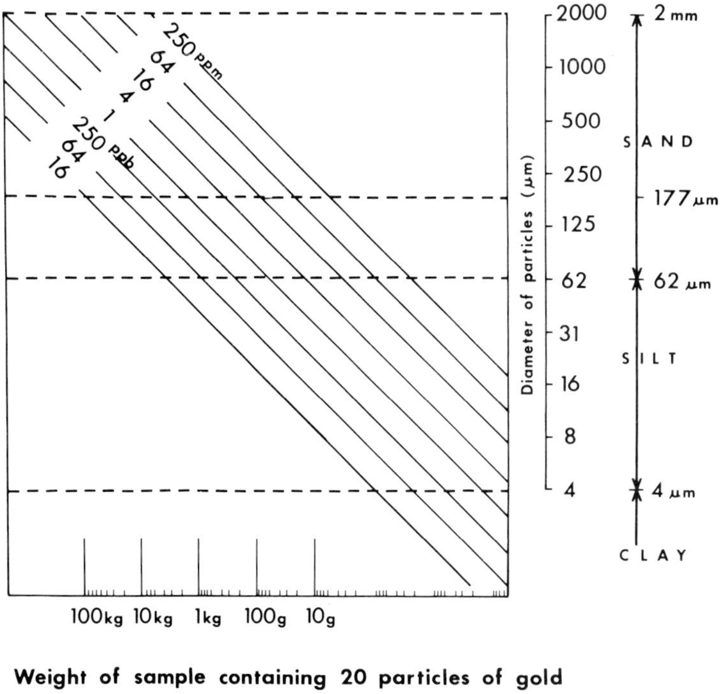

Figure 3 Weight of sample containing twenty particles of gold as a function of gold particle size and content, assuming all gold particles are of uniform size and randomly distributed. (From Clifton *et al.*, 1969)

will contain 20 gold particles per 200 g sample, and a sample containing 64 ppb will contain 20 gold particles per 800 g sample.

If one assumes gold occurs in spheres 62 μm (250 mesh) in diameter and the range of interest is 64–250 ppb then clearly the analysis of 10 g sub-samples will give rise to inadequately representative analyses with potentially misleading results. The situation is illustrated diagrammatically in Figure 4 where two samples of 100 g have the same gold content. In the one case (A) all the gold is contained in a single grain, thus nine of the ten 10 g samples taken for analysis would contain no gold and one sample would report ten times the true value. In the other case (B) the same absolute gold content is distributed in many small gold grains and the analyses of each 10 g sub-sample will give a representative analysis. Alternatively, if one considers samples containing

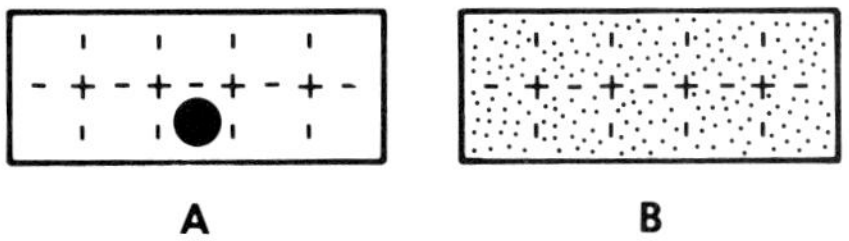

Figure 4 Variation of representivity of sub-sample with particle size of gold grains for a constant gold content

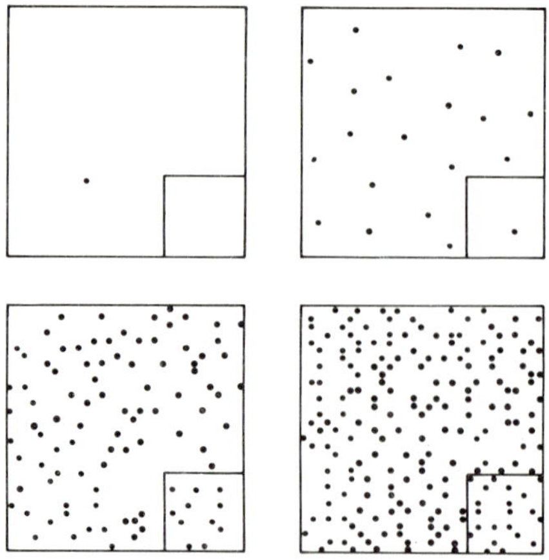

Figure 5 Variation of representivity of sub-sample for analysis according to concentration (constant sized gold particles)

gold particles of the same size, but widely differing absolute gold contents, it is apparent that the size of the sample containing twenty gold particles increases as the absolute gold content decreases (Figure 5).

The relation between relative error and number of gold particles in a sample has been illustrated diagramatically by Clifton *et al.* (1969) (Figure 6). For example, if the particle size/gold concentration of a large sample is such that the sub-sample taken for analysis is expected to contain 1 gold particle, then

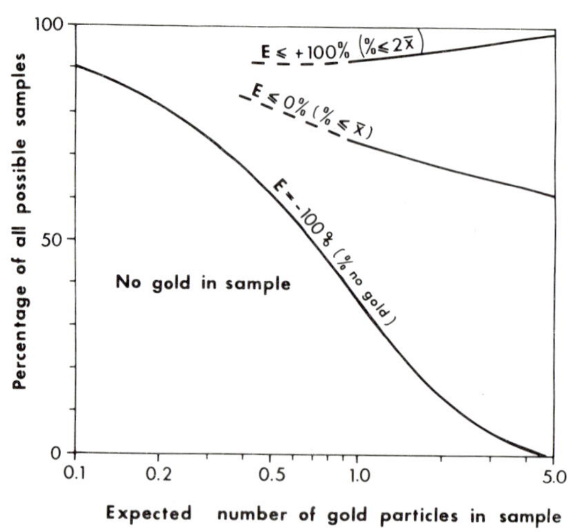

Figure 6 Proportion of analyses at various levels of relative error (E) in relation to expected number of gold grains in the sub-sample to be analysed (From Clifton *et al.*, 1969). (E = –100% shows the percentage of analyses that will show no gold; E⩽0% shows the percentage of values containing up to the true content; and E⩽+100% shows the percentage of analyses reporting up to twice the true value)

37% of replicate analyses will contain no gold, 74% of the analyses will show gold contents equal to or less than the true value and 93% will have analyses equal to or less than twice the true value and, by the same token, 7% of the values will exceed twice the true value.

The validity of these arguments has been demonstrated by the variability of gold analyses obtained by the replicate analyses of 10, 100 and 1000 g portions of a beach sand, Clifton *et al.* (1969). The range of analyses is greatest in the case of 10 g sub-samples with forty-nine of ninety-six analyses reporting gold contents below the detection level of 40 ppb. In contrast, replicate analyses of the 1000 g sub-samples shows a much narrower range. Similar examples have been demonstrated by other investigators.

Whilst the reporting of misleadingly high values is highly undesirable, it will at worst lead to follow-up in unwarranted areas, whereas the reporting of less than true values is much more serious as it may lead to mineralized areas being overlooked. Although this may be less serious in follow-up investigations where interpretation will be directed to detecting patterns in the gold content of a relatively large number of samples and undue significance will not be attached to the value of a single sample, in the reconnaissance phase the recognition of areas for follow-up is based in some cases on the gold values of single samples and adequate sampling and analyses are particularly important in reconnaissance level exploration.

Hence it is clear, from theoretical consideration at least, the analyses of a 10 g sample of minus 177 μm (80 mesh) material for gold is inadequate if the assumptions are made that the gold occurs as discrete particles 62 μm (250 mesh) in diameter and concentrations of interest are up to 250 ppb.

In terms of the occurrence of gold accumulations in surface environments, it is well known that gold placers occur in coarse grained sediments, in which the gold is of a much finer size than the associated quartz of silicate minerals, the variations in sizes of grains of the same shape of different minerals being related to specific gravity. Mechanically-formed sediments are composed of aggregates of different-sized minerals that are hydraulically equivalent to one another, i.e. have the same settling velocity. The settling velocity is related to the specific gravity and the shape of minerals, with round grains settling faster than flakes of a given mineral. As an example, a sphere of quartz 100 μm in diameter settles at the same velocity, (8 mm), as a sphere of gold 31 μm in diameter or a flake of gold having a diameter : thickness ratio of 10 and a diameter of 62 μm, Tourtelot (1968). Thus gold and quartz grains of these dimensions tend to be associated together. The behaviour of particles of different specific gravity in still water is however only a general guide to their behaviour in natural environments. In nature the transport and deposition of particles is also related to velocity of the current and its distribution in a stream profile and cross-section, the proportions of turbulent and laminar flow, the volume of water, the morphology of the channel in which the water flows and the characteristics of sediment load, Tourtelot (1968). Thus in terms of exploration significance, assuming some information is available on the size of gold particles in bedrock mineralization, and that this gold is transported mechanically and incorporated in the drainage sediment, then this information can be applied in focussing attention on collecting the appropriate type of sample, e.g. gravel, sand, silt, clay, etc.

Analytical procedures

There is a variety of analytical procedures available for gold analyses. Historically, the most widely adopted procedure was fire assay involving the concentration of the gold into a silver bead and following removal of the silver, the gravimetric estimation of the gold. This procedure, using 1/2 to 1 assay-ton sample weights, had practical detection limits of 170–340 ppb, which is inadequate for many exploration programmes.

Following the advent of atomic absorption with the capability of detection limits in the range 1–40 ppb, according to analytical conditions, a number of procedures have been adopted involving variously a fire assay start, or different acid digestions followed by estimation of the gold content by atomic absorption. More recently, plasma emission spectrometry and instrumental neutron activation analyses have been adopted as estimation procedures. All these procedures, which are the commonest used in geochemical exploration, involve the analyses of samples up to 30 g in weight. A number of procedures involving the analyses of samples up to 100 g have been reported by Lakin and Nakagawa (1965) and McHaffie (pers. comm.) in order to lower detection limits or increase the representivity of the sample being analysed.

The mostly widely used procedures for the routine analysis of gold in the sub 100 ppb range involve analysis of samples up to 30 g in weight. It has been previously shown from theoretical considerations that a much larger sample must be analysed to give the necessary levels of representivity.

Case history examples

In this section, emphasis is placed on published information relating to the distribution of gold in different sample media that indicates the nature of gold dispersion in relation to mineralization. This information is used to highlight some of the unique features of geochemical exploration for gold deposits.

DISTRIBUTION OF GOLD IN SOIL PROFILES

The distribution of gold within soil profiles has been reported for twelve gold districts in Colorado, Utah, Nevada and Arizona, Lakin *et al.* (1974). The soil profiles sampled are generally poorly developed because they are in areas where frost action or soil creep together with climate inhibit soil development. The gold distribution within the profiles displays concentration variously in the humus and surface soil horizons or towards the base of the soil profiles, Figure 7.

The high concentration of gold in humus and surficial soil horizons is

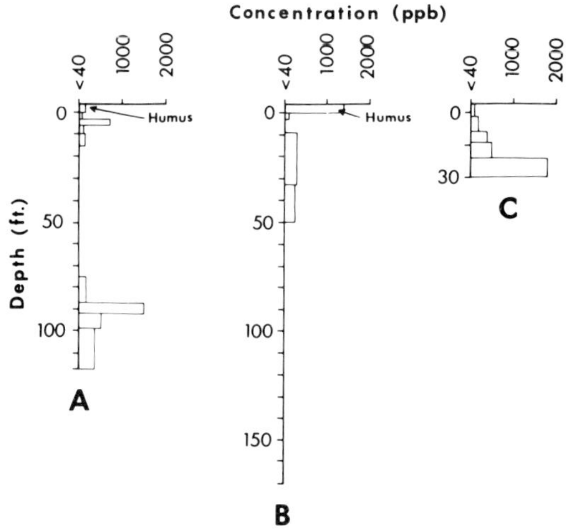

Figure 7 Distribution of gold within soil profiles. (From Lakin *et al.*, 1974)

related primarily to the behaviour of gold in the biogeochemical cycle. To a lesser extent the concentration of gold in the surface soil horizons may be due to surface run-off and wind erosion. The high concentration of gold near the bottom of the profiles may be due to an *in situ* disintegration of gold-bearing material and/or the downward migration of gold within the profile. An instance is cited where gold-bearing pebbles with up to 11 ppm Au occur at a depth of 30–46 cm (12–18 in) from the surface, at which depth the associated soil contains no detectable gold. Lower down in the profile, pebbles containing 1.0 and 3.8 ppm gold at depths of 200–400 cm (79–94 in), however, contributed gold to the soils, indicative of the breakdown of material due to abrasion during soil creep and/or chemical weathering being more active lower in the profile.

The precise relationship of the soil profiles to gold mineralization is not quoted, nor is the extent of variation in the distribution of gold within soil profiles over an area of the size that might be considered typical of that for detailed exploration. However, the data clearly indicate marked variations in soil depth and distribution of gold within soil profiles, which emphasizes the need for particular care at the sampling stage to ensure that samples are taken from the appropriate soil horizon.

GLACIAL OVERBURDEN

In recent years there has been increasing recognition of the importance of considering the nature of glacial stratigraphy in geochemical exploration programmes being carried out over areas of glacial overburden. In brief, glacial overburden comprises a mixture of materials differing in both direction, distance of transport and degree of sorting from their source. Most geochemical exploration involving the sampling of glacial overburden is carried out in detailed scale exploration where the objective is to identify geochemical anomalies of local derivation. Such locally derived anomalies mostly exist in basal or lodgment till and to a lesser extent ablation till. In addition to till, however, glacial overburden comprises a variety of types of stratified drift which mostly have been transported considerable distances and bear no compositional relationship to local bedrock or mineral deposits.

Typically a stratigraphic column through glacial sediment will consist of basal or lodgment till at the base overlain by units of various types of stratified drift, in some cases interspersed with younger till units. The key to exploration success in this situation, as far as geochemical exploration is concerned, is to obtain material from locally-derived basal till. The validity of these considerations has been well recognized in base metal exploration in glaciated terrain in Canada and Scandinavia for some time, and this situation has led to the development of a variety of deep overburden drilling procedures.

Information on the dispersion of gold in glacial material is less common but is becoming increasingly available. The distribution of gold within different components of glacial overburden is well illustrated by data from northern Scotland, Plant and Coleman (1973). The overburden section comprises surficial alluvium and soil underlain sequentially by varved clays and interbedded sand, coarse fluvioglacial detritus, varved clays and interbedded sand and a basal till layer, (Figure 8). Concentrates of up to 500 g from samples initially of 360 kg were prepared and analysed for gold, and the gold content in the original sample calculated according to the concentration factor (weight of original sample/weight of concentrate). Detectable contents of gold occur only in the basal till and coarse fluvioglacial detritus. The latter occurrence is considered to be related to melting of a local snow field and accumulation of gold with hydraulicaly equivalent coarser-grained but lighter minerals. These data dramatically illustrate the marked variation in gold content amongst different types of glacial sediments. In addition, the data indicate the fairly consistent patterns of gold that can be identified by collecting a large sample initially and then concentrating the gold in the heavy mineral fraction.

Whilst recognizing that the identification of anomalous gold contents in till has made significant contributions to the discovery of some gold deposits, e.g. the Aquarius and Watabeag prospects in northern Ontario, Gray (1983), the question arises in how many cases has the geochemical technique employed failed to identify a significant mineral occurrence.

The results of a very informative examination of the nature of gold dispersion in glacial overburden associated with 'vein-type' mineralization in the Beardmore–Geraldton area of northwestern Ontario have been described by Closs and Sado (1979).

Three gold occurrences were examined: (a) The Northern Empire area where the occurrence consists of quartz, arsenopyrite, pyrite, pyrrhotite, visible gold and tourmaline; (b) the Archie Lake area where quartz-carbonate material contains arsenopyrite, low-gold values but no visible gold; and (c) the Knox Lake area where visible gold is rare in the mineralization. On this basis

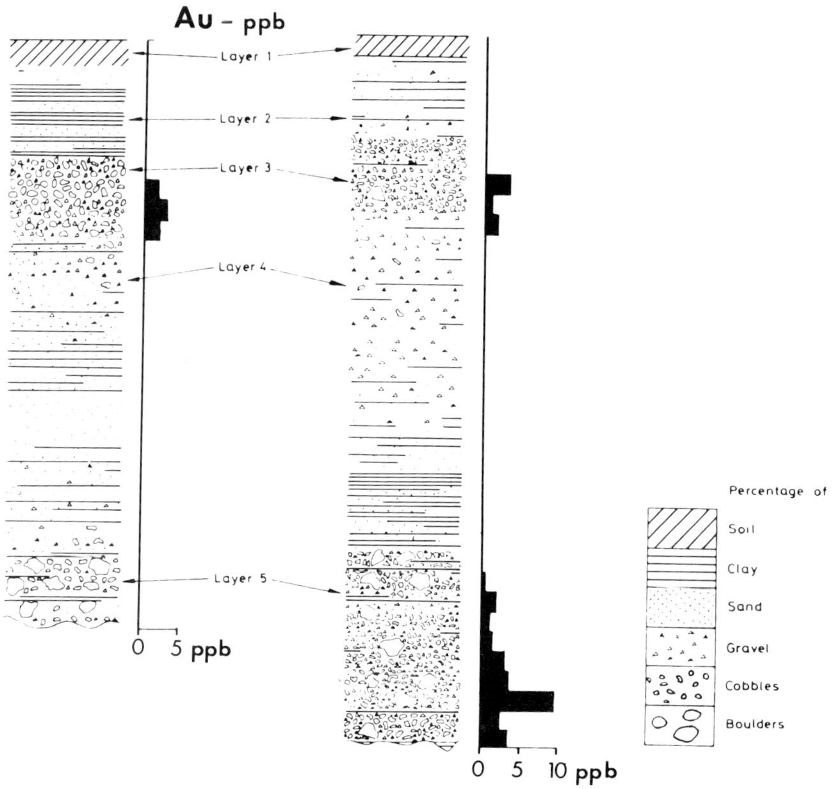

Figure 8 Variation of gold content with different types of glacial sediment. (From Plant and Coleman, 1973)

the mineral association and size of the gold of different vein-type deposits appears to vary within a short distance of one another. Sampling involved the collection of upper B horizon soils and glacial overburden the latter varying from 'till, possible till or non-till material'. Conventional sized soil samples were collected together with 2.5–4 kg glacial sediments. The minus 177 μm (80 mesh) fraction of the soils, the minus 62 μm (250 mesh) fraction of the glacial material and the minus plus 62 μm (250 mesh) fraction of the heavy mineral separate of the glacial material were prepared and analysed for Ag, As, Cu, Cr, Zn, Pb, Sn, B, Zr and Au. Suffice it to say that no consistent element dispersion pattern related to mineralization was noted in any sample medium, indicative at least in part of the different mineral associations of the three occurrences.

With specific reference to gold in the Northern Empire area an anomalous dispersion of gold is indicated in the B horizon soils and the heavy mineral separates but not in the minus 62μm (250 mesh) fraction. The restriction of anomalous dispersion patterns of gold to the B horizon soils and the minus 250 plus 62μm (−60+250 mesh) fraction of the heavy mineral concentrates of

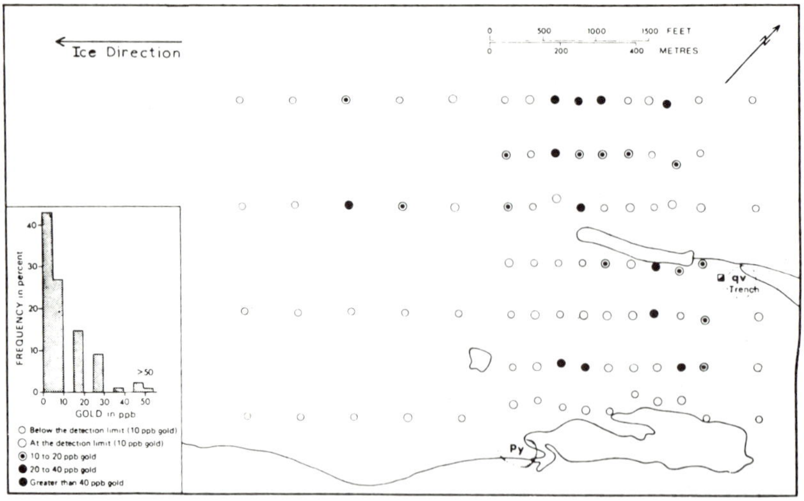

Figure 9 Distribution of gold in B horizon soils associated with gold occurrence in the Knox Lake area, Ontario. (From Closs and Sado, 1979.) (Note 'qv' represents quartz vein location and 'Py' pyrite occurrence)

the glacial materials is interpreted as being related to the restriction of gold to the fine sand or coarser fraction, a feature borne out by the appearance of visible gold in the Northern Empire occurrence.

In the case of the Archie Lake area only very spotty anomalous gold values are apparent in the soils with no clear relationship to the mineral occurrence. With regard to the Knox Lake area, the main area of mineralization gives rise to anomalous down-ice dispersion of gold in the B horizon soils, minus 62 μm (250 mesh) fraction of the glacial sediments and the minus 250 plus 62 μm (−60+250 mesh) heavy mineral concentrate, (Figure 9). However, along strike westwards from the main mineralization, anomalous gold values are present only in the glacial sediments and significantly absent in the soils. The absence of anomalous gold in the soils is attributed to the soils being formed from glacial sands rather than glacial till. The occurrence of anomalous gold in the minus 62 μm (250 mesh) fraction of the glacial sediments is attributed to the occurrence of gold blebs and gold stringers within pyrite, and numerous occurrences of gold-like material down to 2 μm in diameter as was recognized in a heavy mineral concentrate of the minus 62 μm (250 mesh) fraction of glacial sediments. As noted previously, gold in the main occurrence at the Knox area occurs as blebs associated variously with pyrite quartz, and silicates in the 2–20 μm size range, thus supporting the mechanical derivation of gold in the minus 62 μm (250 mesh) fraction of the till.

The basic conclusion to be drawn from this study is the existence of variable anomalous dispersion trains in overburden associated with mineral occurrences. The reasons for this include: (a) variations in the nature of the gold occurrences in terms of mineral associations and the size of the gold itself;

(b) variations in the provenance of the glacial overburden and thus its suitability for sampling as an indicator of local mineralization; and (c) variations in the provenance of the parent material of the soils, giving rise to the limited suitability of surficial soils as a sampling medium.

One of the over-riding features of the gold distribution in all three sample types is the marked variability on a local scale, symptomatic of poor sample representivity. In the case of the Northern Empire area, anomalous gold values were detected in minus 177 μm (80 mesh) fraction of the B horizon soils and the minus 250 plus 62 μm (−60+250) mesh fraction of the heavy mineral separates of glacial overburden, but not in the minus 62 μm (250 mesh) fraction of the glacial overburden, a feature which was ascribed to the occurrence of only visible gold in the bedrock mineralization. If one considers the gold distribution in the soils this displays anomalous values in the range 10–50 ppb with some one percent of the values greater than 50 ppb. If one assumes that the gold occurs in particulate form and is of the same size as the primary mineralization $>$62 μm (250 mesh) then one can determine the size of sample to be analysed where gold occurs as 62 μm (250 mesh) spheres and the content of a sample is 16 ppb. Twenty particles of gold occur in some 3 kg, Clifton *et al.* (1969). If this assumption is valid then there would be a considerable number of values below the detection limit, with a few erratically high values. Thus a major cause of the 'noise' in the data could be due to the poor sample representivity.

DISTRIBUTION OF GOLD IN HUMUS

The association of gold with organic matter has been observed in a number of instances together with the enrichment of gold in the humus horizon of soils derived from colluvium and glacial drift associated with gold mineralization in the Empire district, Colorado, Curtin *et al.* (1968). Investigations carried out in areas near Kirkland Lake, Noranda and Val d'Or overlain by permeable till have indicated that the best anomalous response in areas covered by permeable till is that in the humus horizon. In areas where well-developed podsols occur, marked anomalies occur in the humus, related in some cases to gold-bearing zones below up to 37 m (120 ft) of glacial cover, Gleeson and Boyle (1979). The existence of anomalous zones of gold content in humus overlying gold mineralization has been noted at the Arntfield and the Powell–Rouyn mines in Quebec (Figure 10) Hoffman and Brooker (1983).

A feature of gold in humus is the much more homogeneous distribution than in the case of glacial overburden. The application of humus geochemistry as an indicator of gold mineralization has been clearly demonstrated. This feature has been related variously to the high adsorptive and absorptive capacity of organic matter and the presence of reducing conditions, Boyle (1979); or to the absorption of gold by cyanogenic plants, the accumulation of gold in the humus layer following the death of the plant, Lakin *et al.* (1974). In many instances the concentration of gold in the humus layer gives a very sharp and narrow anomaly reflecting very closely the location of underlying

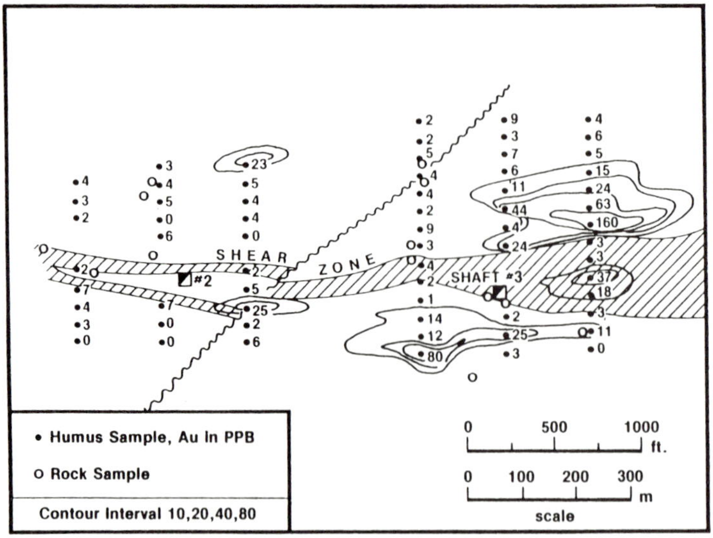

Figure 10 Gold in humus overlying the Arntfield gold mine, northwestern Quebec. (From White *et al.*, 1979)

mineralization. In other instances, anomalous gold contents occur in humus in swamps remote from any mineralization.

This variation in the spatial relation of gold anomalies to gold mineralization indicates a variable behaviour of gold in the surface environment; possibly related to variations in the nature of the primary mineralization, glacial overburden or the physicochemical conditions of the local environment. At the present time the sampling and analysis of the humus horizon is probably the most widely used geochemical exploration technique being applied to gold exploration in Canada. However, there are wide differences in the nature of the humus material being sampled together with considerable variations in the nature of the underlying glacial material. These features, together with the varying spatial relationship between anomalous gold contents in humus and mineralization, make the interpretation of humus geochemistry difficult and unreliable.

Limited studies by Soviet geochemists have indicated a variation in the form of gold associated with organic matter with distance from source mineralization, i.e. adjacent to mineral occurrences gold may occur associated with humin, humic and fulvic acid components of organic matter, whereas remote from mineralization gold appears to be restricted to being associated with the fulvic acid component of organic matter, Roslyakov (1983).

DISTRIBUTION OF GOLD IN STREAM SEDIMENTS

The validity of the theoretical considerations of sample representivity described by Clifton *et al.* (1969) referred to earlier have been vividly

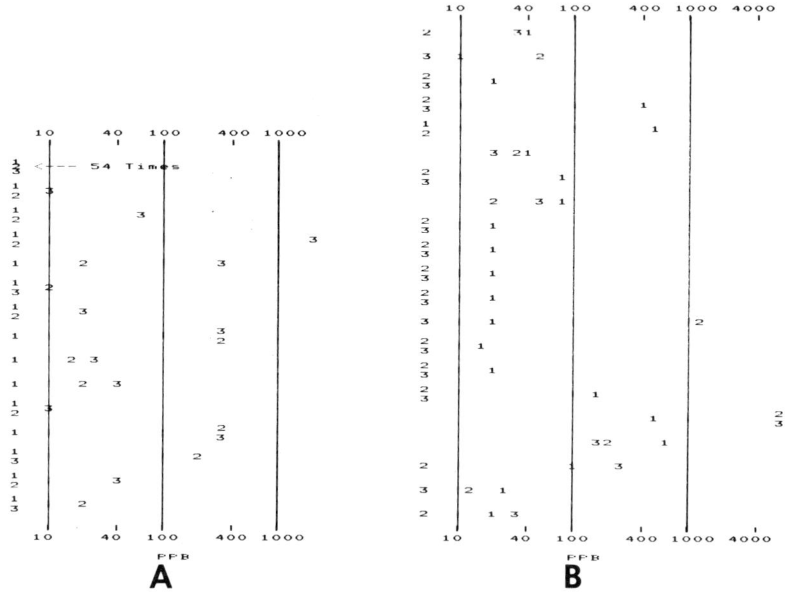

Figure 11 Variation of gold contents in 3 replicate analyses of 5 g sub-samples of stream sediments. (Adapted from Harris, 1982.) (Note: 1, 2 and 3 represent first, second and third analyses respectively: Detection limit is 10 ppb. Determinations reported as <10 ppb plotted as 5 ppb). 'A' represents replicate on first analyses reported gold contents below detection limit and on subsequent analyses in some cases gave anomalous values. 'B' represents replicate analyses of samples which displayed anomalous values on initial analyses and in many cases below detection contents on replicate analysis

demonstrated with regard to gold contents of stream sediments from British Columbia, Harris (1982).

The unreliability of the analyses of 5 g sub-samples of minus 177 μm (80 mesh) material was clearly demonstrated on the basis of replicate analyses (Figure 11). In some cases anomalous gold values identified in initial analyses were not confirmed in subsequent analyses, whereas in other cases samples with non-detectable values on initial analyses indicated anomalous contents on replicate analyses. The relevance of this information lies in the fact that it clearly demonstrates the distinct possibility of not detecting anomalous gold contents in stream sediments in an area of gold mineralization due to unrepresentative sampling and analyses and thereby perhaps overlooking significant mineralization. This situation seriously reduces the reliability of any such programmes.

The reliability that can be placed on the results of a programme was shown to be increased by replicate analyses of samples and then consideration of the mean contents, Harris (1982). In a comparison of replicate analyses of three conventional sized samples and a bulk sample taken within a few feet of one another, Harris (1982) firstly demonstrated the existence of considerable analytical imprecision, due to the non-homogeneous distribution of gold within the individual portions taken for analyses, and secondly the existence of significant variations in the mean metal content of samples taken essentially from the same location, indicative of the existence of poor representivity of the individual samples.

Harris provides comparative data on the replicate analyses of unpulverized and pulverized portions of three samples. The purpose of pulverizing was to homogenize the gold within the sample by reducing the size of the gold particles. The data indicate only a small improvement in analytical consistency attributable to the malleable nature of gold, with the result that the size of the gold particles were not significantly reduced during the pulverizing procedure. However the mean content of the analyses of the pulverized portions was lower than the original unpulverized equivalents attributable to gold having been smeared on the components of the pulverizing equipment.

HEAVY MINERAL CONCENTRATE

An alternative to the analyses of a particular size fraction of a sample, such as the minus 177 μm (80 mesh) fraction, has involved the separation of fractions according to specific gravity or magnetic susceptibility. The principle behind this approach is that it allows the relative concentration of mineral(s) of interest by removing the large quantities of lighter fractions composed of quartz or common rock forming silicates which cause a diluting effect. The subsequent mineralogical or chemical analyses of these fractions has in a number of instances been shown to indicate the existence of mineral deposits that was not evident from the analysis of the traditional minus 177 μm (80 mesh) fraction, Overstreet and Marsh (1981).

The applicability of gold concentration of heavy mineral concentrates as an indicator of gold mineralization was demonstrated in Canada over a 4900 km^2 (1900 sq mile) area of the Yukon, Boyle and Gleeson (1972). In this investigation all known gold deposits were identified and also there were indications of others. An outstanding example of the longer and more consistent anomalous contents of gold mineral concentrates of stream sediments relative to equivalent minus 177 μm (80 mesh) fractions is illustrated by data from British Columbia (Thomson, pers. comm.), (Figure 12). The procedure involved the collection of an 8 kg sample and the sieving to pass 841 μm (20 mesh) at the sample site. The minus 841 μm (20 mesh) fraction is then sized and separated in the laboratory and the plus 3.3 s.g. fraction concentrated using methylene iodide, of which the non-magnetic fraction is analysed for gold. The analysis of the conventional minus 177 μm (80 mesh) fraction displays only a single anomalous value whereas the analyses of the heavy mineral concentrate shows a consistent anomalous train extending at least three miles downstream from gold occurrence with values of 300 ppm relative to an arbitrary threshold of 30 ppm. The existence of the much more extensive anomaly in the gold contents of the heavy mineral

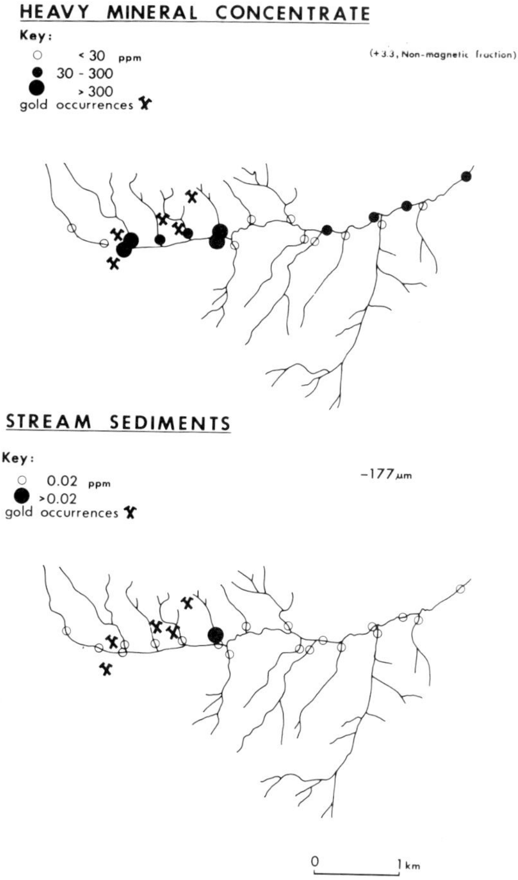

Figure 12 Comparison of gold contents of heavy mineral fraction and minus 177 μm (80 mesh) fraction of stream sediments, British Columbia (Thomson, pers. comm., 1983). (Note: orientation is omitted for proprietory reasons)

concentrate allows the adoption of a much greater sample interval and at the same time a much higher level of confidence can be attached to the results. Whilst the cost of obtaining a heavy mineral concentrate is much higher than a minus μm (80 mesh) fraction, the higher level of unit cost for the heavy mineral concentrate is more than offset by the lower sample density that will suffice.

In an investigation of the nature of dispersion in the minus 841 plus 177 μm (−20 + 80 mesh) heavy mineral concentrates and the minus 177 μm (80 mesh) fraction of stream sediments associated with a number of gold and silver deposits in British Columbia, a marked enhancement of gold in absolute terms occurs in the heavy mineral concentrate but only in some cases were the anomalous contents unique to the heavy mineral concentrate, Barakso and Tegart (1982) (Figure 13). The gold, as noted by Barakso and Tegart, in part

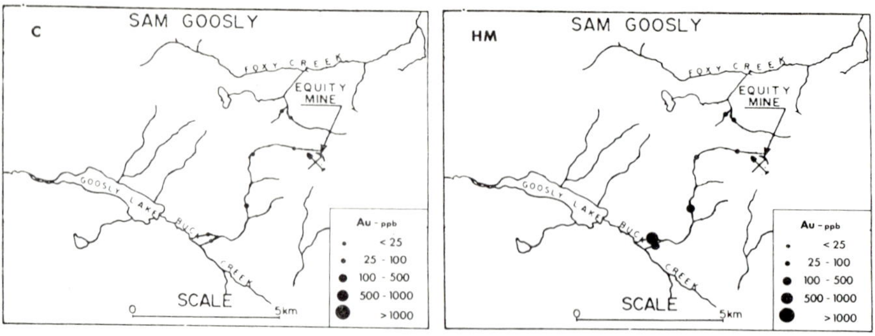

Figure 13 Comparisons of gold content of minus 177 μm (80 mesh) 'C' and heavy mineral fraction 'HM' of stream sediments in the Sam Goosly area, British Columbia. (From Barakso and Tegart, 1982)

occurs as chemically precipitated flour gold and is not concentrated in the heavy mineral fraction.

In an attempt to reduce variability of replicate analyses of stream sediment samples, Harris (1982) obtained heavy mineral concentrates of fifteen samples and analysed them for gold. However, gold values in the heavy mineral concentrate were significantly higher than the minus 177 μm (80 mesh) fractions in only 40% of the samples. This situation was attributed to much of the gold being held as inclusions in silicates and not liberated.

One of the most important aspects for the successful use of heavy mineral concentrate geochemistry in gold exploration is the selection of appropriate sample sites. From theoretical considerations discussed previously, gold grains of a particular size and shape occur with lighter minerals of considerably greater dimensions, Tourtelot (1968). The evidence of gold being associated with unsorted glacial till as opposed to fluvioglacial deposits and varved clays in northern Scotland may be applied to present-day drainage environments. Thus meaningful results can only be obtained when care is taken to collect samples from sites characteristic of the appropriate depositional environment for the accumulation of gold.

Discussion and conclusions

It should be apparent from the consideration of theoretical aspects of sample representivity, analytical methods for gold analysis, and the nature of gold dispersion in different sample media that reliable geochemical exploration for gold requires careful attention to the design of the programme. The fundamental problem relates to sample representivity and that concentrations of gold in the sub-100 ppb range are frequently of interest.

A number of possible solutions to the problem have been described by Harris (1982), involving alternatively modifications to the sampling, sample preparation or analytical procedures.

With regard to selection of appropriate sites or material, this feature is highlighted by the theoretical considerations of Tourtelot (1968) that draw attention to the size-relationship of hydraulically equivalent grains of different specific gravities. According to the grain size of the gold in a stream sediment sample so the sample has to be collected from a site where gold of this size would accumulate together with coarser grained lighter minerals.

With regard to the sampling of basal till, it is important that the sample is taken from till as opposed to other types of glacial sediments of different provenance. To ensure this requirement is fulfilled is not always straightforward when the sampling is undertaken by reverse circulation drilling and positive identity of the material being sampled is difficult.

With regard to the collection of larger samples, this is absolutely critical. It has been shown that to meet precision requirements of ±50% at the 95% confidence level at a content of 64 ppb gold with particles 62 μm in diameter an 800 g sample must be analysed. This means inevitably that a sample of several kilograms must be collected in the first instance. The size of the sample necessary to constitute a representative sample will vary inversely with the size of the gold particles and content features that vary from case to case.

The collection of duplicate samples is a further procedure to increase the reliability of the estimation of gold content at a particular location. However, although information on between site variability of gold content is necessary to aid interpretation, a more satisfactory procedure is surely to collect a single appropriately large sample that is a composite from several sites within a short distance of one another.

The initial sieving of samples at the sample site may be advantageous as a means of reducing the weight of the sample to be transported. However, wet sieving of stream sediments is not recommended as any fine gold will be lost.

Regarding sample preparation stages, care should be taken to ensure that the entire sample is prepared (e.g. sieved, crushed, etc.) and thoroughly homogenized, rather than sieving only sufficient for analysis, such a portion possibly being unrepresentative of the entire sample.

The reduction of the size of the gold particles is a further alternative, but as discussed previously gold is malleable and not prone to easy grinding and in addition gold may be smeared on to the walls of the pulverizer, which may then serve as a source of contamination to subsequent samples.

Perhaps the most satisfactory alternative is to adopt a sample preparation procedure that concentrates the gold in the heavy mineral fraction. However, particular care needs to be exercised that the gold is retained in the heavy mineral concentrate and not floated off in the field-panning process.

At the analytical stage, representivity can be improved by replicate analyses and the calculation of averages. However, it is surely more satisfactory to obtain a more representative sample in the first instance. Analysing larger samples is a further alternative but has an adverse effect on productivity.

In summary, the diversity of the nature of gold deposits and the weathering environment create a range of dispersion characteristics of gold dispersion associated with gold deposits. Careful design of geochemical exploration methodology has never been more critical for the success of an exploration programme than in the case of gold.

References

Antweiler, J.C. and Campbell, W.L. (1977). Application of gold compositional analyses to mineral exploration in the United States, *JGE*, **8**, 17–29.

Barakso, J.J. and Tegart, P. (1982). Revised stream sediment geochemistry at selected precious metal deposits in British Columbia, Western Miner, **April**, 53–65.

Boyle, R.W. (1979). The goechemistry of gold and its deposits (with a chapter on geochemical prospecting for gold), *GSC Bull.*, **280**.

Boyle, R.W. (1979). The geochemistry of gold and its deposits (with a chapter on geochemical Keno Hill area, Yukon Territory, *GSC* Pap., **71-51.**

Clarke F.W. (1980). The data of geochemistry; *Bull. US Geol. Surv.*, 330.

Clifton, H.E., Hunter, R.E., Swanson, F.J. and Phillips, R.L. (1969). Sample size and meaningful gold analysis, *US Geol. Surv. Prof. Pap.*, **625-C.**

Closs, L.G. and Sado, E.V. (1979). Geochemical drift prospecting studies near gold mineralization Beardmore-Geraldton area, Northwest Ontario, Canada. In: *Geochemical Exploration 1978* (Eds. J.R. Watterson and P.K. Theobald), pp. 459–477; Assoc. of Expln. Geochemists.

Curtin, G.C., Lakin, H.W., Neuerburg, G.J. and Hubert, A.E. (1968). Utilization of humus-rich forest soil (mull) in geochemical exploration, *US Geol. Surv. Circ.*, **562.**

Denisov, S.V., Koshman, P.N. and Yugay, T.A. (1966). Trace elements in gold from some districts of the Amur region, *Geochim. Int.*, **3**, 904–908.

Gay, N.C. (1963). A review of the geochemical characteristics of gold in ore deposits, *Univ. Witwatersrand. Econ. Geol. Res. Unit. Information Circ.*, **12.**

Gleeson, C.F. and Boyle, R.W. (1979). Consider geochemistry when seeking gold, *Northern Miner 64*, **52,** Cl.

Gray, R.S. (1983). Overburden drilling as a tool for gold exploration, 85th Annual Meeting CIM, Winnipeg.

Guindon, D.L. and Nichol I. (1983). Speciation of free gold in glacial overburden, *Ontario Geol. Surv. Open-File Rep. 5444.*

Gy, P. (1954). Erreur commise dans le prevelement d ún echantillon sur un lot de minerai (Errors committed in taking a sample from a lot of ore). *Revenue de l'Industrie Minerale*, **35,** 311–345.

Harris, J.F. (1982). Sampling and analytical requirements for effective use of geochemistry in exploration for gold. In: *Precious Metals in the Northern Cordillera*, pp. 53–67; The Assoc. of Expln. Geochemists, Spec. Vol. 10.

Hodgson, C.J., Chapman, R.S.G. and MacGeehan, P.J. (1982). Application of exploration criteria for gold deposits in the Superior Province of the Canadian Shield to gold exploration in the Cordillera. In: *Precious Metals in the Northern Cordillera*, pp. 173–206; The Assoc. of Expln. Geochemists. Spec. Vol. 10.

Hoffman, E.L. and Brooker, E.J. (1983). Biogeochemical prospecting for gold with reference to some Canadian Gold Deposits. Paper presented at Organic matter, biological systems and mineral exploration symposium UCLA.

Lakin, H.W., Curtin, G.C. and Hubert A.E. (1974). Geochemistry of gold in the weathering cycle; *US Geol. Surv. Bull.*, **1330.**

Lakin, H.W. and Nakagawa, H.M. (1965). A spectrophotometric method for the determination of traces of gold in geologic materials, *US Geol. Surv. Prof. Pap.*, **525C**, 168–171.

Landsweert (1869). Metallic deposits—Gold Nuggets Theory as to their formation, *Mining and Scientific Press, San Francisco*, **17,** 82, 274 and 306.

Lewis, A. (1982). Gold geochemistry, *Eng. Min. J.*, **(December),** 56–60.

Ottley, D.J. (1966). Gy's sampling slide rule, World Mining, **19, 9,** 40–44.

Overstreet, W.C. and Marsh, S.P. (1981). Some concepts and techniques in geochemical exploration. In: *75th Aniv. Vol. Econ. Geol.* (Ed. B.J. Skinner), pp. 775–805.

Plant, J.A. and Coleman, R.F. (1973). Application of neutron activation analysis to the evaluation of placer gold concentrations, *London IGES,* 373–381

Roslyakov, N.A. (1983). Zoning of forms of gold in the surficial environment as a criterian for buried gold deposits (abs.), 10th Into. Geochem. Explor. Symp., Helsink; pp. 66–67.

Springer, J. (1983). Invisible gold. In: *The Geology of Gold in Ontario.* (Ed. A.C. Colvine), pp. 240–250; Ont. Geol. Survey Misc. Paper 110.

Tourtelot, H.A. (1968). Hydrologic equivalence of grams of quartz and heavier minerals, and implications for the study of placers, US Geol. Surv. **594 F.**

Warren, H.V. and Thompson, R.M. (1944). Minor elements in gold, *Econ. Geol.,* **39,** 457–471.

White, M.V., Brooker, E.J. and Eagles, T.E. (1979). Some geological and geochemical criteria for the selection of exploration targets for base metals and gold (Unpubl).

Detection of Concealed Mineral and Energy Resources by Vapour Geochemistry

W. T. MEYER, J. S. LOVELL and M. HALE

Introduction

Gases and vapours, being inherently mobile, appear to offer an attractive means by which concealed mineralization or hydrocarbon accumulations may communicate their presence to the surface. Various gases and vapour techniques for the detection of hydrocarbons have been discussed since the 1930s, and for mineral deposits since the 1950s, with the earliest reports being principally in Russian literature. Research and development has continued, and analytical techniques in particular have greatly improved. However, with the possible exception of offshore hydrocarbon 'sniffing', vapour geochemistry has not yet achieved routine use in exploration and has remained a fringe technique showing a promise that is often not fulfilled. In view of the mobility which renders the gaseous species so attractive in the first place, it is perhaps not surprising that vapour geochemical anomalies tend to be transitory and difficult to repeat. Nevertheless, it is the purpose of this paper to suggest that vapour geochemistry does have a potential role to play in exploration and that the continued development of methods towards greater cost effectiveness, sensitivity and reliability will bear fruit in the 1980s.

The basis for vapour geochemistry

The relationship between mineral and energy resources and their gaseous dispersion patterns ranges from direct to conjectural. In the case of petroleum and gas reservoirs, the gaseous hydrocarbons up to and including butane may leak into the overlying sediments; even the liquid hydrocarbons up to quite high molecular weights have appreciable vapour pressures. Some hydrocarbon accumulations contain a few percent helium, and this gas may experience comparable leakage.

A less clear relationship exists between non-volatile minerals and gaseous species. In the case of uranium mineralization, both radon and helium are the products of the radioactive decay series of uranium. Mercury, being a chalcophilic element, is a common associate of sulphide mineralization. Although sulphur is ubiquitous in sulphide mineralization, the mechanism of its transformation into gaseous species remains obscure. There have been a number of descriptions of inorganic sulphur gas anomalies in the overburden above concealed sulphide mineralization. However, the majority of these surveys have not been substantiated by later work. Recent work by Taylor *et al.* (1982), has confirmed earlier work by Byman (1977) and has shown that, contrary to the prediction of thermodynamics (Rose *et al.*, 1979), the principal gases produced by oxidizing wet pyrite are not sulphur dioxide or hydrogen sulphide, but carbonyl sulphide (COS) and carbon disulphide (CS_2).

The relationship of the carbon dioxide and oxygen content of soil air to sulphide mineralization is indirect. The use of these gases as a mineral exploration tool was first suggested in Russia in 1955–1956 (Glebovskaya and Glebovskii, 1960). The suggestion was based on the belief that the oxygen consumed by the oxidation of the sulphide minerals was derived from the interstitial air of the overburden and that the acidic environment that is generated should decompose carbonate minerals with a consequent release of carbon dioxide into the surroundings. It can be shown that the oxidation of a single ton of pyrite to limonite could produce a fall of 1% oxygen (1%ΔO_2) in 350 000 m^3 of overburden and add approximately 0.5% CO_2 to this same volume (Lovell, 1979). Reports in the Russian literature have described the successful application of this technique (Kulikova, 1960; Glebovskaya, 1969) but, until recently, there has been little work in the West in this area.

The association between hydrocarbons and sulphide mineralization is related to the genesis of the ore itself. Deposits which may include an organic component at the time of formation include Mississippi Valley-type lead-zinc mineralization (Connan, 1979), roll-front and Athabasca unconformity uranium deposits (Granger and Warren, 1974) and kimberlites (Petersilie, 1960).

Dispersion

The dispersion of a gas into its surroundings is controlled by two principal mechanisms. First, a gas may diffuse due to a partial pressure or concentration gradient; and secondly, it may move by mass flow along an absolute pressure gradient. A third possibility is mass flow as a result of gravitational or pressure gradients whilst the gas is in solution.

The dispersion of gases from near-surface mineral deposits in the aerated environment is perhaps simple to visualize but, in fact, is still poorly understood. Radon, for example, has been shown to travel much farther than can be accounted for by diffusion alone (Tanner, 1964), suggesting that mass flow is an important transportation method, although the mechanisms by which mass flow is induced in the near-surface have not yet been adequately defined.

Hydrocarbon deposits are found at depths that are generally one or two orders of magnitude greater than ore bodies and the principal objections to the use of gas geochemistry in hydrocarbon exploration are related to the concepts of leakage and vertical migration. The source of the gas is readily apparent, but the route by which it reaches the surface is still a matter of some considerable conjecture. Hydrocarbon gases must escape from the reservoir through an overlying cap rock which, by definition, is considered to be impermeable. A leaking trap is generally not prospective for hydrocarbon accumulations. However, it is probable that, on a molecular scale at least, there is no such thing as a totally impervious rock. In addition, any fracturing that develops within the sedimentary column is likely to open in the direction of minimum stress, i.e. the fracturing will be vertical and will result in an anisotropy that will plainly favour vertical movement.

The movement of gases through rocks by diffusion alone is a very slow process and time periods of the order of millions of years are required for the gases to reach the surface. However, it has been reported that surface geochemical anomalies above producing oil fields have disappeared as the field is depleted (Horvitz, 1969). This has been confirmed by the experience of two of the authors (Meyer and Lovell) and it appears that the removal of the oil and gas from a reservoir may alter the surface geochemical expression in periods of as little as 25 years. Clearly, a mass transport mechanism must exist to bring the gases to the surface.

A possibility exists that this mass movement within a sedimentary basin may be promoted by the twice daily response to earth tidal movement. The stress changes brought about by this diurnal variation are likely to find expression in the flexing of the strata, the opening and closing of fractures and, probably, the mass transfer of fluids. Sugisaki (1981) demonstrated that the ratio of the deep source gases ^{40}Ar and He being emitted from a hydrothermal spring varied in direct relationship to the calculated earth tidal strain and showed no correlation with diurnal temperature or barometric changes. The concentration of helium in natural gas reservoirs has also been shown to vary with earth tides (Tugarinov and Sardarov, 1975).

It is surprising that there is so much resistance to the concept of near-surface expressions of oil and gas leakage developing onshore, as a great many of the oil fields of the world are associated with either macroscopic oil seepages or the development of 'paraffin dirt' by the action of micro-organisms on leaking gases (Link, 1952). Offshore, the search for hydrocarbon leakage is facilitated by the existence of a biologically inactive, anaerobic zone, virtually down to the sediment surface, and by the general absence of groundwater movement. A further advantage is that the presence of leakage can often be ascertained by the escape of visible quantities of bubbles from the sea floor. However, if offshore fields are leaking gases, then there is no apparent reason for doubting that onshore fields should also leak, although the search will be made more difficult by microbial oxidation of the hydrocarbons, groundwater movement and the generally higher energy of the near-surface onshore environment.

Methods

Reliable collection and analysis of trace quantities of gases requires sensitive analytical techniques. Indeed, the history of vapour geochemical research has generally been governed by the availability of suitable analytical techniques. Thus, helium geochemistry required the development of sensitive, inexpensive mass spectrometers; interest in mercury closely followed the introduction of flameless atomic absorption techniques; radon relied upon the development of portable scintillation counting methods and alpha particle-sensitive films; and sulphur and hydrocarbon gases have been greatly dependent upon the introduction of more sensitive, reliable and rapid gas chromatography.

The sampling techniques may be divided into two principal types. Firstly, the free interstitial soil air may be sampled by the use of a hollow probe penetrating to various depths (Jones and Drozd, 1983). The sample depth is generally of the order of 1 m, but hydrocarbon sampling from depths of up to 50 m has been reported. The soil air may be pumped directly to a portable field instrument, for example, for the on-site measurement of carbon dioxide and oxygen concentrations (Lovell *et al.*, 1980) or may be returned to the laboratory in a suitable container for chromatographic analysis.

The second method of sampling relies upon the determination of gases absorbed either onto the soil itself or onto passive absorbents such as activated charcoal or molecular sieves. The gases are normally released for instrumental determination by heat or chemical treatment. The main advantage of using absorbents, whether natural or emplaced, is that they should help to avoid the problems which result from rapid short term fluctuations related to diurnal or meteorological changes, and should thus give a more representative sample. Integration periods of up to four weeks may be used for emplaced absorbents or alpha particle-sensitive films. A recent development in this area has been announced by Klusman and Voorhees (1983). An activated charcoal trap cemented to a Curie point wire is emplaced in a cavity in the soil for up to two weeks. After collection, the charcoal is thermally desorbed by radio frequency heating and the effluent is carried into a quadrupole mass spectrometer. Pattern recognition techniques are used to distinguish the fingerprints of the volatiles collected above oil and gas from those gathered in background areas.

Case histories

There are numerous case histories within the literature and within the experience of the authors. Those included in this section of the paper have been chosen to demonstrate a variety of techniques and to indicate the scope of gas and vapour geochemistry.

PATRICK DRAW OIL FIELD, WYOMING

The Patrick Draw Field lies some 120 miles to the east of Rock Springs in southern Wyoming. It has produced over 100 million barrels of oil from a stratigraphic pinch-out in the Upper Cretaceous Almond Formation at a depth of 1250–1400 m. The field is cut by a number of vertical extensional faults with an east-northeast to easterly trend.

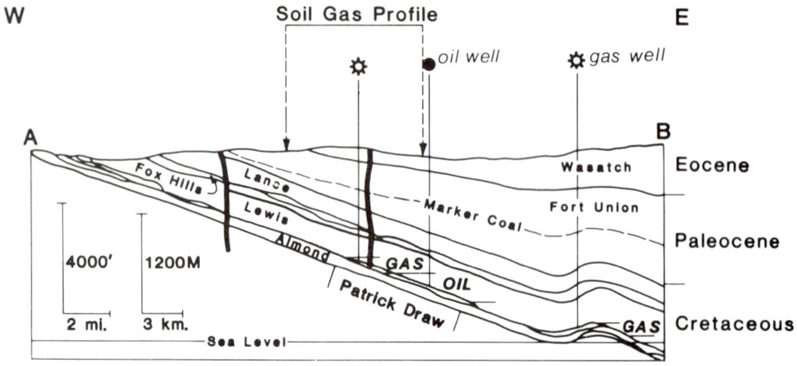

Figure 1 Geological cross-section (after Weimer, 1966) and location of soil gas profile, Patrick Draw oil and gas field, Wyoming

Soil air samples were collected from a depth of 75 cm at 320 m intervals along a line across the field and the gas cap and extending into background areas (Figure 1). Gas chromatographic analysis of the samples revealed highly anomalous concentrations of hydrocarbons in the soil air. These anomalous values have a maximum of 1.5% methane and 1% ethane over the gas cap, but the values across the oil cannot be distinguished from background (Figures 2 and 3). The extreme levels of hydrocarbons in the soil air would appear to account for the vegetational stress that overlies the gas cap (Richers *et al.*, 1982). Helium gave a less well-defined expression of the gas but a maximum value of 100 ppb ΔHe (5340 ppb total He) occurred immediately above the gas cap (Figure 3).

DRAGOON OIL FIELD, COLORADO

The Dragoon Field lies approximately 35 miles to the east of Denver in northeastern Colorado. Oil is trapped at a depth of about 2300 m in a pinch-out of the D_1 (Lower Cretaceous) sand (Figure 4). The D_1 zone represents bar sands that have generally good porosities and are superimposed on deltaic sediments developing from the northeast. The D_1 sands are potentially prolific producers, but generally present difficult exploration targets.

Soil samples were collected along a line across the southern half of the field at 80 m intervals from both the surficial A_0 soil horizon and at a depth of approximately 50 cm. Soil gas samples for helium analysis were collected at 320 m intervals at a depth of 75 cm. The samples were analysed for total volatile sulphur and hydrocarbons using a system of controlled heating and direct injection into a gas chromatographic detector (Lovell, 1979). The results of these analyses show a good expression of the oil field in the surficial A_0 soil horizon; this expression is absent from the deeper soils (Figure 5).

Helium values across the field (Figure 6) show an extraordinarily strong expression of the hydrocarbon reservoir with the highest values being concentrated over the eastern portion. This reflects the thickest and up-dip

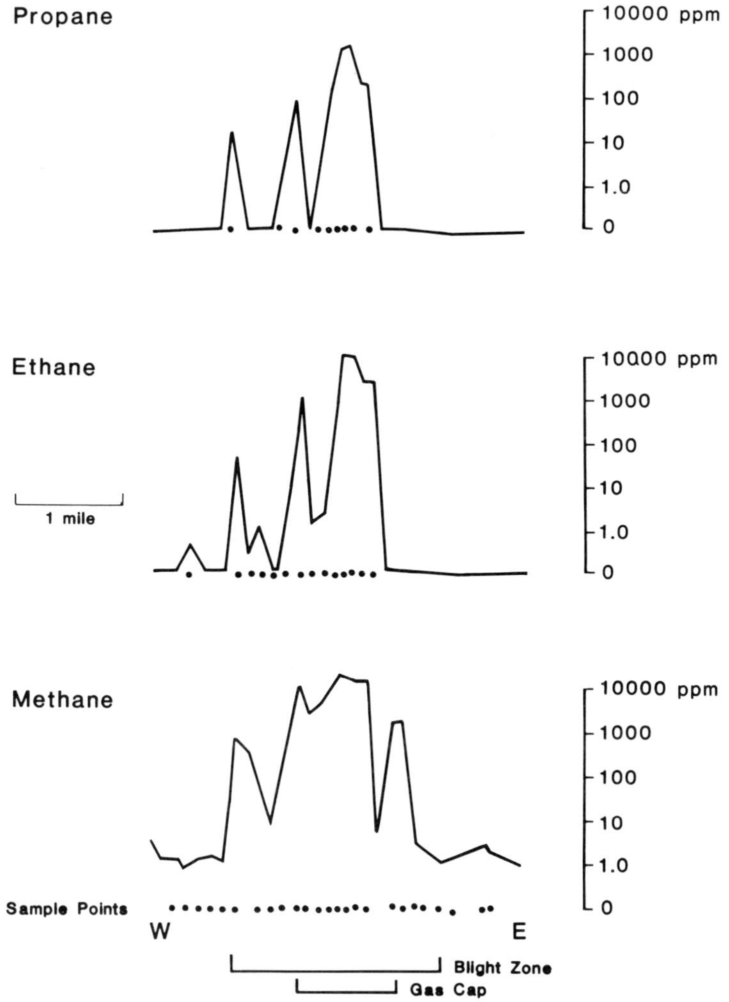

Figure 2 Methane, ethane and propane in soil gas over the Patrick Draw oil and gas field, Wyoming

area of the sand. The maximum value was 10 670 ppb ΔHe (15 910 ppb total He) which is an exceptional anomaly contrast. Beyond the boundaries of the field the helium values are generally below the normal threshold of 60 ppb Δ He (5300 ppb He).

RABBIT HILLS FIELD, MONTANA

Jurassic oil is produced from a depth of 1050 m at the Rabbit Hills Field, Montana. Surficial A_0 soil horizon samples were collected on a 12.5 km line

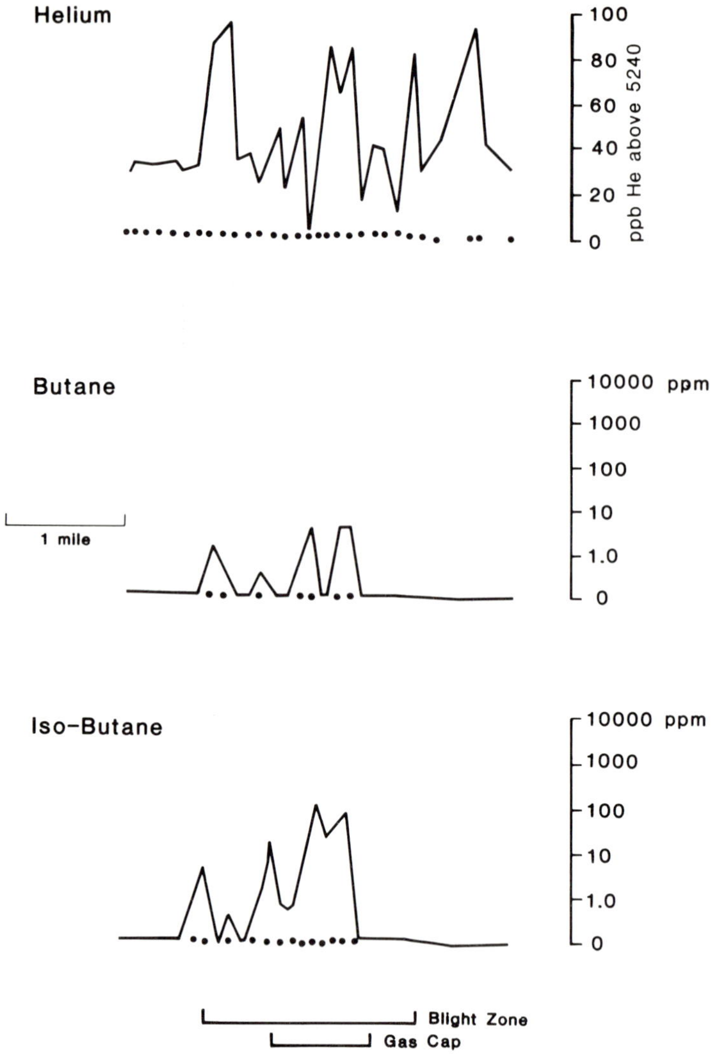

Figure 3 Helium, butane and iso-butane in soil gas over the Patrick Draw oil and gas field, Wyoming

with a sample interval of 320 m. These samples were analysed for evidence of methane, ethane and propane leakage. It is very clear that the field has a good response, particularly in the propane data, with a much less well-defined expression in methane (Figure 7). This is generally diagnostic of oil accumulations although methane is still the most common gaseous hydrocarbon present in the reservoir. Over gas reserves, particularly if they are dry biogenic gas, the best expression is generally provided by methane with the propane expression either absent or poorly defined.

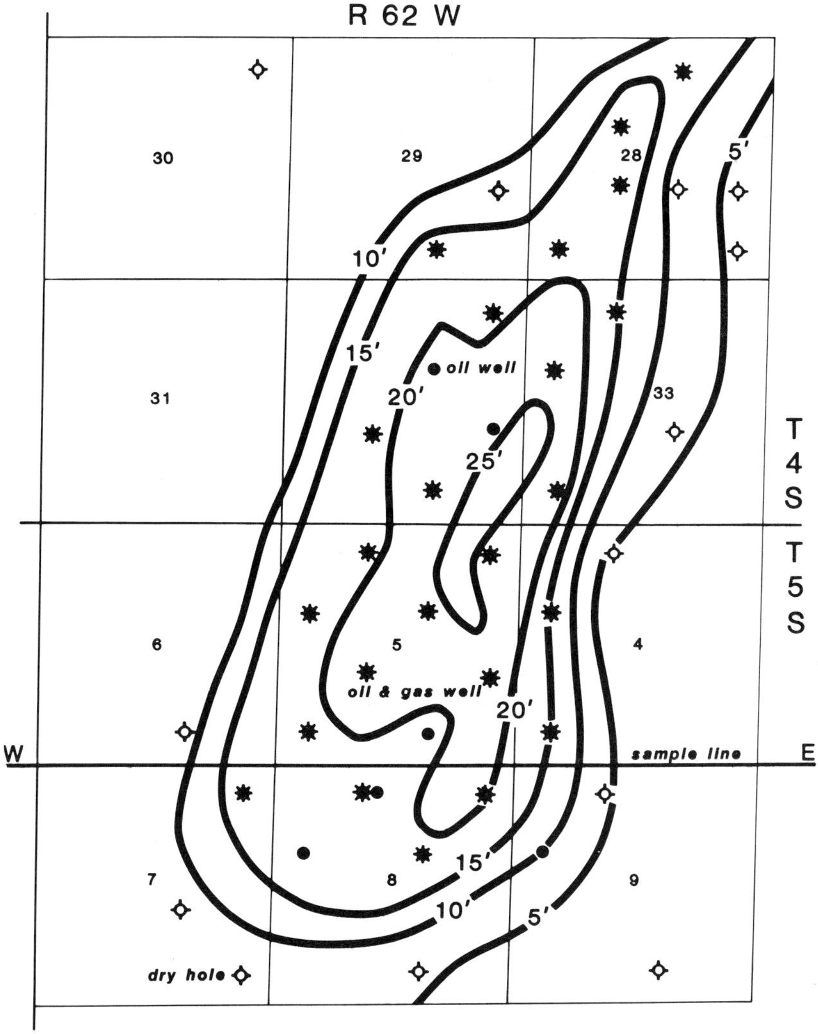

Figure 4 Total D sand isopach and location of sampling traverse, Dragoon oil and gas field, Colorado. Each numbered section represents one square mile

DRILL-SITE TEST PROGRAMME

To use a producing oil field as a geochemical test area can raise many potential objections. Is the expression of the field a result of leakage from the casings or is it due to the drilling activity? A producing reservoir is, in effect, a trap which is venting to the surface on a very major scale through controlled mechanical rupture of the seal. A producing field may give a very different surficial geochemical expression from an undisturbed accumulation.

Therefore a series of trials was carried out (on behalf of Barringer Resources

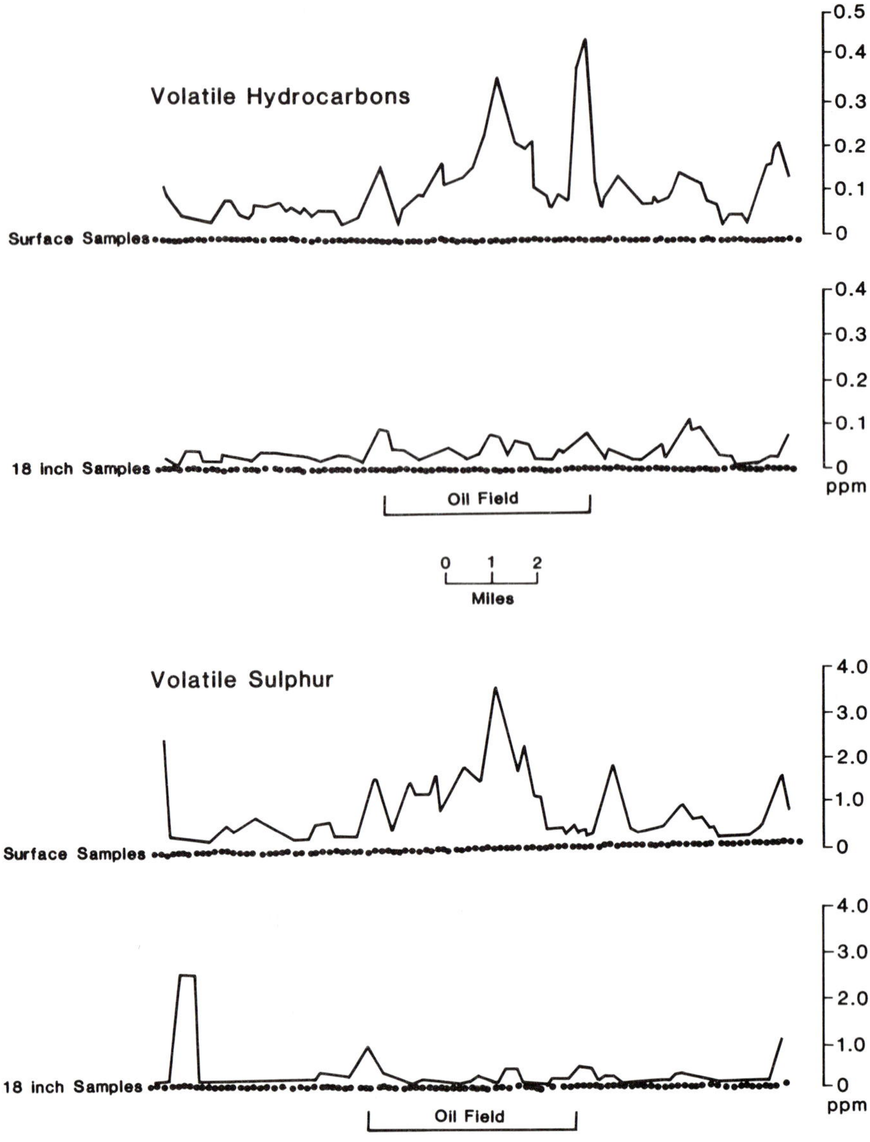

Figure 5 Volatile hydrocarbons and sulphur in surface and 50 cm (18 inch) deep soils overlying the Dragoon oil and gas field, Colorado

Inc.) over a total of 49 proposed drill sites covering a variety of surface and subsurface environments in Kansas and Colorado. Test sites made available to Barringer were selected by mean of conventional geological and seismic methods. Soil samples were collected on star-shaped patterns with arms of 400 m, centred on the proposed drill sites, and analysed prior to any surface disturbance or drilling.

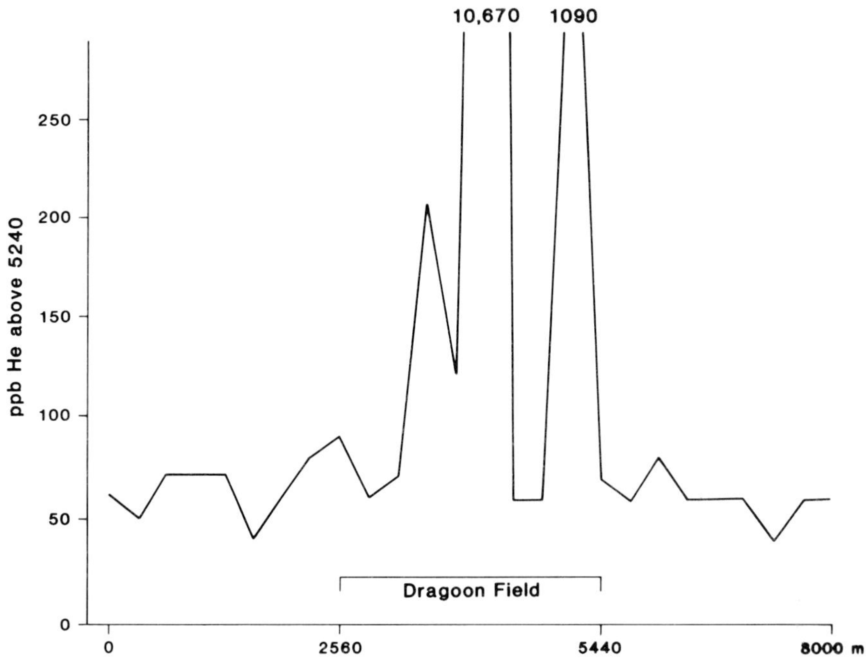

Figure 6 Helium in soil gas collected over the Dragoon oil and gas field, Colorado. Baseline for the profile is the normal content of helium in the atmosphere

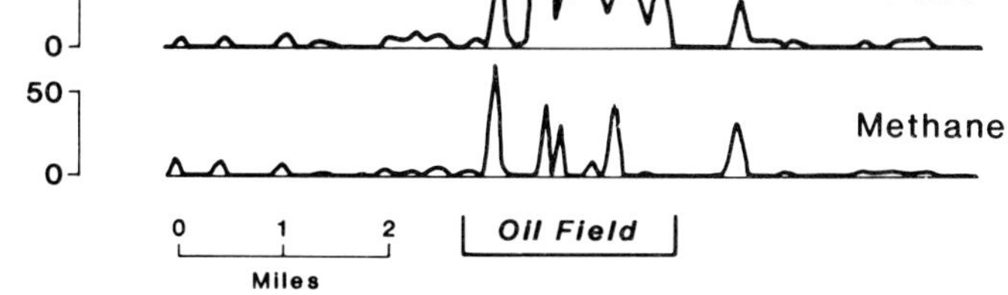

Figure 7 Propane, ethane and methane in surface soils from a traverse line across the Rabbit Hills oil field, Montana. Data scaled to maximum reading equals 100 units

Geochemistry did not play a part in the selection of the 39 locations that were drilled and only 3 wells were productive. The methane, ethane and propane values for the 9 samples at, and immediately surrounding, each proposed location were averaged allowing a reasonably accurate estimate of the leakage at the site. These were then plotted as histograms of the entire data

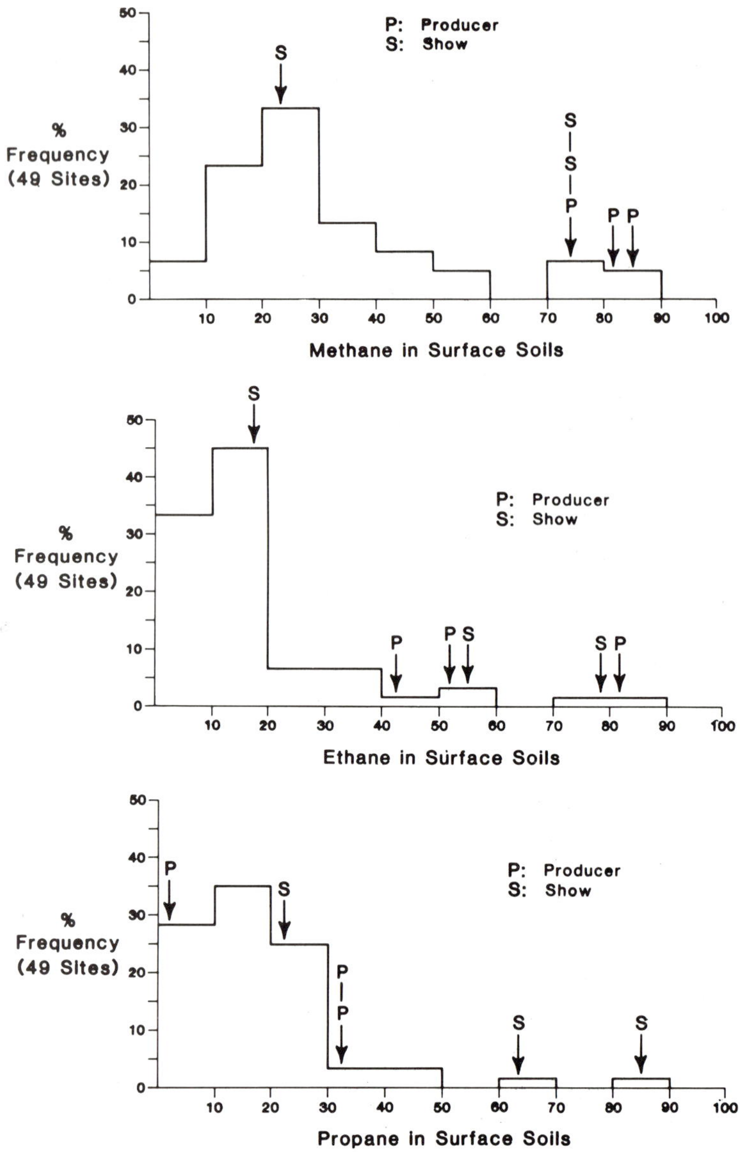

Figure 8 Frequency distribution histograms of (a) methane, (b) ethane and (c) propane in surface soils averaged at each drill site and presented on a scale with 100 representing the maximum value recorded. The average values of soils over producing and 'show' wells are indicated

set (Figures 8a, 8b and 8c). It is readily apparent that the productive wells are strongly anomalous when compared to the rest of the sites, particularly in methane and ethane. One of the producers is not anomalous in propane, but analysis of the natural gas showed it to be very dry, with less than 1% ethane and only traces of the higher molecular weight hydrocarbons.

These results indicate that leakage of hydrocarbons does occur from undisturbed reservoirs and this may lead to a detectable anomalous expression at the surface.

JOHNSON CAMP, ARIZONA

Research into three vapour geochemistry techniques for mineral exploration has been carried out at Johnson Camp, between the towns of Benson and Wilcox, southeast Arizona. In this area, metamorphosed limestones host three (or more) discontinuous tabular sheets of mineralization, thought to have been introduced during the last stages of emplacement of the nearby Texas Canyon quartz monzonite stock (Cooper and Silver, 1964). The principal sulphide minerals are sphalerite, chalcopyrite and bornite, and typical grades are 0.8% Zn and 0.8% Cu. The mineralized zones dip to the northeast, and their suboutcrops are concealed beneath pediment gravels and alluvium which thicken to the northeast. At surface there is an alkaline sandy soil with variable development of caliche, and a sparse vegetation of sagebrush, creosote brush and yucca.

Soil air samples were taken from a depth of approximately 1 m at sites 10–30 m apart along an 1800 m traverse, extending in the line of dip across the Zone I and Zone II mineralization. Portable oxygen and carbon dioxide analysers connected in series were used to record the concentration of the respective gases at the sample site. The results are expressed as absolute CO_2 enrichment and as relative oxygen depletion (ΔO_2) from the 21% O_2 level in the atmosphere. The oxygen depletion is seen as a reflection of oxygen demand during active sulphide oxidation, and the carbon dioxide enrichment as a manifestation of dissolution of carbonates by the resulting acidic

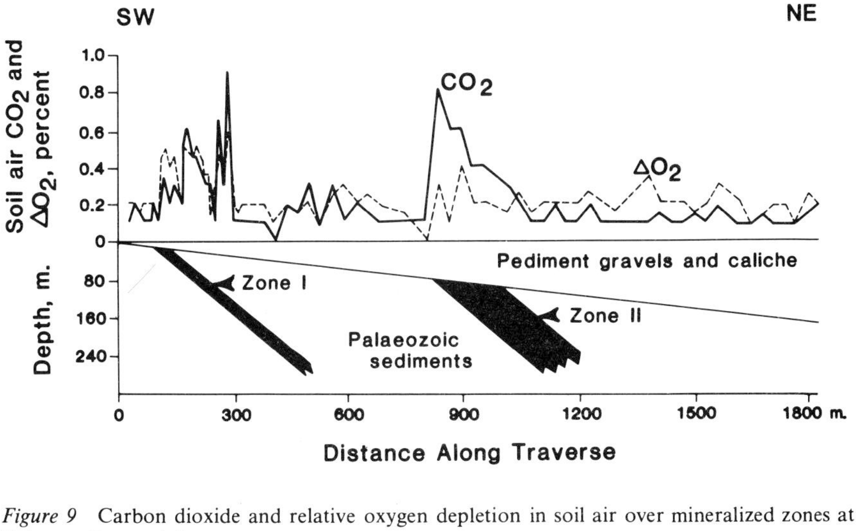

Figure 9 Carbon dioxide and relative oxygen depletion in soil air over mineralized zones at Johnson Camp, Arizona

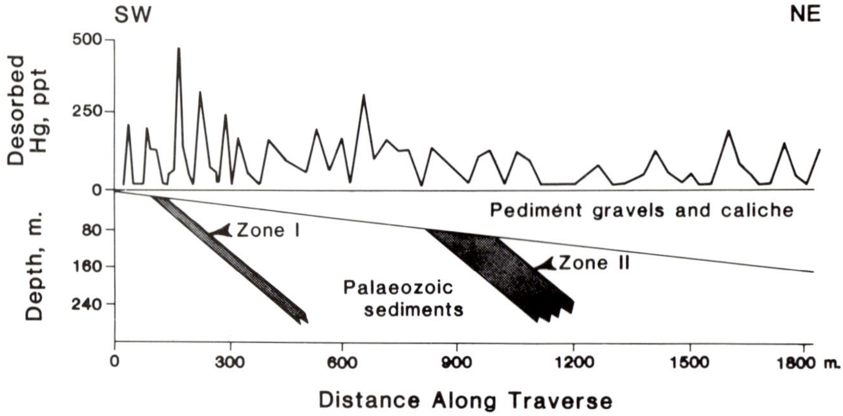

Figure 10 Thermally desorbed mercury from surface soils at Johnson Camp, Arizona

solutions. Concentrations up to 0.9% CO_2 and 0.75% ΔO_2 occur in the soil air overlying the concealed suboutcrop and immediate downdip extension of Zone I mineralization (Figure 9). Values of 0.8% CO_2 and 0.4% ΔO_2 occur over the concealed suboutcrop of Zone II. These anomalous levels compare with a background range of 0.1–0.3% for both gases. The peaks of each of the soil air anomalies are apparently due to the oxidation of mineralization 80 –100 m below surface. This can therefore be considered to be the depth at which most active oxidation is taking place at Johnson Camp, and compares reasonably well with a report of oxidation to a depth of 60 m at the nearby Peabody Mine (Scott, 1916). The anomalous CO_2 and ΔO_2 measurements falling between threshold and peak values probably reflect weaker mineralized sources such as sulphide remnants in the oxide section of Zone I, and preferential oxidation in fractures in the more deeply buried sulphide section of Zone II.

Samples of the A_0 horizon of the soils along the same traverse were collected and analysed for sorbed mercury (Hg) vapour. This form of Hg was selectively stripped by subjecting samples to a vacuum reduction step, and introducing the desorbed Hg into a flameless atomic absorption spectrophotometer (Ruan, 1981). Anomalous Hg vapour concentrations of up to 500 ppt are found above the downdip extension of Zone I immediately below the suboutcrop (Figure 10), compared with background concentrations of up to 200 ppt. The peak value is, however, displaced updip relative to the CO_2 and ΔO_2 peaks in this section of the traverse. There is no unambiguous expression of Zone II in sorbed Hg. A single point anomaly 150 m southwest of the Zone II suboutcrop may be due to Hg vapour migration along northeast dipping bedding planes in the pediment gravels and alluvium. Hg vapour may, therefore, be more prone to some degree of lateral migration along preferred channelways in the overburden than CO_2 or O_2.

The surface samples were also analysed for their total volatile sulphur content released by thermal desorption. The uncalibrated data are interpreted

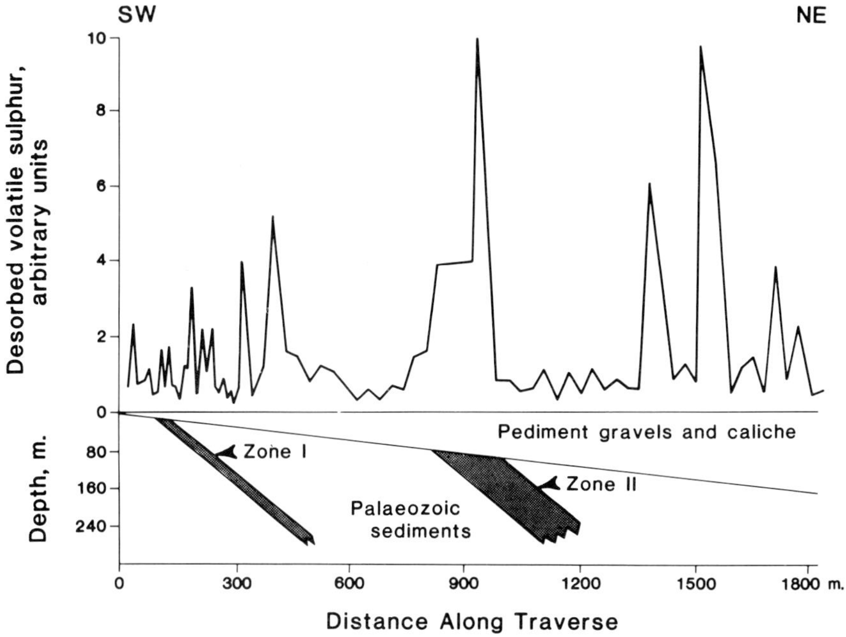

Figure 11 Desorbed volatile sulphur from surface soils over concealed sulphides at Johnson Camp, Arizona

as three anomalous halos of up to 10 units compared with a background variation below 3 units (Figure 11). Two of the anomalies have an approximate spatial coincidence with the CO_2 and ΔO_2 anomalies in soil air, but in detail each of the volatile sulphur peaks are about 100 m further to the northeast than the corresponding CO_2 and ΔO_2 anomaly peaks. A source further downdip in Zone I and Zone II mineralization is therefore suggested. The third desorbed volatile sulphur halo, towards the northeastern end of the traverse, does not coincide with any of the other vapour geochemical anomalies described above. This is probably a spurious anomaly, but could be derived from an extension of Zone III which is known to suboutcrop approximately 150 m to the northwest.

Speciation of the total desorbed volatile sulphur from Johnson Camp soils showed the main constituent of the anomalies to be carbonyl sulphide (COS). Further research then provided a rapid and extremely sensitive thermal desorption/gas chromatography method for the determination of COS in soils. This was applied to a grid of 186 surface samples collected at 30 m intervals along six southwest/northeast trending lines 60 m apart. The area covered by the grid and the COS pattern determined from the samples are shown in Figure 12. Background values of less than 200 pg g^{-1} COS are found over most of the area. In the vicinity of the suboutcrop of Zone II, there are discontinuous COS anomalies exceeding 300 pg g^{-1}. The strongest anomaly extends over the suboutcrop of the contact between the mineralization and the

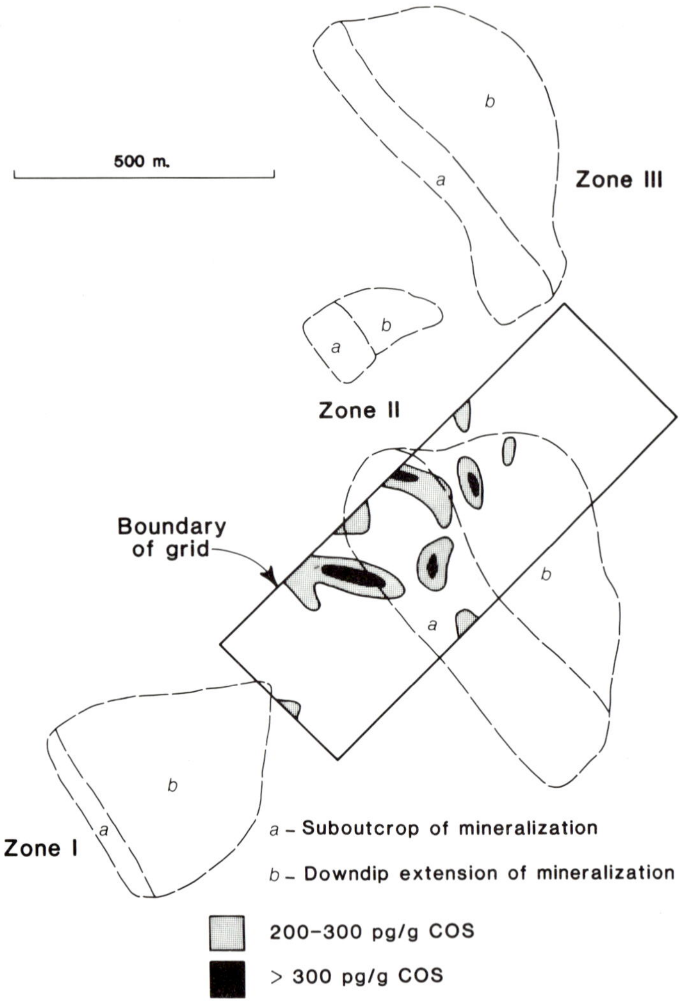

Figure 12 COS desorbed from surface soils collected on a 30×60 m grid covering Zone II mineralization, Johnson Camp, Arizona

footwall rocks. Zone II mineralization is at its shallowest about 80–90 m below surface, and at this depth is likely to be experiencing the most active oxidation. Extension of this COS anomaly to the southwest may again reflect vapour dispersion along northeasterly dipping bedding planes in the overburden. Other COS anomalies occur above the suboutcrop of Zone II, and there is one anomaly of more than 300 pg g^{-1} above the downdip extension.

The determination of COS in the surface soils formed on exotic overburden has since been applied (along with geophysical methods) to exploration for concealed mineralization in the Johnson Camp area (disclosure of the exact location is withheld for commercial reasons). An exploratory drillhole close to

COS and IP anomalies subsequently intersected sulphide mineralization at a depth of 75 m.

Discussion and conclusions

This paper is not intended to be an exhaustive review of either sampling or analytical methods. The case histories have been chosen to demonstrate that concealed hydrocarbons or mineralization may give significant vapour geochemical anomalies at the surface. Additional examples of vapour geochemistry that could have been included are: mercury in soil air; radon for uranium, oil and geothermal exploration; and helium for uranium prospecting.

As vapour geochemistry is essentially an indirect search technique for minerals, it should be employed as a useful adjunct to other geological, geochemical and geophysical programmes in areas of exotic overburden. In mineral exploration it may, perhaps, find its most useful role in testing geophysical anomalies and in extending known mineral reserves beneath transported cover. On the other hand, surface geochemical methods can form an integral part of both regional and detailed programmes in hydrocarbon exploration. There remains, of course, a continuing need for further refinement and development of these techniques and a need to demonstrate their application in both mineral and energy exploration programmes.

ACKNOWLEDGEMENTS

The case histories reported in this paper are drawn from work carried out by Barringer Resources Inc., Denver, USA, and the Applied Geochemistry Research Group, Department of Geology, Imperial College, London. Vapour geochemistry research for mineral exploration was sponsored at the Applied Geochemistry Research Group by a Mineral Industries Research Organization syndicate comprising Charter Consolidated p.l.c., Consolidated Gold Fields p.l.c., General Mining Union Corporation (UK) Ltd. and RTZ Services Ltd. The work at Johnson Camp reported here was carried out with the help of J.A. Kantor and M.E. Hinkle, and with the permission of New Beginnings Resources Incorporated and B.A. Black and Associates. The authors are greatly indebted to many of their colleagues whose work has contributed to the contents of this paper, and in particular Mr B.W. Oakes and Mr Ruan Tianjian, both formerly of the Applied Geochemistry Research Group, for their research on COS and Hg, respectively, at Johnson Camp.

References

Byman, J. (1977). Investigation of sulphur vapour techniques in exploration geochemistry, *M. Phil. Thesis, University of London* (unpubl.).

Connan, J. (1979). Genetic relation between oil and ore in some Pb–Zn–Ba ore deposits, *Geol.*

Soc. S. Afr., Spec. Publ., **5,** 263–274.

Cooper, J.R. and Silver, L.T. (1964). Geology and ore deposits of the Dragoon Quad., Arizona, *US Geol. Surv. Prof. Paper 416.*

Glebovskaya, V.A. (1969). Gaseous surveying in prospecting for ore deposits. In: *Geochemical Methods of Prospecting for Oil and Gas Deposits,* Proceedings of the Conference on Geochemical Methods, Moscow 1969, 161–162 (in Russian).

Glebovskaya, V.S. and Glebovskii, S.S. (1960). The possibility of application of gas surveys in prospecting for sulphide deposits (in Russian). Translation announced in GSC Paper 74–58, Item 63, as GSC Translation No. 603.

Granger, H.C. and Warren, G.C. (1974). Zoning in the altered tongue associated with roll-type uranium deposits. In: *Formation of Uranium Ore Deposits,* IAEA Proceedings, Athens 1974.

Horvitz, L. (1969). Hydrocarbon geochemical prospecting after 30 years. In: *Unconventional Methods in Exploration for Petroleum and Natural Gas,* (Ed. W.B. Heroy), pp. 205–218; Southern Methodist University, Dallas, Texas.

Jones, V.T. and Drozd, R.J. (1983). Prediction of oil or gas potential by near surface geochemistry, *AAPG Bull.,* **67,** no. 6, 932–952.

Klusman, R.W. and Voorhees, K.J. (1983). A new development in petroleum exploration technology, *Mines Magazine,* Colorado School of Mines, Golden, March 1983.

Kulikova, N.N. (1960). On the use of a gas-core survey for detecting blind ore zones in the region of the Balei deposits. Izvestiya of Higher Educational Institutions, Geology and Exploration, Moscow 1960, no. 12, 78–82.

Link, W.K. (1952). Significance of oil and gas seeps in world exploration, *AAPG Bull.,* **36,** 1505–1541.

Lovell, J.S., Hale, M. and Webb, J.S. (1980). Vapour Geochemistry in mineral exploration, *Mining Magazine,* **September,** 229–239.

Lovell, J.S. (1979). Applications of vapour geochemistry to mineral exploration, *PhD Thesis, University of London* (unpubl.).

Petersilie, A. (1960). Combustible gases and bitumens of alkali rocks on the Kola Peninsula. Petrographic provinces and metamorphic rocks, Internat. Geol. Congress, XXI session, reports of Soviet geologists, Problem 13, Moscow 1960.

Richers, D.M., Reed, R.J., Horstman, K.C., Michels, G.D., Barber, R.N., Lundell, L. and Marrs, R.W. (1982). Landsat and soil-gas geochemical study of Patrick Draw oil field, Sweetwater County, Wyoming, *Am. Assoc. Pet-Geol. Bull,* **66**, no. 7, 903–932.

Rose, A.W., Hawkes, H.E. and Webb, J.S. (1979). *Geochemistry in Mineral Exploration.* (Second ed.); Academic Press, London.

Ruan, T. (1981). Some new approaches in vapour geochemistry. Research Report, Imperial College.

Scott, W.A. (1916). Mining operations at Johnson Camp, Arizona, *Mining Engineering World,* **45,** 141–143.

Sugisaki, K. (1981). Deep seated gas emission indicated by the earth tide: A basic observation for geochemical earthquake prediction, *Science,* **212,** 1264–1266.

Tanner, A.B. (1964). Radon migration in the ground. In: *The Natural Radiation Environment* (Eds. J.A.S. Black and W.M. Lowder) pp. 161–178; University of Chicago Press.

Taylor, C.H., Kesler, S.E. and Cloke, P.L. (1982). Sulfur gases produced by the decomposition of sulfide minerals: application to geochemical exploration, *J. Geochem. Explor.,* **17,** no. 3, 165–185.

Tugarinov, A.Z. and Sardarov, S.S. (1975). Variations in the deep streams of radiogen gases as a consequence of elastic deformations in the earth's crust, *Proc. Acad. Sci. USSR,* **223,** no. 4, 668–671.

Weimer, R.J. (1966). Time-stratigraphic analysis and petroleum accumulations, Patrick Draw Field, Sweetwater County, Wyoming, *AAPG Bull.,* **50,** 2150–2175.

Regional Geochemistry in the Detection and Modelling of Mineral Deposits

J. A. PLANT, M. D. FORREST, J. F. HODGSON, R. T. SMITH and A. G. STEVENSON

Introduction

Following the pioneering work of H.E. Hawkes and J.S. Webb in the 1950s, many millions of items of regional geochemical data have been accumulated covering large areas of the world. The data include chemical analyses of soils, waters, plants, rocks and, perhaps most important of all, stream sediment samples. It was the realization by John Webb that stream sediment samples reflect the average chemical composition of the catchment area upstream of the sampling point which enabled large regions to be covered rapidly and with a high probability of selecting the best areas for further exploration. Another of John Webb's important contributions was his early appreciation of the value of computers in the treatment and analysis of multi-element geochemical data. Therefore the theme which we have selected for this special volume concerns the application of regional geochemical maps based on stream sediment samples to mineral exploration using sophisticated computer techniques, and in particular, image analysis systems.

Regional geochemical mapping continues to be regarded as one of the most valuable methods for the direct discovery of ore deposits. However, as easily identified orebodies have been exhausted and large low-grade deposits have become less economic, exploration geochemistry has moved towards an approach based on conceptual models for specific high-value ore deposits such as granite-related Sn–W mineralization and high-grade deposits such as volcanogenic massive sulphides. Increasingly, emphasis is on the collection of precise, highly standardized, data on a wide range of elements, including heat-producing and key petrogenetic elements and those indicative of hydrothermal activity. At the same time, advances in computing and image processing are enabling sophisticated exploration criteria to be used for the rapid screening of large, computer-readable data sets including multi-element geochemical, geophysical and remotely sensed data, for exploration and resource evaluation. Regional geochemistry has a major role to play in this new area of exploration technology both by providing a view of geochemical variation comparable to that provided for geology by remotely-sensed data and as a basis for the identification of the geochemical characteristics of orebodies and their host rocks.

The geochemical maps of northern Scotland prepared by the Institute of Geological Sciences (IGS, 1978–83), on which the discussion that follows is based, show the distribution in stream sediments of 26 elements, including B, Ba, Be, Co, Cr, Cu, Li, Mo, Ni, Pb, Sn, V, U, Zn (based on sampling at a minimum density of 1 per 2 km^2) (Plant and Moore, 1979). The geochemical maps are evaluated first as a basis for the *direct* identification of areas for further exploration. A search area of approximately 3 km around recorded mineral occurrences is used, and highly anomalous values are defined as those greater than the ninety-fifth percentile for the area of the Great Glen 1: 250 000 map sheet, which contains a subset of the main geological units of the Caledonides of northern Scotland. The geochemical data are also used, together with regional geophysical, geological and remotely sensed data, in interactive studies based on the NERC I^2S interactive image analysis system to develop metallogenic models and exploration criteria. Three main types of metalliferous mineral association are considered:

1. Stratabound Ba–Zn–Pb, Fe–Cu–Ni and Fe–Cu–Zn mineralization in the Middle Dalradian metasediments of the Caledonian cover sequence.
2. Mo–Cu and Mo–(W) mineralization and Sn–U–Be–B–F enrichments associated with 'newer' Caledonian granite (*sensu lato*) plutons.
3. U–Mo enrichments and U (Ba and Pb) mineralization in post orogenic Old Red Sandstone basins.

The geochemical data used are those published in the Institute of Geological Sciences Regional Geochemical Reconnaissance Atlas Series (IGS, 1978–83) and they are available in computer-readable form from the UK National Geochemical Data Bank. The methods of sampling, analysis and data handling are described elsewhere (Plant and Moore, 1979). The mineral occurrences are based on the compilation of Gallagher *et al.* (1971), supplemented by data from the Institute's Mineral Reconnaissance Report Series.

Sampling as a basis for the detection and conceptual modelling of metalliferous mineralization

It is now widely recognized as a result of stable isotope and fluid inclusion studies of rocks (e.g. Sheppard and Taylor, 1974; Roedder, 1977), investigations of extant geothermal fields (e.g. Fairbank *et al.*, 1981), and hydrothermal activity associated with ocean ridge systems (e.g. Mottl, 1983), that many types of metalliferous mineral deposit form by the interaction of large volumes of epizonal water* with rocks. Hence descriptive classification systems, for example that of Smirnov (1976), are giving way to conceptual models of ore deposition as a basis for exploration. The model for Kuroko

*Including metamorphic, formational and meteoric water.

massive sulphide deposits associated with calc-alkaline volcanism on the sea floor in island arc tectonic settings (Ohmoto and Rye, 1974; Sakai and Matsubaya, 1974) provides one such example. These deposits and associated sediments, like those of Cyprus type (Spooner and Fyfe, 1973), are accounted for by a model of heat and mass transfer in a sub-sea floor geothermal system involving large scale convective interaction of seawater and volcanics. The metals are derived by leaching of the volcanic rocks, which also supply heat to drive convective hydrothermal cells. Kuroko-type deposits probably correspond to single-pass convective systems with very high water/rock ratios (Fyfe and Kerrich, 1976); such hydrothermal mineralizing systems generally involving the distributed recharge of fluids while discharge, and hence metal precipitation, is highly focussed.

It is implicit in such models that metals are redistributed from primary silicates or accessory minerals into ore, gangue and secondary mineral phases. For example, the redistribution of U in granites as a result of mineralization has been described by Simpson *et al.* (1979). The uranium, which is held in zircon or other accessory mineral inclusions in primary biotite crystals in unaltered granite is remobilized into secondary minerals and late microfractures during mineralization; the mean U content is unchanged. Mineralization is generally found to produce little change in the average metal content over mineralized systems, although the standard deviation increases (Garrett, 1971; Simpson *et al.*, 1979) reflecting the redistribution of metals. Moreover, some metals in ore deposits, for example Au, may be concentrated by factors of the order of 10 000 : 1 compared to their source rocks (Anhauser *et al.*, 1975; Fyfe and Kerrich, 1976). A summary of the principal host minerals of some trace elements in metalliferous and mineralized rocks* is given in Table 1.

Stream sediments reflect anomalous concentrations of metals due to the occurrence of ore bodies, in addition to the primary elemental levels in rocks, and secondary redistribution as a result of hydrothermal alteration. This information is becoming increasingly important in understanding and modelling metallogenesis.

For example, the primary geochemistry of certain rock types reflected by petrogenetic elements such as K, Ba, Rb, Sr and REE may indicate their mineralization potential. Thus, highly fractionated REE patterns with marked negative Eu anomalies have been found to characterize acid volcanics (Campbell *et al.*, 1981) and granites (Plant *et al.*, 1980) associated with Cu–Zn and Sn–U mineralization, respectively. The chondrite-normalized REE patterns of granites persist in stream sediments (Howarth *et al.*, 1981). Other primary characteristics of granites associated with Sn–U mineralization and Mo and Mo–Cu porphyries are given by Plant *et al.* (1980 and 1983) and

*The term 'metalliferous' is used throughout to describe rocks containing high primary concentrations of elements predominantly in silicate minerals. 'Mineralized' rocks are those in which metals are in secondary concentrations in ore minerals.

Table 1 **Comparative table of Host Minerals for elements in mineralized and non-mineralized lithologies**

Element	Mineralized rocks	Non-mineralized rocks		
		Ultrabasic rocks	Intermediate rocks	Acid rocks
Lithium	Lithium micas (pegmatites), spodumene, amblygonite		hornblende, biotite, feldspar	biotite, feldspar
Rubidium	Secondary micas and clays, low temperature potassium feldspar (adularia)		biotite and other micas, feldspars	micas, K-feldspar
Beryllium	Beryl, bertrandite, phenacite, chrysoberyl		plagioclase, muscovite	plagioclase, muscovite
Fluorine	fluorite, topaz, villiaumite, apatite		amphiboles, micas	micas
Manganese	secondary oxides in surficial environment	magnetite, ilmenite pyroxenes, olivines	pyroxenes, magnetite, ilmenite	
Copper	sulphides	pyroxenes, amphiboles, magnetite	pyroxenes, amphiboles, biotite, magnetite	biotite in Cu—Mo porphyries
Zinc	sulphides	amphiboles, magnetite	amphiboles, biotite, magnetite	biotite
Lead	sulphides		feldspars, U/Th rich phases	feldspars, U/Th rich phases

Nickel	sulphides	olivines, pyroxenes	pyroxenes, biotite	biotite
Niobium	Columbite, pyrochlore, niobotantalates, loparite	ilmenite, zircon	ilmenite, zircon, sphene, biotite	ilmenite, zircon, sphene, biotite
Tantalum	tantalite, microlite, niobotantalates	ilmenite, zircon	ilmenite, zircon, sphene, biotite	ilmenite, zircon sphene, biotite
Molybdenum	molybdenite, wolfenite		magnetite, ilmenite, sphene, plagioclase	sphene, K-feldspar, plagioclase
Tin	cassiterite, tourmaline		biotite, muscovite	biotite, muscovite
Cesium			biotite, muscovite	biotite, muscovite
Tungsten	tungstates		biotite	biotite
Thorium	monazite, thorite	zircon, sphene, monazite, apatite, biotite, xenotime,		allanite
Uranium	uraninite, pitchblende, niobo-tantalates, phosphates, hematite, martite	zircon, sphene, monazite, apatite, biotite, xenotime,		allanite, grain boundaries
Rare-earths	monazite, allanite, xenotime, niobo-tantalates, complex REE silicates	garnet, amphiboles (HREE)	monazite, allanite, sphene, plagioclase (Eu), biotite	apatite, zircon, micas (LREE)
Barium	baryte, celsian, cymrite	pyroxenes	micas, plagioclase	micas, K-feldspar

Mutschler *et al.* (1981), respectively (see below).

The detection of mineral deposits beneath barren sedimentary cover by the use of regional data may, in some circumstances, be improved where alteration associated with hydrothermal activity is sought, rather than the mineralization itself. The former will generally be much more extensive (Fyfe, 1975), and elements such as B, Li, Be and F, changes in K/Rb and Mg/Li ratios which reflect the occurrence of secondary silicates, carbonates and borates (e.g. Beus and Grigorian, 1977) and fluid inclusions, are of particular value. The rates of change of Li and B values in regional geochemical data have been shown to provide a powerful method of detecting buried porphyry Cu systems (Botbol *et al.*, 1981).

Modern conceptual models require the interplay of several factors to account for the formation of ore deposits. These include: a source of fluids with an appropriate flow regime in which porosity and fracture systems are important; and a source of energy. The latter may be magmatic heat, heat of crystallization, radiogenic heat or tectonism. The preparation of sound metallogenic models thus requires information on the geological and tectonic setting of ore bodies in addition to geophysical and geochemical data.

The geological framework of northern Scotland

The region contains two principal tectonic units, the Hebridean craton in the extreme northwest and the Caledonian orogen to the south and east. The Hebridean craton has remained tectonically stable since early Proterozoic times and consists largely of Archaean gneisses and granulites which had been deeply eroded by mid-Proterozoic times. It is overlain by a discontinuous sedimentary cover up to a few kilometres in thickness. The eastern margin of the craton is defined by the eastward-dipping Moine thrust-zone at the Caledonian orogenic front.

The Caledonian province in northern Scotland is underlain by a gneissose basement similar to that of the adjacent craton (Bamford *et al.*, 1977). During the Caledonian cycle, the province formed part of the unstable eastern margin of the American–Greenlandic continental plate which is believed to have been welded on to the European plate, the suture passing through southern Scotland. The two main cover units—the Moine and Dalradian—comprise a late Proterozoic to lower Palaeozoic sequence many kilometres in thickness consisting predominantly of metasediments. The cover sequence was folded and metamorphosed during a series of tectonothermal events culminating in the Grampian orogeny (*ca.* 480 Ma) and was subsequently intruded by granites and other igneous rocks (Table 2).

The terminal phase of the Caledonian orogeny which was reached in the late Ordovician–early Silurian, marked a turning point in the style of tectonic activity. 'Newer forceful' granite emplacement, rapid cooling and uplift, and large scale displacements along transcurrent faults (such as the Great Glen fault set) coincided with the cessation of ductile deformation and metamorphism. Subsequently, large granites were emplaced discordantly near to the surface in sub-volcanic complexes at about the same time as vertical movements blocked out the intermontane troughs in which the molasse-like Old Red Sandstone accumulated. Calc-alkaline volcanicity was concentrated near some of the faults which controlled the development of the Old Red Sandstone basins.

The application of geochemical data to mineral exploration in northern Scotland

STRATABOUND MINERALIZATION IN THE MIDDLE DALRADIAN

Stratabound Ba–Zn–Pb, Fe–Cu–Ni and Fe–Cu–Zn mineralization in the Dalradian assemblage is associated with the late Precambrian to Lower Cambrian sedimentary development of the Caledonides preceding the mainly Ordovician tectonothermal events. The existence of an extensive zone of pyrite enrichment in Middle Dalradian rocks of the southwest Highlands had been known since the region was first surveyed geologically (see Wilson and Flett, 1921), but the base metal content of the sulphide-bearing horizon was found to be generally low, except in south Knapdale (see below). It was not until geochemical reconnaissance, conducted by the Institute of Geological Sciences, identified other important mineral localities, notably the Foss Ba–Zn–Pb deposit at Aberfeldy, Perthshire (Coats *et al.*, 1978, 1980, 1981) in the same Middle Dalradian horizons, that the extent of the mineralization was recognized. The mineralized zone at the type locality near Aberfeldy has a thickness of about 100 m and a strike length of 7 km; total reserves of Ba are about 2 million tons. Reserves for other localities are not known.

The direct detection of the mineralization

A map of mineral occurrences is shown in Figure 1 and a summary of anomalies which directly reflect the ore mineral assemblage, and of anomalies for other chemical elements, is given in Table 3. It is apparent that the detection of mineralization by stream sediment samples is good, particularly for the metals represented in the ore assemblage. In the case of the Aberfeldy deposit, for example, sediment samples from streams draining the mineralized zone have levels of Ba of up to 3300 ppm, Zn up to 1250 ppm, and of Pb up to 180 ppm compared with median values for northern Scotland as a whole of 830 ppm, 90 ppm and 20 ppm, respectively. Copper mineralization at Meall Mor is characterized by Cu (+Sb +Zn) anomalies (Smith *et al.*, 1977); Cu–Ni mineralization at Coille–bhraghad and Craignure by Cu (+Ni) anomalies (Coats *et al.*, 1982), while the Pyrite Zone over most of its length has anomalous values of Zn.

There are, however, some limitations to the direct detection of such mineral occurrences by stream sediment sampling. Firstly, anomalous element concentrations occur several kilometres downstream from the mineralized zones and, since many of the streams flow across the regional strike, high

Table 2 **Stratigraphic units (after Watson and Plant, 1979)**

HEBRIDEAN CRATON		
	cratonic cover	
fault basins associated with development of Atlantic margin	Tertiary marine sediments (confined to basins in the Minch and Continental shelf) Palaeocene–Eocene plateau basalts: acid–basic intrusive complexes and dyke swarms. Mesozoic (mainly Triassic–Jurassic fluviatile marine and estuarine sediments) ? U. Palaeozoic	
unconformity	Cambro-Ordovician transgressive marine succession (quartzites, dolomites)	
unconformity	Torridon Group (*ca.* 800 Ma) *unconformity* Stoer group (*ca.* 1000 Ma)	Feldspathic sandstone, minor conglomerates siltstones shales pyroclastic debris in Stoer group mainly fluviatile
basement		
major unconformity		
Lewisian complex metamorphic, mainly gneisses and granulites modified by repeated meta morphism and deformation (2800–1800 Ma)	2700 Ma	Laxfordian stage: inhomogenous deformation, movement on shear zones amphibolite-facies metamorphism, emplacement of granitic and pegmatitic veins emplacement of basic dyke swarm late Scourian state: inhomogeneous deformation, initiation of shear zones, emplacement of granitic and pegmatic veins, early Scourian stage; gneiss-forming event giving gneisses and granulites, parent materials included metasediments, metavolcanics, layered basic intrusives and abundant intrusive tonalitic material.

geochemical values may appear to extend over unmineralized horizons of the Dalradian. Secondly, levels of Ba and Pb in panned concentrates generally give better contrast, and are better indicators of minor mineral occurrences than stream sediments (Coats *et al.*, 1982), although the detection of Zn and Cu is improved by the use of stream sediment samples, particularly over the Pyrite Zone.

Table 2 continued

NW PORTION OF THE CALEDONIAN OROGEN

	cratonic cover	
fault basins associated with development of North Sea	Tertiary and Mesozoic marine sediments (mainly in North Sea and adjacent coastal tracts) Permo-Triassic mainly terrestrial sediments	
unconformity post-orogenic	Upper and Middle Old Red Sandstone: terrestrial and lacustrine sediments, local andesitic volcanics	
unconformity late orogenic gently folded	Lower Old Red Sandstone: terrestrial conglomerates, sandstones, volcanics, latest Caledonian granites	
	products of Caledonian cycle	
major unconformity	Caledonian granites	460–390 Ma
folded cover formations in Caldonides, metamorphic events *ca.* 750, 550–450 Ma	late Proterozoic to early Ordovician	Dalradian Supergroup: lower units mainly shallow-marine, upper units mainly turbidites
	late Proterozoic	Moine Supergroup: upper units, mainly deltaic sediment *ca.* 800 Ma?
	unconformity?	
		Moine Supergroup: lower units, mainly deltaic sediments *ca.* 1000 Ma?
	basement	
major unconformity		
Lewisian complex metamorphic, mainly gneisses and granulites modified by repeated metamorphism and deformation (2800–1800 Ma)	2700 Ma	Laxfordian stage: inhomogenous deformation, movement on shear zones amphibolite-facies metamorphism, emplacement of granitic and pegmatitic veins emplacement of basic dyke swarm late Scourian state: inhomogeneous deformation, initiation of shear zones, emplacement of granitic and pegmatitic veins. early Scourian stage; gneiss forming event giving gneisses and granulites, parent materials included metasediments, meta-volcanics, layered basic intrusives and abundant intrusive tonalitic materials.

Mineralization model

Recorded occurrences of stratabound mineralization are plotted in Figure 1 and are shown in relation to the stratigraphic succession of the Dalradian in Figure 2. The association of the mineralization with particular horizons of the Middle Dalradian (the Easdale Subgroup), and such evidence as the

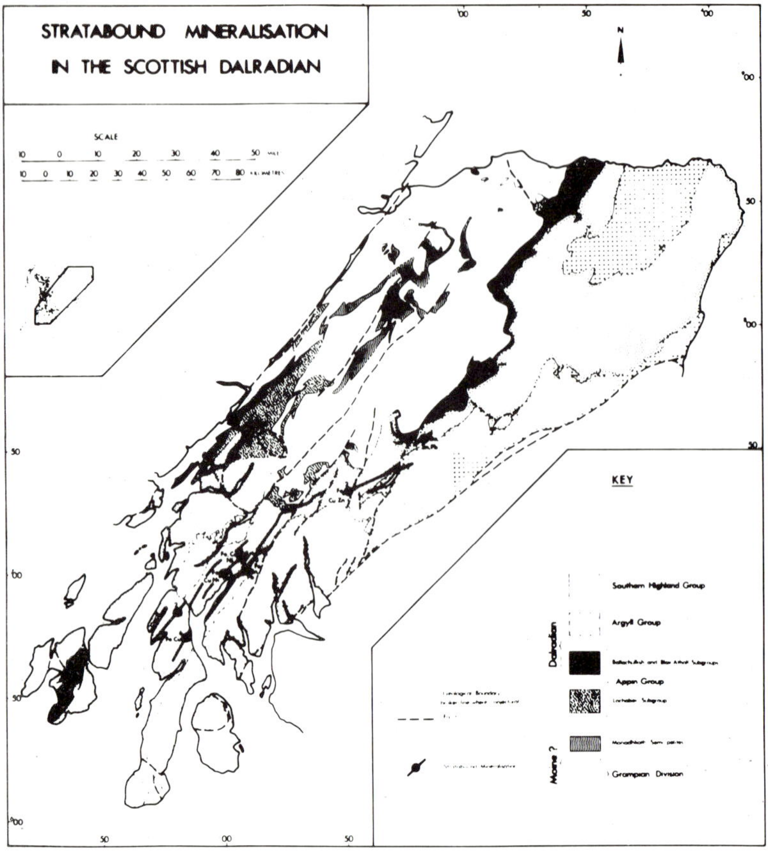

Figure 1 Map showing the occurrences of stratabound mineralization in the Scottish Dalradian. (Locality names: 1—Aberfeldy; 2—Pyrite Zone (Loch Tay and Loch Fyne); 3—McPhun's Cairn; 4—Coille-bhraghad; 5—Craignure; 6—Pyrite Zone (Meall Mor))

concordance of the mineral banding with the adjacent schists, indicates that the mineralization was contemporaneous with sedimentation.

Local controls

In the Middle Dalradian Easdale Subgroup there is a transition from localized Ba–Zn–Pb mineralization, for example at Aberfeldy; to Ni-pyrrhotite at Loch Fyne; to pyrite with Cu, Zn and Pb sulphides in the more extensive Pyrite Zone at the top of the subgroup (Figure 2). In the type locality of the Ba–Zn–Pb mineralization at Aberfeldy, the main mineralized zone consists of interbanded baryte, quartz–celsian and sulphide-bearing carbonate rock; graphitic, calcareous and dolomitic schists are also abundant. The mineralized zone is preceded by about 250 metres of unmineralized sediments which contain increased levels of Ba. Localized footwall brecciation, similar to that at Rosh Pinah, southwest Africa (Page and

Table 3 **Geochemical anomalies associated with Dalradian stratabound mineralization**

Locality	Mineralization	Host minerals	Regional geochemical anomalies
Aberfeldy	Ba–Zn–Pb	Barite, celsian, cymrite, sphalerite, galena	+Zn +Pb +Ba –Cr –Cu
Coille-bhraghad	Fe–Cu–Ni	Nickeliferous pyrrhotite, pyrite, chalcopyrite, pentlandite	+Pb +Zn +Cu
Craignure	Fe–Cu–Ni	Nickeliferous pyrrhotite, pyrite, chalcopyrite, pentlandite	+Pb +Zn +B + elements of basic association
McPhun's Cairn	Fe–Zn–Pb–Cu	Pyrite, sphalerite, galena, chalcopyrite, pyrrhotite	+Zn +Cu +Ba +B +Ti
Pyrite Zone (Meall Mor)	Fe–Cu	Pyrite, chalcopyrite, sphalerite, galena	+Cu +Zn +Pb
Pyrite Zone (Loch Fyne)	Fe–Zn	Pyrite, sphalerite, chalcopyrite	+Zn +Cu +B +Pb +Bi
Pyrite Zone (Loch Tay)	Fe–Cu	Pyrite, chalcopyrite	+Zn +MgO

Watson, 1976), which results from sub-surface boiling of mineralizing fluids, is also a feature of the Aberfeldy deposit. The distribution and grade of baryte, Ba-enriched silicate, carbonate cherts and Pb–Zn–Fe sulphides, have been attributed to the proximity of hydrothermal vents located on growth faults and the topography of the sea floor on which dense mineral brines became ponded (Moles, 1982). Isotopic evidence for the source of sulphur indicates that the bulk of baryte sulphate was formed by rapid mixing of hot brine ($T \geqslant 200°C$) with Vendian seawater, the deposition of sulphides reflecting the reduction of seawater at high T ($\geqslant 350°C$) in the underlying sedimentary pile (Willan, 1982). Precipitation of cymrite instead of barite in some horizons (Coats *et al.*, 1981; Fortey and Beddoe-Stephens, 1982) may have been due to locally reducing conditions (Large, 1981).

The Aberfeldy deposit occurs at, or close to, the top of the Ben Eagach schist, which was deposited during the climax of the third major basin-deepening episode in the Dalradian (Harris *et al.*, 1978). This sequence of rocks is everywhere overlain by a sequence (represented by the Ben Lawers Schist at Aberfeldy and the Ardrishaig (=Craignish) Phyllites at Loch Fyne and Knapdale (see Figure 2)) which was deposited during a prolonged period of basin shallowing when sediment influx outpaced basin subsidence, culminating in periods of emergence (Harris and Pitcher, 1975; Harris *et al.*, 1978). To the northwest of Loch Fyne the Ardrishaig Phyllites host the stratabound Ni-pyrrhotite deposits of Coille-braghad and Craignure. The mineralization, which is believed to form part of a continuous zone (Wilson and Flett, 1921; Coats *et al.*, 1982) consists of nickeliferous pyrrhotite and chalcopyrite, with subordinate pyrite and pentlandite. The Pyrite Zone, which

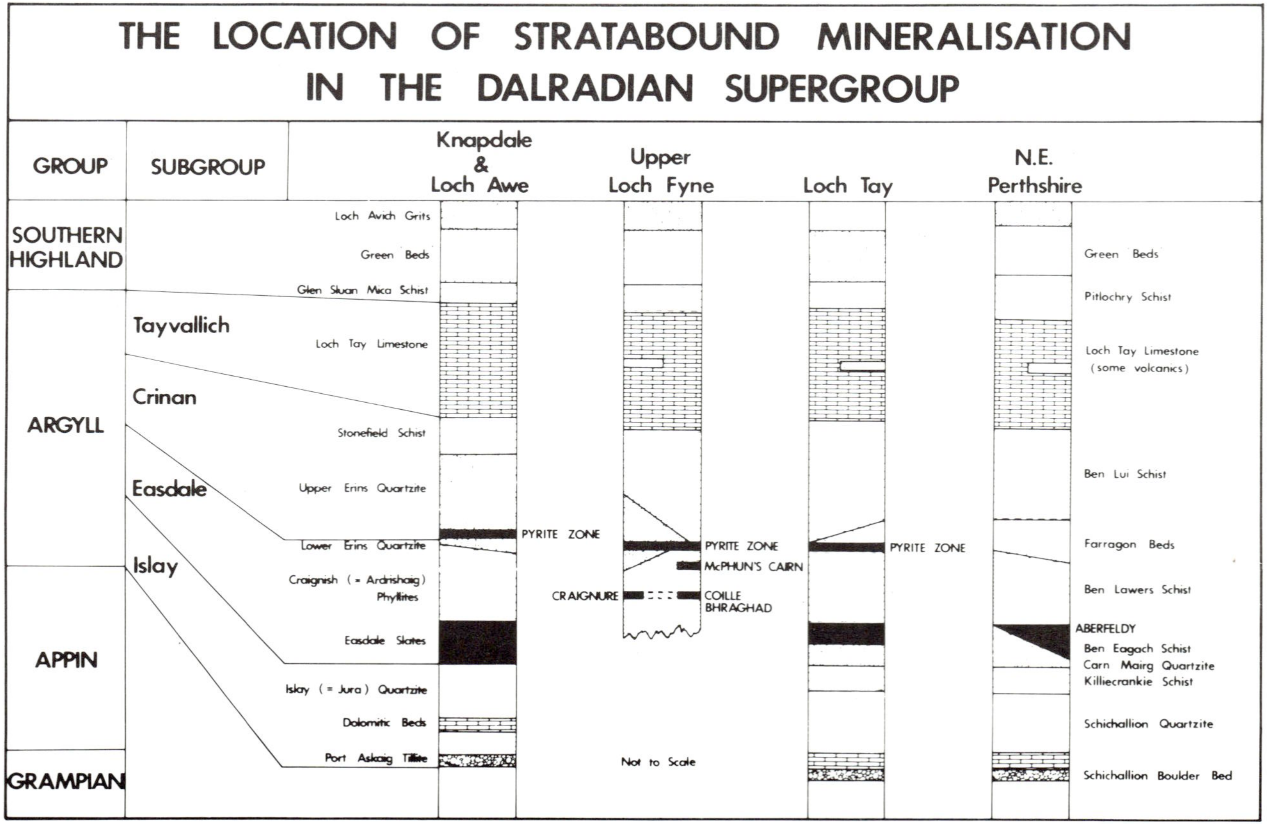

Figure 2 Diagram showing the stratigraphic location of stratabound mineralization in the Dalradian Supergroup (modified after Willan, 1980; stratigraphic correlation from Harris and Pitcher, 1975)

occurs at the top of the Ardrishaig Phyllites (and their lateral equivalents) extends for 190 km from Knapdale to Glen Shee (Willan and Hall, 1980). Its formation coincided with the climax of shallowing of the Easdale basin, probably during a hiatus in tectonic activity. The zone has sharp boundaries and contains up to 5%, and locally up to 20% pyrite; chalcopyrite also occurs, for example at Glen Lochay, where it is present in the upper part of the zone.

Recent work on the primary sulphide species of exhalative ore deposits (e.g. Finlow-Bates *et al.*, 1977; Plimer, 1978; Plimer and Finlow-Bates, 1978) suggests that sulphur activity and Eh (reflecting environmental factors such as water depth and circulation) may be as important as fluid temperature in controlling the iron ore mineral assemblage of such deposits. Hence a decrease in sulphur content and Eh as a result of increased seawater depth will tend to favour pyrrhotite deposition, while shallow water in which Eh and sulphur concentration are generally higher will favour pyrite precipitation. The pyrrhotitic (Cu–Ni) ores at Coille-bhraghad and Craignure are thus interpreted as a high-T deposit formed in conditions of low aS^{2-} and low Eh at medium ($\geqslant$500 m) depth as proposed by Willan and Hall (1980) for the similar mineralization at McPhun's Cairn, Argyllshire. In contrast, the Pyrite Zone is thought to reflect hydrothermal activity in shallow water conditions comparable with those in which the pyritic Fahlbands of the Norwegian Caledonides were deposited (Gammon, 1966). The temperature of the hydrothermal fluids may have been relatively low, reflecting the hiatus in tectonic and magnetic activity represented by the infilling of the Easdale basin.

Regional controls

The local controls of the Dalradian stratabound mineralization are thus fairly well understood, and the main application of geochemical maps is in providing information on the regional setting in which mineralization took place. In particular, comparisons between the Middle Dalradian horizons which host the stratabound deposits (and important vein deposits which may represent remobilized stratabound material), and other cover formations including the Moine and Lower and Upper Dalradian Groups are critical.

Some of the most important lithological, geochemical and geophysical differences in the Caledonian cover sequence are between the Moine and Dalradian metasedimentary assemblages (Plant *et al.*, 1984). The Moine generally comprises a monotonous sequence of quartz–feldspar psammites with subordinate pelites while the Dalradian comprises a varied sequence of interbedded psammites, pelites, black pelites and limestones; and an upper (Southern Highland) Group dominated by turbidites and including basic volcanics (see Harris and Pitcher, 1975) of which the most important are the Tayvallich lavas in the southwestern Highlands. Other major differences between the Moine and Dalradian include isotopic evidence that a pre-750 Ma tectonothermal event affected extensive areas of the Moine while the Dalradian sequence is affected only by Grampian events *ca.* 480 Ma (Piasecki

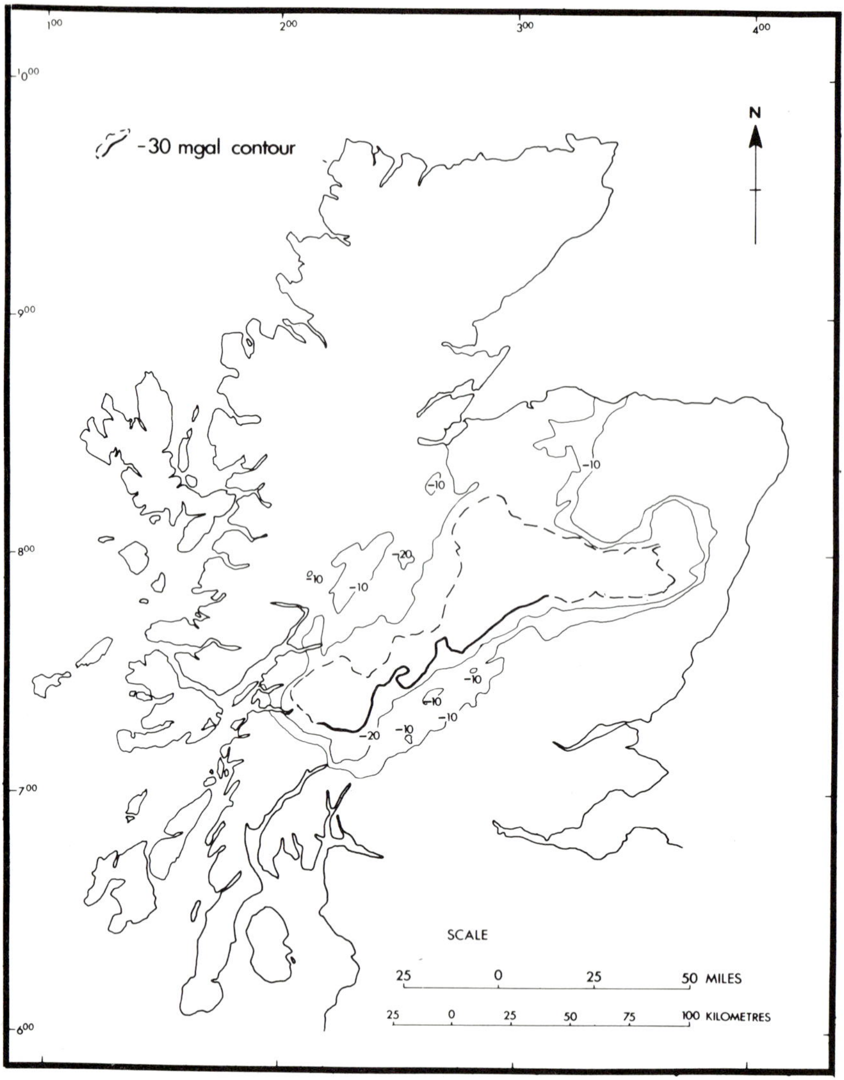

Figure 3 Map showing the Bouguer gravity anomaly associated with the Moine/Dalradian junction (based on Hussain and Hipkin, 1981, first published in Plant *et al.*, 1983)

and van Breemen, 1982); and geophysical evidence which shows a steep Bouguer gravity gradient coincident with the Moine/Dalradian junction, connecting lower values over the Moine with higher values over the Dalradian (Figure 3). The regional geochemistry shows a marked contrast in the levels of chemical elements between the Moine and Dalradian over their entire outcrop extending to Shetland (Johnstone *et al.*, 1979; cf. whole-rock data of Lambert *et al.*, 1981).

In general B, Li and Co, Cr, Cu, Fe, Mg, Mn, Ni, Pb and Zn increase and Be,

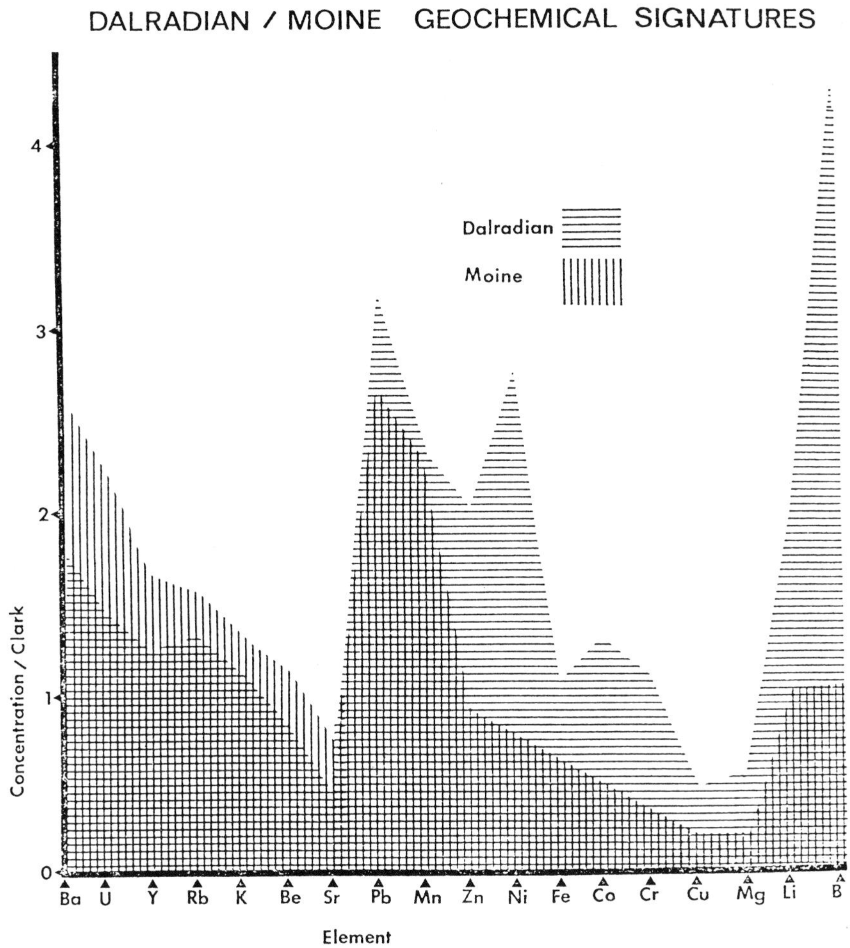

Figure 4 Mean element concentration for the Moine and Dalradian, normalized to the mean continental crust values. (From Plant *et al.*, 1983). The average concentration of each element was obtained for the Moine and Dalradian respectively using the Regional Geochemistry (and Geological) database by means of the NERC I^2S image analysis system

Ba, K, La, Rb, Sr, U, Y and Zr decrease over the Dalradian relative to the Moine (Figure 4), differences which are attributed to a reduction in the component of detrital minerals such as feldspar and an increase in marine clay minerals in the Dalradian relative to the Moine (Plant *et al.*, 1984). The generally higher levels of basic elements, for example Cu (Figure 5), in the Dalradian are thought to record an input from volcanic sources. In a restricted zone near the top of the Argyll group (below and at the level of the Tayvallich lavas, but extending over other mapped rock units, for example, the Crinan Grits), the elements normally associated with ultrabasic-basic

Figure 5 Regional geochemical map of Cu over the Scottish Highlands

rocks, particularly Mg and Co (Plate 1), increase; and at the base of the Southern Highland Group (Upper Dalradian) there is a further increase in Li and B (Plate 2) over the turbidite divisions.

These lines of evidence are all consistent with geological indications that the Moine was deposited on a relatively stable continental margin of normal crustal thickness, whereas the Dalradian was deposited over thinned continental basement which was progressively stretched and affected by tectonism during its deposition (Anderton, 1982). It has been suggested (Plant *et al.*, 1984) that the present Moine–Dalradian junction marks the edge of an 'Old' Moine continental slab which had been affected by a pre-750 Ma tectonothermal event; and that the Dalradian, with its distinctive geochemistry, accumulated in post-Morar times in an extensional marginal basin on a thinned Lewisian-like basement (Figure 6).

The mechanism by which new crust is emplaced beneath marginal basins is uncertain, but it is generally considered that basaltic material is intruded above convective (Karig, 1971; Matsuda and Uyeda, 1971; Mitchell and Garson, 1972) or mechanically driven (Sleep and Toksoz, 1971) cells in the upper mantle. Heat flow above marginal basins is generally high as a result of the transfer of magmas, volatiles, or both upwards through the mantle. The regional geochemical data suggest that the Dalradian sedimentary prism underlying the mineralized layers might have provided adequate sources of Ba and Pb and other trace elements for mineralization. The onset of hydrothermal activity, implying a rise of crustal temperature and increasing tectonic activity, is thought to reflect rifting in the extended continental crust which culminated in the submarine basic volcanicity represented by the Tayvallich lavas and associated tuffs and sills (Figure 6).

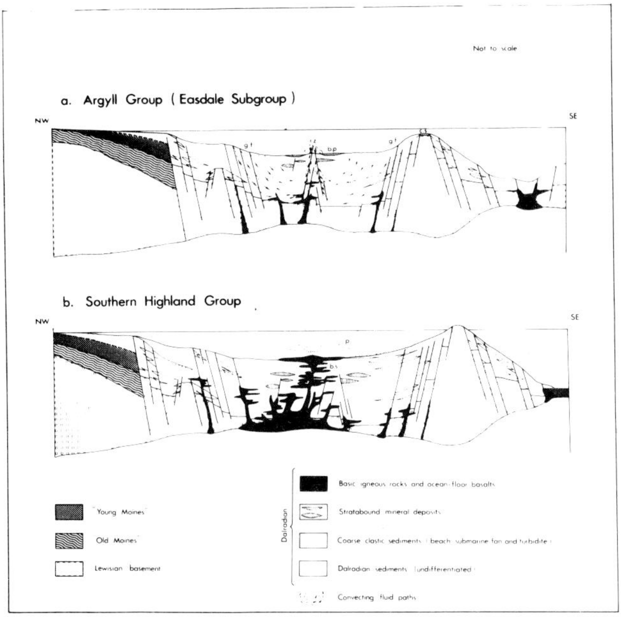

Figure 6 Inferred structural setting of the Dalradian during: 1—Middle Dalradian (Easdale Subgroup); and 2—Upper Dalradian (Southern Highland Group) times (From Plant *et al.*, 1984)

A major basic magmatic contribution to the Aberfeldy mineralization is unlikely, because of the paucity of Cu or other elements of basic association in the deposits, and the rising magma is envisaged principally as a heat source driving the hydrothermal system. Elsewhere, for example during deposition of the Pyrite Zone and associated mineralization in South Knapdale, hydrothermal fluids may have had more direct access to basic volcanic rocks as reflected by an increase in the Cu content of the mineralization. The distinctive regional Mg and Co anomaly (Plate 1) mentioned above, which occurs over volcanics and other rock types stratigraphically above the deposits, may indicate exhalation of fluids which had interacted with juvenile magmas, or the incorporation of basic detritus in sediments. The stratabound Ba–Zn–Pb mineral deposits, tholeiitic volcanics and the chemistry of the Dalradian generally, particularly in the upper Middle-Upper Dalradian sequence, are thus related to the opening of an oceanic rift within an ensialic basin during the late Precambrian/lower Cambrian in which new oceanic crust may have formed (Figure 6).

It is likely that this was preceded by thinning of the lithosphere and partial melting of the upper mantle, associated with a linear zone of anomalous heat flow, and tectonism giving rise to hydrothermal activity and mineralization. The style of stratabound mineralization thus reflects the interplay of magmatism and the temperature of intermittent sub-sea floor to sea floor geothermal systems, with tectonic factors reflecting basin shallowing and deepening. The local evidence on the processes involved in metallogenesis fits well into such a regional model.

MINERALIZATION OF THE GRANITE AND PORPHYRY ASSOCIATIONS

The application of regional geochemical maps to the detection and modelling of metalliferous mineralization associated with granites (*sensu lato*) has been described elsewhere (Simpson *et al.*, 1979, 1982; Plant *et al.*, 1980, 1983) and only a brief summary is given here.

Granite magmas were emplaced into the crust of the Scottish province of the Caledonian orogen during several major episodes of magmatism, some of which were related to movements on deep faults during the late stages of orogenesis (Leake, 1978). Granites emplaced before and during the main Caledonian metamorphic/deformation event (older granites and migmatites) are distinguished from those that are later (the newer granites) (Read, 1961). The latter are subdivided into those emplaced at a relatively deep structural level (the 'forceful granites' of Read) and those emplaced at a shallow level, the late discordant granites of Dewey and Pankhurst (1970). Metalliferous and mineralized intrusions mainly include the youngest of the 'forceful' suite, and also the late discordant intrusions which were emplaced during Lower Devonian times and which have metal associations ranging from Cu–Mo, or Mo–(W) to Sn–U. The 'forceful' intrusions are generally unmineralized, with the exceptions of Grudie (Mo–F), Helmsdale (U–Mo–F), Abriachan (U–F) and Ballachulish (Cu–Mo), which are probably the latest of the suite and are temporally associated with (mainly transcurrent) faulting.

The direct detection of metalliferous mineralization

Occurrences of metalliferous minerals are shown in Figure 7, and geochemical anomalies associated with Caledonian granites are summarized in Table 4. It is clear that metalliferous and mineralized granites are marked by the occurrence of distinctive geochemical anomalies, while barren granites have levels of chemical elements which .are close to background levels. The metalliferous intrusions also tend to have 'juvenile' isotopic compositions, and most have negative Bouguer anomalies and/or large amplitude aeromagnetic anomalies centred on the intrusions. Four regional groups may be identified on the basis of their trace element patterns (Plant *et al.*, 1980, 1983) as follows:

1. Northern Highlands. Late granites of the 'forceful' suite (*ca.* 410 Ma) emplaced in zones in which major transcurrent fault displacement accompanied or followed closely on intrusion (Helmsdale, Grudie, Abriachan).
2. Grampian Highlands. Late, large volume discordant sub-alkaline Sn–U granites (*ca.* 400 Ma) which extend from the Monadhliath intrusion in the west to Mount Battock in the east (Figures 8 and 9).

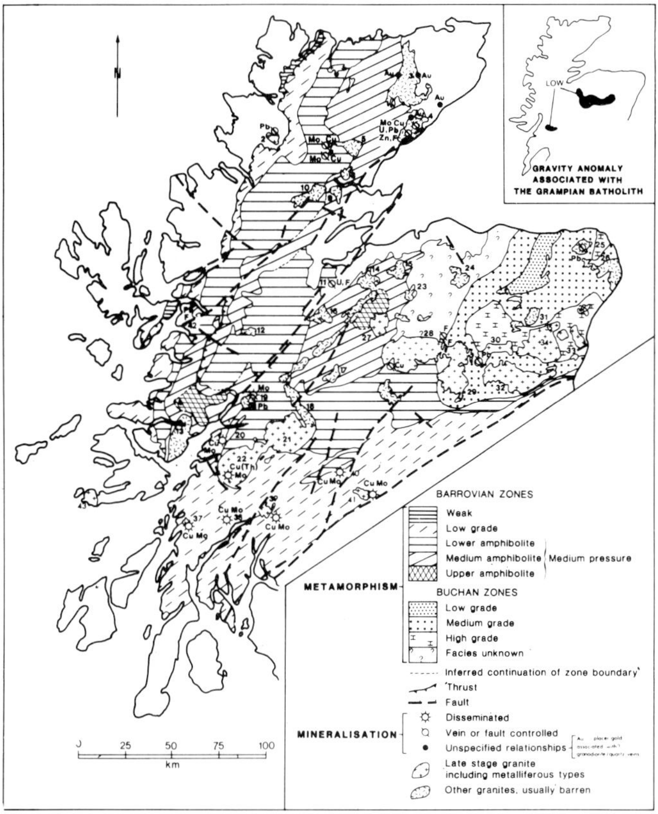

Figure 7 Mineralization associated with granites in the Scottish Highlands. (Granite names: 1—Ben Loyal; 2—Borrolan; 3—Halladale; 4—Helmsdale; 5—Lairg Rogart; 6—Grudie; 7—Loch Shin; 8—Migdale; 9—Fearn; 10—Carn Chuinneag; 11—Abriachan; 12—Cluanie; 13—Strontian; 14—Moy; 15—Ardlach; 16—Foyers; 17—Strathspey; 18—Strath Ossian; 19—Ben Nevis; 20—Ballachulish; 21—Moor of Rannoch; 22—Etive; 23—Grantown; 24—Ben Rinnes; 25—Strichen; 26—Peterhead; 27—Monadhliath; 28—Cairngorm; 29—Lochnagar; 30—Braemar; 31—Bennachie; 32—Mount Battock; 33—Auchlee; 34—Hill of Fare; 35—Skene Complex; 36—Ardlethen; 37—Kilmelford; 38—Garbh Achadh; 39—Garabal Hill; 40—Tomnadashan; 43—Ross of Mull (From Plant *et al.*, 1983)

3. Southwest Highlands. Late discordant (*ca.* 400 Ma) Mo (W?) intrusions (e.g. the Etive complex).
4. Southwest Highlands late discordant porphyry Cu-style intrusions (*ca.* 400 Ma), for example, Kilmelford, Garabal Hill (Arrochar) and Comrie.

Groups (3) and (4) are broadly those recognized by Mutschler *et al.* (1981) who distinguished granodiorite (Cu–Mo) systems from granite (Mo) systems.

The northern group (1) and southwest Highlands groups ((3) and (4)) generally show good agreement between the suites of metals in which intrusions are enriched, as indicated by regional geochemical maps, and

Table 4 **Metalliferous mineralization and geochemical anomalies associated with granites**

Intrusion	Mineralization	Geochemical anomalies
1. Ben Loyal	U–Th–Pb–F–Ba–REE	+Ti +V +Be +U +U (water) +Ba +Pb +Zr –Ni
2. Loch Ailsh-Borrolan	Cu–Pb–Zn in basics—skarn	+Be +U +Pb +elements of basic association
3. Halladale (migmatite)	Placer gold	+Ba +Sr +Mo +Be +Zr
4. Helmsdale	U–Pb–Ba–F	+U (water) +Pb +Ba +Mo +Be +Ca +Mn +B
5. Grudie	Cu–Mo–Pb–F–Ba	+Cu +Mo +Pb +Ba +Be +U
6. Loch Shin	Mo–Cu	+Cu +Mo +Pb +U
7. Carn Chuinneag riebeckite gneiss	Sn–Fe and Mo–Th–F–Cu with faults	+Mo +Sn +U +U (water)
8. Abriachan	U–F	+U +U (water)
9. Strontian	Pb–Zn–Ba–Sr (but associated with Permian lamprophyre dykes)	+Zn +Sr +U +Pb +La +Ca +Cr +K +Ni +Ti +Rb +Li +Y +Zr +Be +V +Co +Ba +Cu
10. Ben Nevis	Mo	+Cu +B +K +Mo +U +Li +Sr
11. Ballachulish	Mo–Cu	+Cu +Mo +Pb +Zn +Sr +K +Rb +elements of basic association
12. Etive	Mo–Cu–Th	+Cu +Ba +Zn +B +K +Bi +Mo +V +Be +Zr +U +Li +Rb +Ti +Sr +La +Ca
13. Peterhead	?F	+V +Pb +Co +Rb +Ni +Bi +Zr +U +U (water) +La +Cu +Zn +B
14. Monadhliath	—	+Li +Be +U +Rb +Y +Pb –elements of basic association –Ba
15. Cairngorm	Sn–Nb–F–Pb–Zn–Li–Ag	+K +Li +Be +B +U +Rb +Y –elements of basic association –Ba –Sr –Zr
16. Lochnagar		
(a) northern part	—	+Y +V +K +Co +Mg +Rb +Zn +Ca +Cu +Bi +Li +La +Ba +U +Pb +Zr
(b) southern part	—	*
17. Braemar	Pb	+Ca +K +Ni +Mg +Be +Li +U +U (water) +Mo +Y +V +Pb +Co +B +Rb +Ba +Zn +Zr
18. Bennachie	U	+Li +Rb +La +U +U (water) +K +Be +Ti
19. Mount Battock		
(a) northern part	F	+Y +Sr +K +Zn +Co +Be +U +U (water) +La +Li +Rb +Ca +Cu +Pb +Bi +B +Ba
(b) southern part	—	*
20. Hill of Fare	U	+Li +Rb +La +Zr +U +Be +Pb
21. Kilmelford	Cu–Mo	+Cu +Zn +Pb +Ba +B +Sr
22. Garbh Achadh	Cu–Mo	+Pb +Zn +Cu +Co +B +Rb
23. Garabal Hill	Cu–Mo	+Pb +Sr +Mo +Zn +B +Rb +Li +Be +Ba
24. Tomnadashan	Cu–Mo	+Pb +Zn +B +Ti +Rb +Li +Be +Bi
25. Comrie	Cu–Mo	+Cu +Zn +B +K +Bi +Co +Li
26. Ratagan (including basics)	Pb–F, Mo–Cu	+elements of basic association +Mo +U +Ba +Pb
27. Ross of Mull	—	+Pb +Zn +La +Ba

*Sampling and analysis are complete and the data will be available shortly through the NGBD

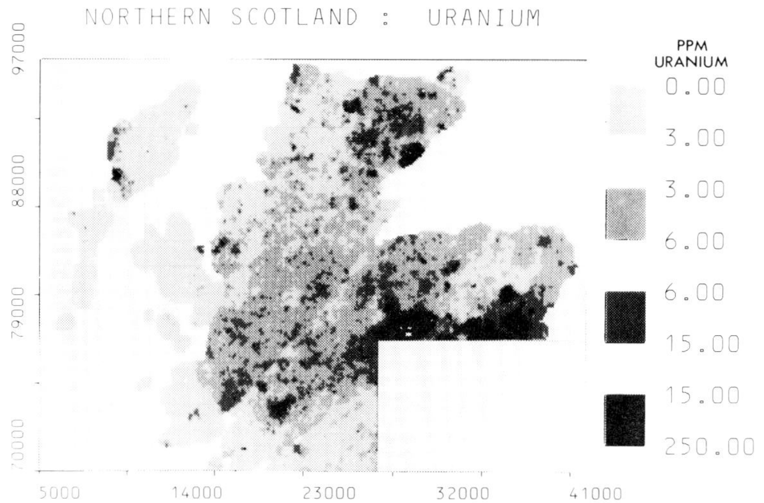

Figure 8 Regional geochemical map of U over the Scottish highlands

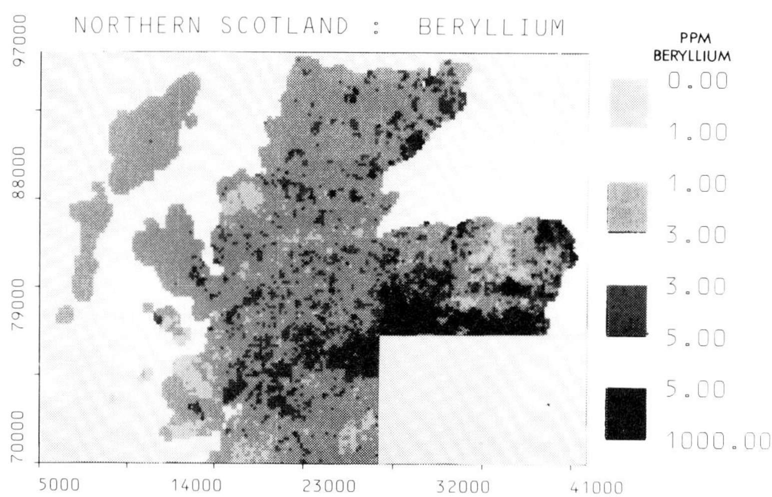

Figure 9 Regional geochemical map of Be over the Scottish highlands

recorded occurrences of metalliferous minerals. In the Kilmelford intrusion, for example, increased levels of Cu and Mo are associated with Cu–Mo mineralization. Nevertheless, the reported quantities of minerals are sub-economic and the intrusions generally lack the full sequence of hydrothermal alteration typical of porphyry Cu deposits elsewhere (Ellis *et al.*, 1977). The disparity between levels of metals indicated by geochemical maps and the quantity of metalliferous minerals is greatest for the group (2) intrusions. These contain high levels of Sn–Li–Be and U, comparable with those of the

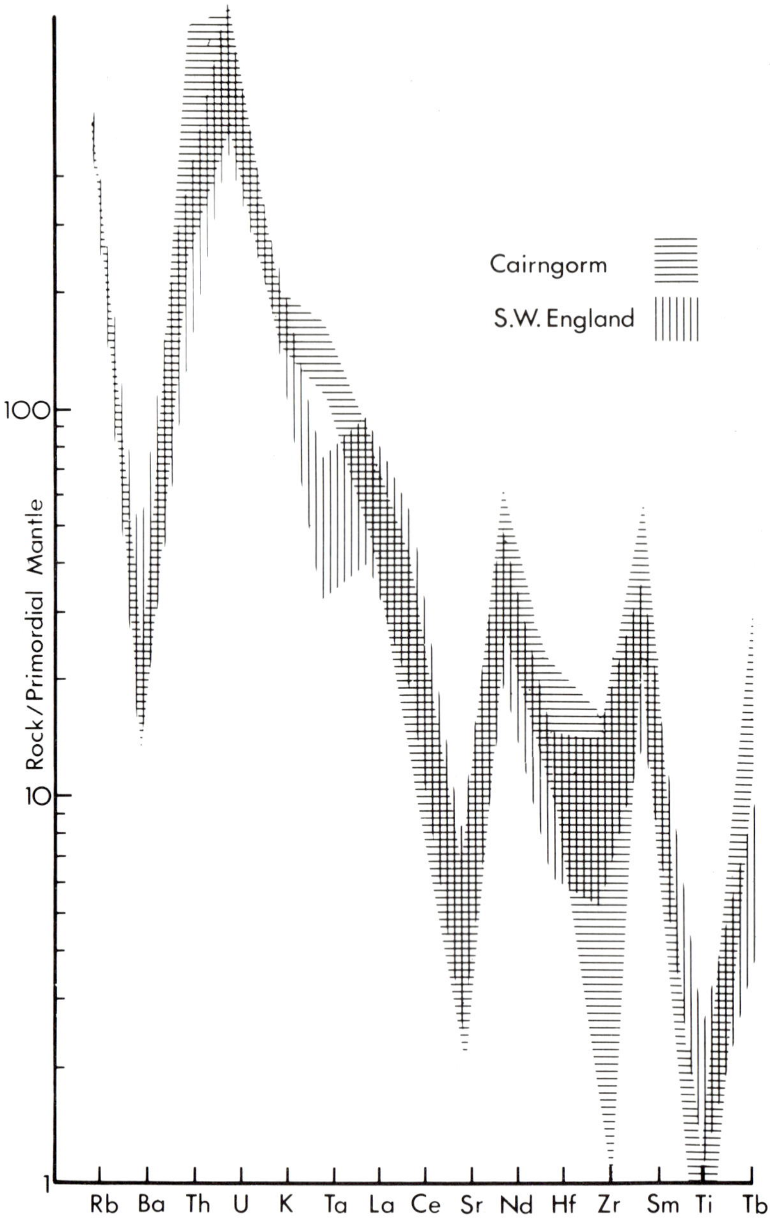

Figure 10 Mean element concentration (normalized to the primordial mantle) for the Cairngorm and Cornubian granite batholiths (Simpson, pers. comm.)

Hercynian Cornubian Batholith of southwest England (Figure 10), but only minor low-temperature mineral veins (for example, of fluorite), and a few grains of cassiterite in heavy mineral concentrates. The Cornubian batholith in contrast, is associated with many substantial concentrations of ore minerals. Comparison of the regional geochemical maps of the group (2) granites of Scotland with the Cornubian batholith (Webb *et al.*, 1978) not only confirms the distinctive and closely similar primary geochemistry of the two suites of intrusion, but also indicates that redistribution of metals has occurred in the Cornubian batholith with the highest values of metals located around centres of mineralization.

Metallogenic models

Exploration for metalliferous minerals associated with granites is frequently based on the 'S' and 'I' type models of granite genesis (Chappell and White, 1974). Hence Cu–Mo porphyries are attributed to an igneous (I-type) origin at destructive plate margins, the mineralization being related to formational/meteoric water–magma interaction while 'two-mica' S-type Sn and U granites are ascribed to crustal anatexis (Tischendorf, 1973; Moreau, 1976) in arc–continent or continent–continent collisions (Mitchell and Garson, 1976; Beckinsale, 1979). The Sn and U mineralization associated with the granites is attributed to palingenesis of wet sediments (Tischendorf, 1973; Moreau, 1976).

Studies based on regional geochemical maps of the Scottish Caledonian province in relation to the crustal and tectonic setting of mineralized intrusions have led to the development of a different model for mineralization associated with granites (Plant *et al.*, 1980, 1983). The metalliferous and mineralized Caledonian granites were emplaced into cool crust up to 85 Ma after the metamorphic climax. Their large volume, discordant structural trends, and lack of major fault displacement indicate that emplacement took place in a tensional tectonic regime, an observation which is supported by the temporal and, in the case of the Etive and Lochnagar complexes, spatial association of the intrusions with dykes. The metalliferous granites, including U–Sn alkaline granites, are thought to be related to subcrustal processes at destructive plate margins (Mitchell and Garson, 1972) involving a large ion lithophile element enriched juvenile source, such as metasomatized lithospheric mantle as first proposed by Simpson *et al.* (1979). The granites were emplaced through deep faults at the end of the Caledonian orogenic cycle when melting occurred locally as a result of a reduction in pressure (Leake, 1978). For granite magmas to rise in the crust they must be water-deficient (Brown and Fyfe, 1970) and mineralization is thought to occur initially as a result of hydration of the granite by host rock fluids at high (metamorphic) temperature (Figure 11). Reduction of silicate and primary accessory minerals will also commence at this stage, evolving into deuteric alteration of the granite magma and finally, as the granite cools and cracks, hydrothermal convection with flow of formational and then meteoric water through fracture systems. Hence, in the Scottish Caledonides mineralization is associated with metalliferous granites (a) in zones that became sites of major fault systems when the granites were areas of high heat flow and fluid circulation and (b) in

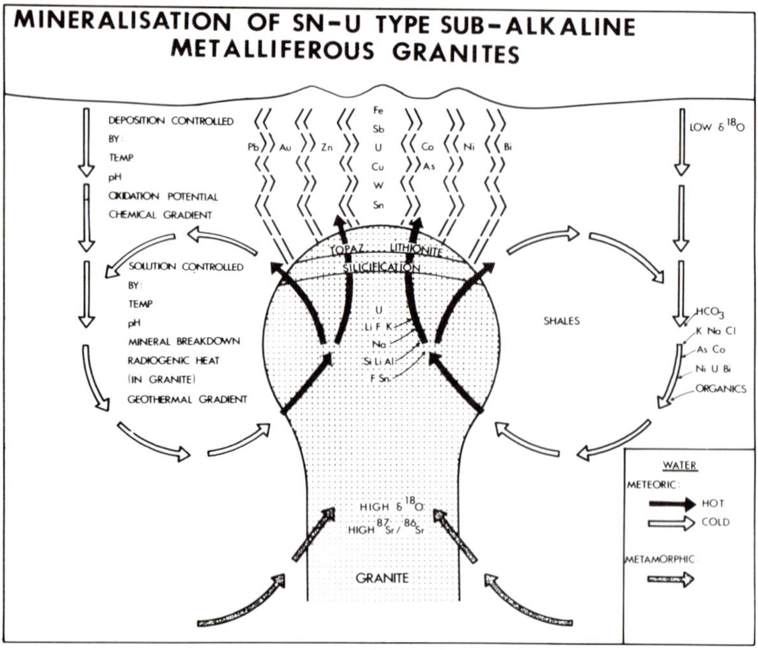

Figure 11 Diagrammatic model for mineralization of Sn–U type sub-alkaline metalliferous granites. (Modified from Simpson *et al.*, 1979)

zones where the country rocks are at greenschist or epidote–amphibolite grade of metamorphism. The latter include the Mo-bearing Ballachulish and large volume Etive intrusive complex (for which mineralization is also favoured by its location in a major fracture zone, the Cruachan line) and the small Kilmelford, Garabal Hill (Arrochar) and Comrie Cu porphyry-style intrusions.

The metalliferous Cairngorm–Mount Battock granite batholith with high levels of Sn–U–Li–Be is largely unmineralized, except where the Ballater intrusion cuts garnet grade pelitic or calcareous sediments along the Moine–Dalradian Junction. In contrast, the geochemically similar Cornubian batholith (which was emplaced in greenschist facies sediments and which was affected by contemporaneous major faults) is thought to represent the eroded centre of a Hercynian geothermal field similar to that of the present day Warakeii springs, New Zealand (Simpson *et al.*, 1979; Moore, 1982). There is evidence (Darnley, 1982) that hydrothermal convection has been re-established at several intervals in Cornubia, particularly at times of increased mantle convective heat flow (such as the Tertiary) when the local geothermal anomaly related to the concentration of heat-producing elements would be enhanced. Moreover, recent hydrogeochemical evidence (Burgess *et al.*, 1982) suggests that deep circulation of meteoric water to depths of 1 km is continuing in the granite at the present time. Fluid inclusion studies of the southwest England batholith (Rankin and Alderton, 1982) are consistent with

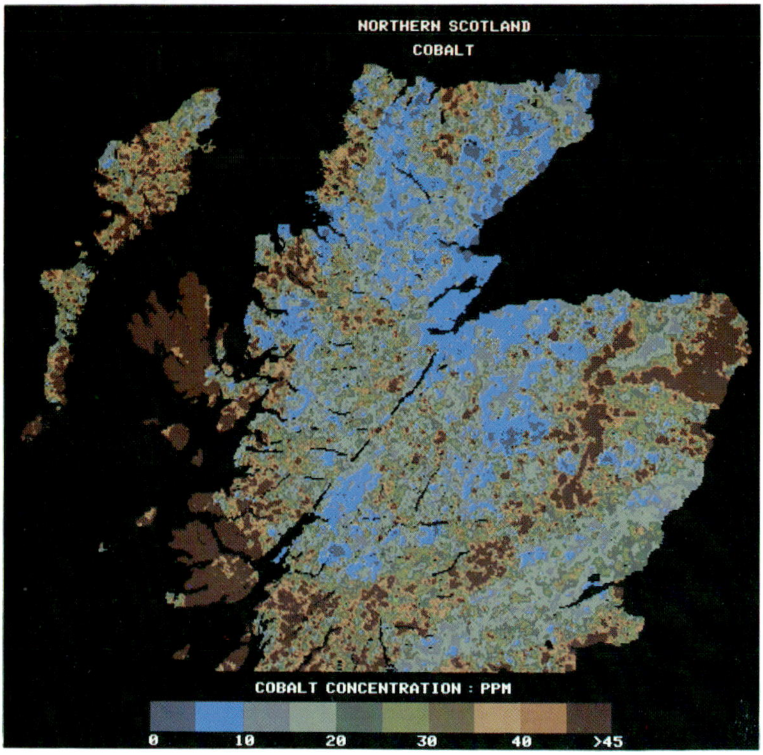

Plate 1 Regional geochemical map of Co over the Scottish highlands, prepared using the NERC interactive image analysis system (see Plant *et al*., this volume)

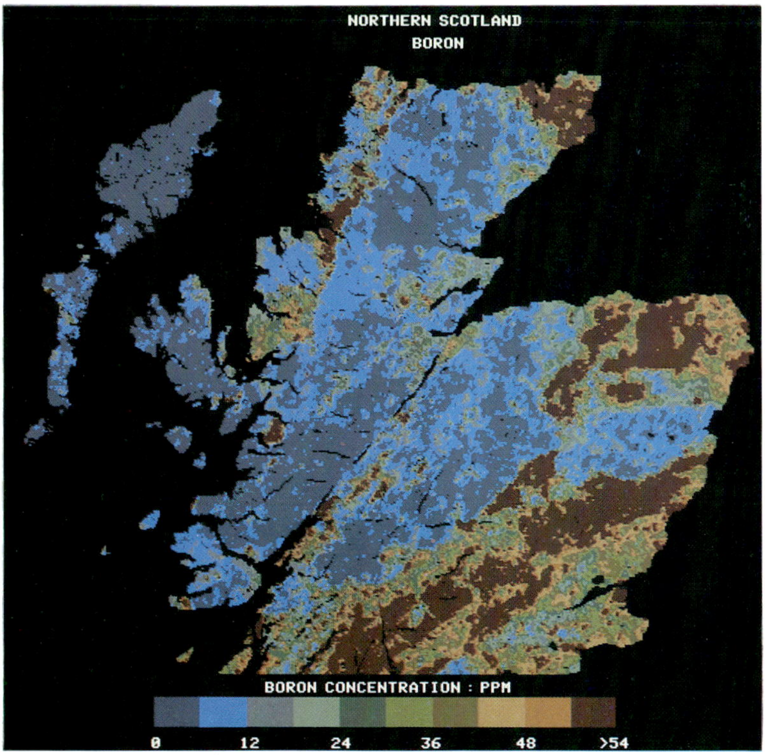

Plate 2 Regional geochemical map of B over the Scottish highlands, prepared using the NERC interactive image analysis system (see Plant *et al*., this volume)

such a model. The exploration criteria derived from this model are principally: (a) metalliferous (as indicated by regional geochemical or lithogeochemical maps) discordant granites; (b) low (garnet–amphibolite) grade pelitic–semi-pelitic host rock lithologies; and/or (c) faulting associated temporally with and continuing after granite intrusion. The combination of these criteria identify all of the mineralized granites in the Scottish Highlands (see Figure 7).

The occurrence of discrete metalliferous mineral concentrations associated with granites in the southern province of the Caledonides and in the Hercynian Province of southwest England is also predicted.

In contrast, classification according to the S- and I-type model gives misleading results. For example, the S-type biotite-muscovite granites such as those of Buchan are barren, whereas the granites that contain high levels of Sn are emplaced in a tensional regime unrelated to crustal anatexis.

MINERALIZATION IN POST-OROGENIC OLD RED SANDSTONE BASINS

The principal outcrops of the molasse–facies Old Red Sandstone in northern Scotland extend along the northern flank of the Midland Valley Graben and for several hundred kilometres around the Moray Firth, to Caithness and Orkney (Figure 12). The deposits accumulated on the flood plains of rivers, and in shallow lakes along the foot of the emerging Caledonian Mountains.

Three main divisions, all of non-marine origin, are recognized. The Lower Old Red Sandstone, which accumulated during the final stages of the Caledonian disturbance, overlaps in time with the intrusion of the late discordant suite of granites. The more extensive Middle Old Red Sandstone was deposited following a period of warping and faulting (including transcurrent dislocation of the Great Glen fault set) and erosion. During this period a large shallow lake (Lake Orcadia) developed in Orkney and Caithness. The post-Middle Devonian earth movements were relatively minor and the topography in Upper Old Red Sandstone times was subdued.

Most of the recorded occurrences of metalliferous minerals in the Old Red Sandstone were identified by radiometric and regional geochemical surveys carried out by the IGS (Gallagher *et al.*, 1971; IGS, 1978–83; Michie and Cooper, 1979). The occurrences, predominantly of U, Pb, Ba and Mo, are almost entirely confined to the Orcadian basin of Caithness and Orkney. They include numerous, small syngenetic concentrations of metals in the lacustrine facies Middle Old Red Sandstone, and two important zones of U mineralization located in major fault belts. One occurs in the Lower Old Red Sandstone Ousdale arkose on the southwest margin of the Orcadian basin and the second is in the Lower–lower Middle Old Red Sandstone in the southwest of mainland Orkney.

The direct detection of the mineralization

The principal occurrences of metalliferous mineralization are shown in Figure 12, and a summary of associated geochemical anomalies is given in Table 5.

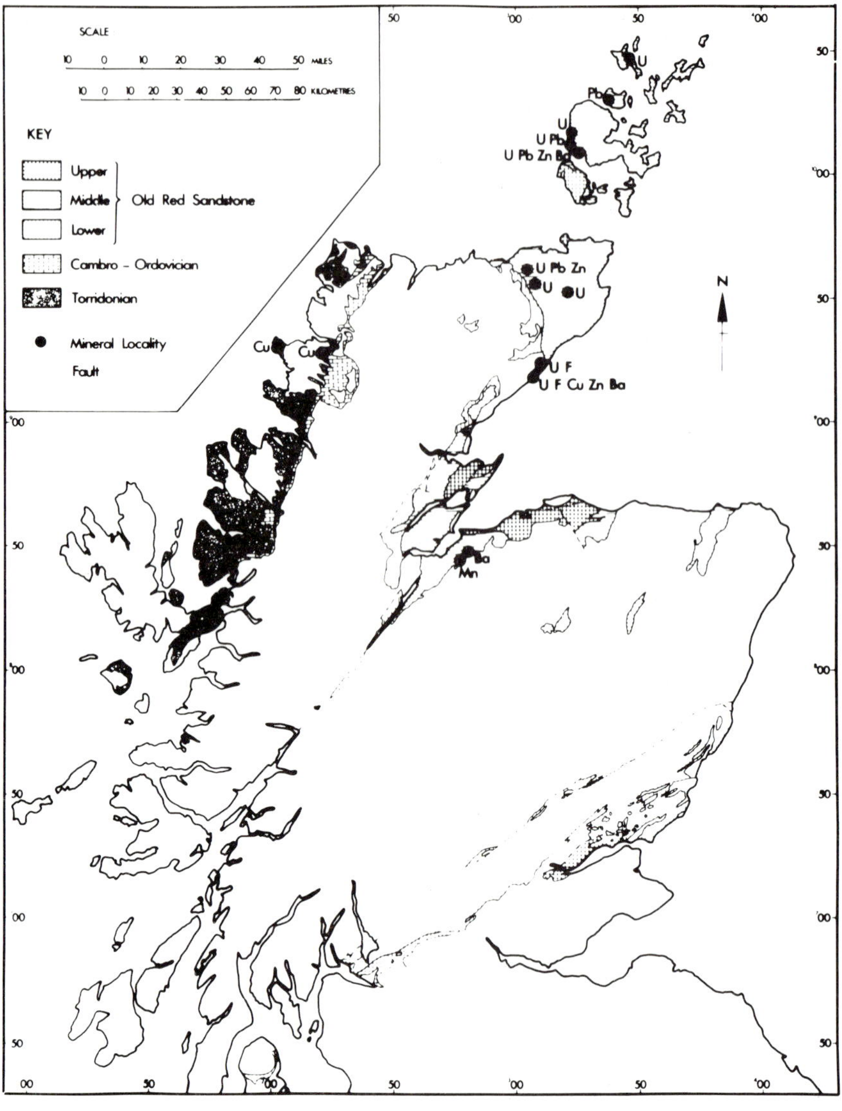

Figure 12 Distribution of sedimentary mineralization in the Old Red Sandstone and older formations of northern Scotland

Uranium is generally the most important constituent of the mineralization but in the high pH and bicarbonate regime of the Old Red Sandstone areas of Scotland it is highly soluble. Stream sediments may not, therefore, reflect local anomalies which are detected more readily by water samples, or radiometric measurement. Associations of anomalous element levels can, however, be used to indicate the sedimentary horizons in which U and other trace elements are concentrated (Table 5). The geochemical data are also valuable at the

Table 5 **Geochemical anomalies associated with sedimentary mineralization in Caithness and Orkney**

Locality	Formation/ facies	Mineralization	Regional geochemical anomalies for metals in mineralized formations	Other geochemical anomalies
Stromness Mainland, Orkney	Lower Stromness Flags (lacustrine facies)	U–Pb–Zn–Ba	+U +Pb +Zn +Ba	+Mn +Mo +B
Orkney (several localities)	Upper Stromness and Rousay Flags (lacustrine facies)	U	+U	+B +Zn
Blar Cnoc na Gaoith	Robbery Head and Lybster Groups (lacustrine facies)	U–Pb–Zn	+U +Pb	+B +Mo
Brawlbin	Helman Head Beds (flaggy lacustrine facies)	U	+U	+B +Mo +Pb +Zn
Houstry of Dunn	Helman Head Beds (lacustrine facies)	U	+U	+B +Ba +Mo +Pb +Zn
Langwell Valley	Lower Old Red Sandstone (basal conglomerate)	U–F	+U	+B
Ousdale	Lower Old Red	U–F–Cu–Zn–Ba	+U +Ba	+Mo +Pb

regional level; they clearly indicate the marked concentration of U (Figure 8), Mo (Figure 13) and Be (Figure 9) and of other elements in the Orcadian basin compared to other outcrops of the Old Red Sandstone elsewhere in Scotland.

Mineralization model

Many of the occurrences of U–Mo mineralization in the Middle Old Red Sandstone Orcadian basin are associated with dark phosphatic horizons, which are rich in organic matter and are regarded as penecontemporaneous with deposition Bowie *et al.* (1973), Michie and Cooper (1979). Remobilization of uranium as a result of faulting has occurred locally and concentrations of up to 740 ppm have been recorded, for example in the

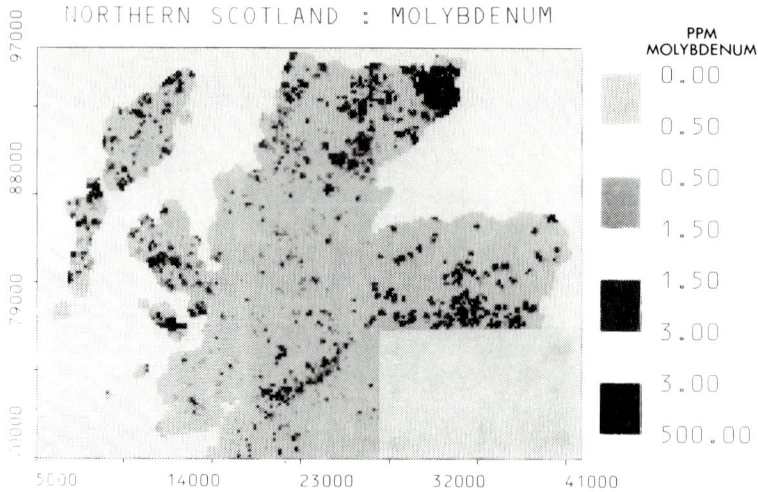

Figure 13 Regional geochemical map of Mo over the Scottish highlands

calcitic cement of a fault breccia of the Brough fault where it cuts the Middle Old Red Sandstone in Caithness (average 5–10 ppm U Bowie *et al.*, 1973). The uranium mineralization in the vicinity of the Ousdale arkose and in the southwest Orkney is considerably more extensive. Both occurrences are close to the basal Old Red Sandstone unconformity on the upthrow side of faults believed to have been active during and after deposition .

The Helmsdale granite which underlies the Ousdale arkose is heavily jointed and is traversed by secondary southeast–northwest faults. Veins of kasolite–galena–fluorite–pyrite and occurrences of baryte, fluorite, quartz, calcite and pyrite in veins and shears have been found locally in the granite itself; disseminations of U and fluorite occur in the coarser horizons of the Ousdale arkose, and thin veins of uraniferous fluorite and pyrite cut this arkose (Bowie *et al.*, 1973). In southwest Orkney U, Pb, Mn and Ba occur in veins, and show anomalously high levels in stream sediment and water samples from the mineralized area (IGS, 1978). The regional geochemical data, when considered in relation to the geological history of the Old Red Sandstone of Scotland, suggest a model which accounts for the enrichment of metals in the Orcadian basin, and for the uranium mineralization at Ousdale and in South Orkney.

The principal outcrop of the Lower Old Red Sandstone in the region is the Strathmore syncline, adjacent to the Highland Boundary fault zone. It comprises coarse immature sediments derived from the emerging Grampian Highlands, interstratified with volcanic rocks from local centres. The Lower Old Red Sandstone of the Orcadian basin only outcrops in narrow tracts of mainly locally-derived arkoses and sandstones, along the fault-controlled western margin of the basin. Post-Lower Old Red Sandstone earth movements raised the Midland Valley to an upland area which received little or no sediment during the Middle Devonian, while in the north, the highest part of the Caledonian Mountain chain shifted westward. During this period, large rivers flowing northeastwards carried sediment on to alluvial flats around the present Moray Firth and into Lake Orcadia. In the extensive Middle Old Red Sandstone of Caithness and Orkney fluviatile sandstones and siltstones, which constitute a marginal red-bed facies, pass laterally into a thick, central lacustrine, sequence of calcareous dolomitic and phosphatic grey siltstones, with a few thin bituminous limestones (see, e.g. Westoll, 1977).

The geochemistry of the Lower and Middle Old Red Sandstone arkosic and fluviatile facies sediments is closely similar to that of the immediate hinterland, supporting evidence that they comprise relatively unmodified detritus from emerging blocks of the Caledonian Mountains. Hence, the Lower Old Red Sandstone of the Strathmore syncline is almost indistinguishable from the Dalradian to the north in its levels of B, Ba, Ca, Cu, La, Li, K, Ni, Pb, Rb, Sr, Ti, Y and Zr (although the linear zone of high Co and Mg in the Dalradian does not appear to have contributed significantly to the Old Red Sandstone sediments). The lacustrine facies of the Middle Old Red Sandstone of the Caithness and Orkney basin, in the principal exception. It contains higher levels of Zn, B and Ni than the adjacent Moine while U, Mo and Be are, with the exception of the late forceful and discordant suite of granites, generally enriched in the basin at significantly higher levels than the Caledonian complex of northern Scotland. The concentrations of elements can be explained partly as a result of deposition of the metal-enriched colloidal and clay fraction of sediment under the low-energy conditions of the internal drainage basin. However, an additional source of enrichment, particularly of Mo, Be and U, is suggested. As discussed above, large granite batholiths had carried concentrations of U, Mo, Be and other metals to a high level of the crust during the Lower Old Red Sandstone, probably in association with acid volcanism; the Lintrathen Porphyry in the Strathmore syncline has been suggested to represent an ignimbrite from the Lochnagar volcanic centre (Paterson and Harris, 1969). The erosion of such igneous centres around the edge of the Moray Firth/Caithness basin in Middle Old Red Sandstone times may have contributed to the concentration of metals in the Caithness basin (Figure 14). The large Cairngorm/Mount Battock batholith which is enriched in U, Mo and Be, could have provided a suitable large volume source, particularly if Middle Old Red Sandstone or post-Lower Middle Old Red Sandstone dextral displacement on the Great Glen fault is assumed (Donovan *et al.*, 1976; Mykura, 1975). Such a model for the enrichment of the elements in the Orcadian basin is consistent with Pb isotope data reported by Parnell and Swainbank (in press).

A model involving the interaction of heated ground water in deep fault zones with magma or granite containing U and F, has been used to explain the Ousdale mineralization in the Lower Old Red Sandstone, Watson and Plant (1979); possibly associated with a high local geothermal gradient which persisted after emplacement of the granite (Figure 14). The extent to which the mineralization was due to a high primary concentration of U or F in the

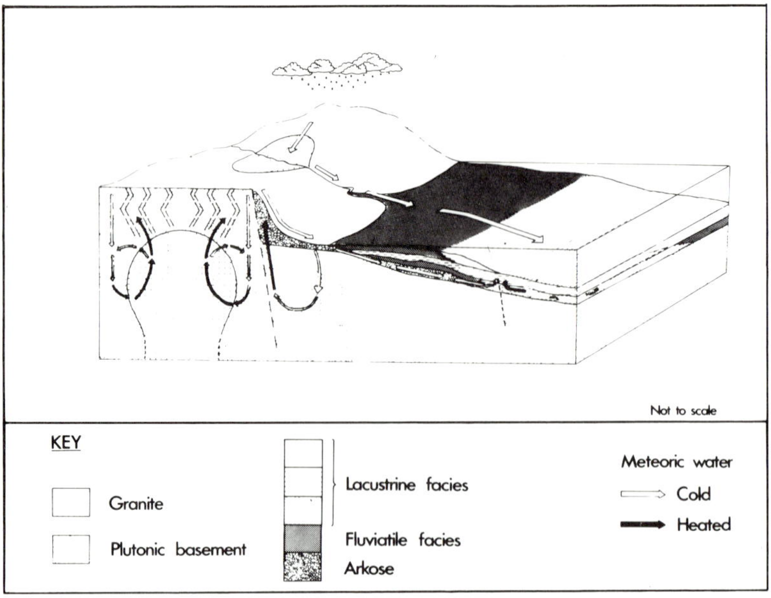

Figure 14 Model for the development of sedimentary mineralization in fault-bounded (Old Red Sandstone) basins

Helmsdale granite, or to concentration in brines interacting with granites containing normal levels of U and F, is not known (Simpson *et al.*, 1982).

In the case of the mineralization in southwest Orkney, enrichment of U associated with the Pb–Ba–Zn mineralization has been attributed to the migration of deep pore fluids up-slope from the central part of the basin (Figure 14); this is thought to have occurred during periods of tectonic activity in Lower and Lower–Middle Old Red Sandstone times driven partly by high residual heat flow (Watson and Plant, 1979). Lead isotope data (Parnell and Swainbank, in press) suggests that a Permian event associated with dyke emplacement also affected the mineralization in southwest Orkney.

Discussion and conclusions

Geochemical maps based on stream sediment samples provide a powerful means of detecting mineralization in the Scottish Caledonian province. They clearly identify stratabound metalliferous minerals in Dalradian metasediments; they distinguish metalliferous and mineralized granites from barren intrusions; and they identify sedimentary formations containing anomalous levels of metals in the Old Red Sandstone as well as specific mineral occurrences.

Vein mineralization has not been considered in earlier sections because it has a wide range of ages, from late Caledonian to approximately 140 Ma (Moorbath, 1962), and in some cases it merely represents remobilization of

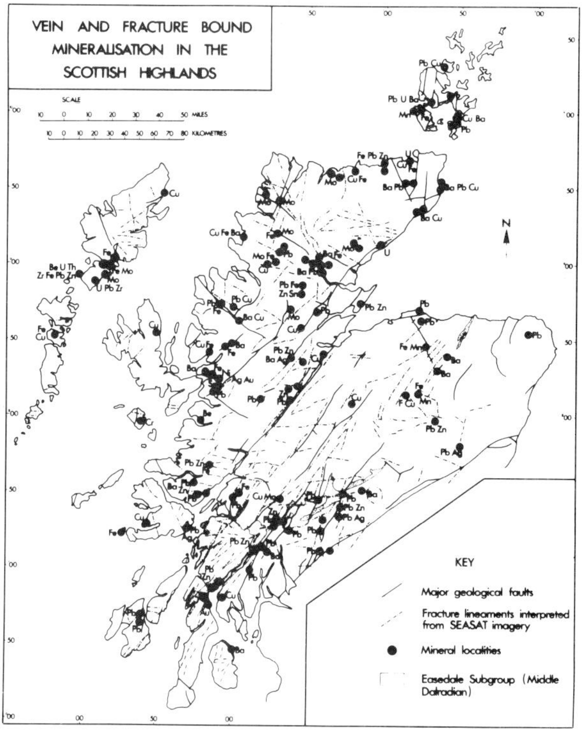

Figure 15 The location of fault and fracture-bound mineralization in the Scottish highlands

pre-existing metal concentrations. Nevertheless, vein-type mineralization is important in northern Scotland and it is therefore included in evaluating the application of regional geochemical maps to the *direct* identification of mineralization. There is a strong association of vein-type mineralization with faults and fracture systems, with 69% of occurrences ≤2 km from a major lineament (Figure 15). Some of the most important veins, for example at Strontian and Tyndrum, have been dated at 240 to 230 Ma, corresponding with Permo–Carboniferous dyke emplacement (Moorbath, 1962). The relationship between occurrences of vein-type metalliferous mineralization, (including minor showings) and geochemical anomalies is shown in Table 6. It is apparent that 96% of all occurrences are identified by an anomalous value for one or more chemical elements; 65% showing an anomaly which directly reflects their metalliferous mineral assemblages.

Failure to detect mineralization occurs only where the mineralized bodies outcrop on the coast or in areas of poor surface drainage. In the Caithness region, it is also likely that the sample density of 1 per 2.3 km^2 was too low. There is a broad correlation between the geochemical association of the vein mineralization and the chemistry of the host rock. For example, Mo and U veins occur mainly in the Middle Old Red Sandstone of Caithness and near to late discordant granite batholiths, suggesting that metals were generally

Table 6 **Percentage of total mineral occurrences identified by geochemical anomalies**

	Total number of occurrences	Total percentage showing geochemical anomalies	Percentage showing anomaly reflecting metalliferous minerals ± other indicator elements	Percentage showing anomalous levels of other indicator elements only
Stratabound mineralization	7	100	71	29
Granites (mineralized intrusions)	17	100	100	—
Old Red Sandstone (mineralized formations)	5	100	100	—
Vein type deposits	115	96	65	31

concentrated from their host rocks. Taking the four types of mineralization which have been considered together, more than 98% of all occurrences are detected by one or more anomalous geochemical values (Table 6). Stream sediments thus identify all of the significant, and most of the minor, recorded occurrences of metalliferous minerals in northern Scotland (although ideally, water samples, radiometric readings and heavy mineral concentrates should be incorporated in geochemical reconnaissance programmes aimed at detecting U and Ba, Pb and Sn mineralization respectively). Regional geochemical maps based on stream sediment samples can also be used to prepare information on the rocks associated with mineral deposits. For example, granites may be classified as barren, transitional, Cu–Mo–granodiorite, Mo(W) granite and Sn–U granites by the use of trace element patterns, while changes in the distribution of metals over intrusions and of key ratios such as B/Ga indicate the occurrence of hydrothermal alteration associated with mineralization (Simpson *et al.*, 1982).

The most powerful application of geochemical maps in exploration, however, is their analysis in relation to geological (particularly tectonic) and geophysical data, using interactive computer graphics devices. It is possible by this means to develop conceptual models for mineralization which incorporate field, isotopic and mineralogical data. Key exploration criteria can then be derived for the automated searching of data held in computer-readable form. For example, it has been suggested that B is an indicator of stratabound mineralization in metamorphic terrains such as that of the Dalradian (Slack, 1982). The analysis presented here, however, suggests that B indicates only that the sedimentary regime of the Dalradian was consistent with that of a marine marginal basin. Hence, while metasomatic tourmaline in rocks may be a useful indicator of hydrothermal activity locally, parameters other than high B levels are required to limit the search area for the metalliferous mineralization to anything less than the Dalradian as a whole. If however, such parameters as: Pb $>$ 40 ppm, Ba $>$ 1650 ppm and Zn $>$ 200 ppm occurring within 20 kilometres, and to the north of, the zone of high Co and MgO, in rocks containing B and Li levels $>$ 30 ppm respectively is used, then all of the recorded occurrences of stratabound mineralization are identified.

Finally, the use of regional geochemistry in relation to tectonic history enables some general features of metallogenesis to be identified. Hence, in the northern Province of the Caledonides, mineralization appears to have occurred in discrete episodes of the order of a few tens of millions of years.* These were principally in Vendian times in the case of the Dalradian stratabound deposits, and in the Lower Old Red Sandstone times in the case of the mineralization associated with granites.

In both cases mineralization was associated with a marked regional or local steepening of the geothermal gradient, associated with melting of the lower crust or lithospheric mantle in a tensional tectonic regime; these events were accompanied by deep fracturing of the crust or movement on pre-existing deep fractures. Hydrothermal activity and mineralization thus reflect deep thermal and tectonic disturbances. The most important sites of mineralization in the Old Red Sandstone, which are located on the upthrow sides of major deep faults, are also thought to be associated with zones of fluid migration and high heat flow. The metals concentrated in occurrences of metalliferous mineralization in the Caledonian province generally appear to have been derived from their host rocks, in a setting in which abundant fluids were available. Mineralization of the granite association is an exception. Large granite batholiths carried heat and metals to a high level in Lower Old Red Sandstone times, but mineralization is thought to have been limited by the anhydrous nature of the Caledonian crust at the time of the granite emplacement. The intrusions and their associated volcanics may, however, have provided large volume sources of metals which became concentrated in sediments of the Caithness–Orkney basin.

It is concluded that computer-readable regional geochemical maps, which were pioneered by John Webb and the Applied Geochemistry Research Group at Imperial College, not only provide a powerful means of detecting mineralization directly but also provide information fundamental to metallogenic studies. The demand for such geochemical maps, which is already considerable, is likely to increase further as the significance of new image processing techniques becomes more widely appreciated by earth scientists, and as geochemical maps are used for interactive studies in relation to geophysical, geological and remotely sensed data. The significance of John Webb's contribution to geochemistry, and hence to geology, is yet to be fully appreciated.

*It should be noted that certain deposits e.g. Au have not been considered.

ACKNOWLEDGEMENTS

The authors thank D. Ostle for critically reading the manuscript.

The paper is published by permission of the Director, the Institute of Geological Sciences, NERC.

References

Anderton, R. (1982). Dalradian deposition and the Late Precambrian-Cambrian history of the N. Atlantic region: a review of the early evolution of the Iapetus Ocean, *J. Geol. Soc. Lond.*, **139**, 421–431.

Anhauser, C.R., Fritze, K., Fyfe, W.S. and Gill, R.C.O. (1975). Gold in "primitive" Archaean volcanics, *Chem. Geol.*, **16**, 129–135.

Bamford, D., Nunn, K., Prodehl, C. and Jacob, B. (1977). LISPB III upper crustal structure of northern Britain, *J. Geol. Soc. Lond.*, **133**, 481–488.

Beckinsale, R.D. (1979). Granite magmatism in the tin belt of South-East Asia. In: *Origin of Granite Batholiths: Geochemical Evidence* (Eds. M.P. Atherton and J. Tarney), pp. 34–44; Shiva Publishing, Nantwich.

Beus, A.A. and Grigorian, S.V. (1977). *Geochemical Exploration for Mineral Deposits*, Applied Publishing Ltd., Illinois.

Botbol, J.M., Sinding-Larsen, R., McCannon, R.T. and Gott, G.B. (1978). Regionalized multivariate approach to target selection in geochemical exploration, *Econ. Geol.*, **73**, 534–546.

Bowie, S.H.U., Ball, T.K. and Ostle, D. (1971). Geochemical Methods in the Detection of Hidden Uranium Deposits, Geochemical Exploration, *CIM Special Volume No. 11.*, pp. 103–111.

Bowie, S.H.U., Simpson, P.R. and Rice, C.M. (1973). Application of fission-track and neutron activation method to geochemical exploration, In: *Geochemical Exploration 1972* (Ed. M.J. Jones), pp. 359–372; IMM, London.

Brown, G.C. and Fyfe, W.S. (1970). The production of granitic melts during ultra-metamorphism, *Contr. Mineral. Petrol.*, **28**, 310–318.

Burgess, W.G., Edmunds, W.M., Andrews, J.N., Kay, R.L.F. and Lee, P.J. (1982). The origin and circulation of groundwater in the Carnmenellis granite: the hydrogeochemical evidence. In: *Investigation of the geothermal potential of the U.K.*; IGS, London.

Campbell, I.H.; Coad, P., Franklin, J.M., Gorton, M.P., Hart, T.R., Scott, S.D., Sowa, J. and Thurston, P.C. (1981). Rare earth elements in felsic volcanic rocks associated with Cu-Zn massive sulphide mineralization, *Ontario Geological Survey Misc. Pap.*, **98**, 45–53.

Coats, J.S., Smith, C.G., Fortey, N.J., Gallagher, M.J., May, F. and McCourt, W.J. (1980). Stratabound barium-zinc mineralization in Dalradian schist near Aberfeldy, Scotland, *Inst. Min. Metall. Trans.*, **89**, B110–B122.

Coats, J.S., Smith, C.G., Gallagher, M.J., May, F., McCourt, W.J., Parker, M.E. and Fortey, N.J. (1978). Stratabound barium-zinc mineralization in Dalradian schist near Aberfeldy, Scotland: Preliminary Report, *Min. Rec. Prog. Rep.Inst. Geol. Sci.*, **26**.

Coats, J.S., Smith, C.G., Gallagher, M.J., May, F., McCourt, W.J., Parker, M.E. and Fortey, N.J. (1980). Stratabound barium-zinc mineralization in Dalradian schists near Aberfeldy, Scotland: Final Report, *Min. Rec. Prog. Rep. Inst. Geol. Sci.*, **40**.

Coats, J.S., Tandy, B.C. and Michie, U. McL. (1982). Geochemical drainage survey of central Argyll, Scotland, *Min. Rec. Prog. Rep. Inst. Geol. Sci.*, **50**.

Darnley, A.G. (1982). "Hot" granites; some general remarks. In: *Uranium in Granites* (Ed. Y.T. Maurice), pp. 1–10; Geol. Surv. Canada, Paper 81–23.

Dewey, J.F. and Pankhurst, R.J. (1970). The evolution of the Scottish Caledonides in relation to their isotopic age pattern, *Trans. R. Soc. Edinburgh*, **68**, 361–389.

Donovan, R.M., Archer, A., Turner, P. and Tarling C.H. (1976). Devonian palaeography of the Orcadian basin and the Great Glen fault, *Nature*, **250**, 550–551.

Ellis, R.A., Coats, J.S., Haslam, H.W., Michie, U. McL., Fortey, N.J., Johnson, C.E. and Parker, M.E. (1977). Investigation of disseminated copper mineralization near Kilmelford, Argyllshire, Scotland, *Min. Rec. Prog. Rep. Inst. Geol. Sci.*, **9**.

Fairbank, B.D., Openshaw, R.E., Souther, J.G. and Stauder, J. (1981). J. Meager Creek Geothermal Project; an exploration case history. *Bull. Geothermal Resources Council 10*: **6**, 3–7.

Finlow-Bates, T., Croxford, N.J.W. and Allan, J.M. (1977). Evidence for, and implications of, a primary FeS phase in the Lead-Zinc-bearing sediments at Mount Isa, *Min. Dep.*, **12**, 143–149.

Fortey, N.J. and Beddoe-Stephens, B. (1982). Barium silicates in stratabound Ba–Zn mineralization in the Scottish Dalradian, *Miner. Mag.*, **46**, 63–72.

Fyfe, W.S. (1975). Hydrosphere solid-earth interaction. In: *Canada's Continental Margins and Offshore Petroleum Exploration* (Eds. C.J. Yorath, E.R. Parker and D.J. Glass); pp. 763–772; Can. Soc. Pet. Geol., Mem., **4**.

Fyfe, W.S. and Kerrich, R. (1976). Geochemical prospection: extensive versus intensive factors, *J. Geochem. Explor.*, **6**, 177–192.

Gallagher, M.J., Michie, U. McL., Smith, R.T. and Haynes, L. (1971). New evidence of uranium and other mineralization in Scotland, *Inst. Min. Metall. Trans.*, **80**, B150–B173.

Gammon, J.B. (1966). Fahlbands in the Precambrian of southern Norway, *Econ. Geol.*, **61**, 174–188.

Garrett, R.G. (1971). Molybdenum, tungsten and uranium in acid plutonic rocks, southeast Yukon, *Can. Min. J.*, **92(4)**, 37–40.

Harris, A.L., Baldwin, C.T., Bradbury, H.J., Johnson, H.D. and Smith R.A. (1978). Ensialic basin sedimentation: the Dalradian Supergroup. In: *Crustal Evolution in Northwestern Britain and Adjacent Areas* (Eds. D.R. Bowes and B.E. Leake); pp.115–138; Geol. J. Spec. Issue, **10**.

Harris, A.L. and Pitcher, W.S. (1975). The Dalradian Supergroup. (Eds. A.L. Harris *et al.*) In: *A Correlation of Precambrian Rocks in the British Isles.* pp.52–75. Geol. Soc. Lond., Spec. Rep. **6**.

Howarth, R.J., Koch, G.S., Plant, J.A. and Lowry, R.K. (1981). Identification of uraniferous granitoids in the USA using stream sediment data, *Miner Mag.*, **44**, 455–470.

Hussain, A. and Hipkin, R.G. (1981). *Bouguer Anomaly Map of the British Isles (Northern Sheet)*; J. Bartholomew and Son, Edinburgh.

Institute of Geological Sciences. (1978a). *Geochemical Atlas of Gt. Britain: Orkney Islands*; Inst. Geol. Sci., London.

Institute of Geological Sciences. (1978b). *Geochemical Atlas of Gt. Britain: Shetland Islands*; Inst. Geol. Sci., London.

Institute of Geological Sciences. (1979). *Geochemical Atlas of Gt. Britain: Caithness and South Orkney*; Inst. Geol. Sci., London.

Institute of Geological Sciences. (1983a). *Geochemical Atlas of Gt. Britain: Sutherland*; Inst. Geol. Sci., London, in press.

Institute of Geological Sciences. (1983b). *Geochemical Atlas of Gt. Britain: Great Glen*; Inst. Geol. Sci., London, in preparation.

Institute of Geological Sciences. (1975–83). Min. Rec. Prog. Rep. Inst. Geol. Sci., **1–63**.

Johnstone, G.S., Plant, J.A. and Watson, J.V. (1979). Regional geochemistry of the northern Highlands of Scotland. (Eds. A.L. Harris, C. Holland and B.E. Leake); In: *The Caledonides of the British Isles-Reviewed,* pp. 117–128; Geol. Soc., Spec. Publ. 8 London.

Karig, D.E. (1971). Origin and development of marginal basins in the Western Pacific, *J. Geophys. Res.*, **76**, 2542–2561.

Lambert, R. St. J., Winchester, J.A. and Holland, J.G. (1981). Comparative geochemistry of pelites from the Moinian and Appin Group (Dalradian) of Scotland, *Geol. Mag.*, **118**, 477–490.

Large, D. (1981). The geochemistry of the sedimentary rocks in the vicinity of the Tom Pb–Zn–Ba deposit, Yukon Territory, Canada, *J. Geochem. Expl.*, **15**, 203–217.

Leake, B.E. (1978). Granite emplacement: the granites of Ireland and their origin. In: *Crustal Evolution in Northwest Britain and Adjacent Region* (Eds. D.R. Bowes and B.E. Leake), pp. 221–248; Geol. J. Spec. Issue No. 10, Seal House Press.

Matsuda, T. and Uyeda, S. (1971). On the Pacific-type orogeny and its model—extension of the paired belts concept and possible origin of marginal seas, *Tectonophysics*, **11**, 5-27.

Michie, U. McL. and Cooper, D.C. (1979). Uranium in the Old Red Sandstone of Orkney, *Rep. Inst. Geol. Sci.*, **78/16**.

Mitchell, A.J.G. and Garson, M.S. (1972). Relationship of porphyry copper and circum-Pacific tin deposits to palae o-Benioff zones, *Inst. Min. Metall. Trans.*, **81**, B10–B25.

Mitchell, A.H.G. and Garson, M.S. (1972). Relationship of porphyry copper anc circum-Pacific tin deposits to palaeo-Benioff zones, *Inst. Min. Metall. Trans.*, **81**, B10–B25.

Moles, N.R. (1982). Controls on the distribution of baryte and sulphide mineralization in the Foss deposit, Aberfeldy. *Abs. Mineral deposits studies group*. Joint meeting of geological societies of Great Britain. Glasgow University. 24 Sept. 1983.

Moorbath, S. (1962). Lead isotope abundance studies on mineral occurrences in the British Isles and their geological significance, *Phil. Trans. R. Soc. Lond.*, **(A)254**, 295–360.

Moore, J. McM. (1982). Mineral zonation near the granitic batholiths of southwest and northern England and some geothermal analogues. In: *Metallization Associated with Acid Magmatism* (Ed. A.M. Evans), pp. 229–241.

Moreau, M. (1976). L'uranium et les granitoides: essai d'interpretation. In: Geology, Mining and Extractive Processing of Uranium (Ed. M.D. Jones), pp. 83–102; IMM, London.

Mottl, M.J. (1983). Metabasalts, axial hot springs and the structure of hydrothermal systems at mid-ocean ridges, *Geol. Soc. Am. Bull.*, **94**, 161–180.

Mutschler, F.E., Wright, E.G., Ludington, S. and Abbott, J.T. (1981). Granite molybdenite systems, *Econ. Geol.*, **76**, 874–897.

Mykura, W. (1975). Possible large-scale sinistral displacement along the Great Glen Fault in Scotland, *Geol. Mag.*, **112**, 91–93.

Ohmoto, H. and Rye, R.O. (1974). Hydrogen and oxygen isotopic compositions of fluid inclusions in the Kuroko deposits, Japan, *Econ. Geol.*, **69**, 947–953.

Page, D.C. and Watson, M.D. (1976). The Pb–Zn deposit of Rosh Pinah Mine, South West Africa, *Econ. Geol.*, **71**, 306–327.

Paterson, I.B. and Harris, A.L. (1969). Lower Old Red Sandstone ignimbrites from Dunkeld, Perthshire, *Rep. Inst. Geol. Sci.*, **69/7**.

Piasecki, M.A.J. and Van Breemen, O. (1982). Field isotopic evidence for a c. 750 Ma tectonothermal event in Moine rocks in the Central Highland region of the Scottish Caledonides, *Trans. R. Soc. Edinburgh: Earth Sci.*, **73**, 119–134.

Plant, J.A. and Moore, P.J. (1979). Regional geochemical mapping and interpretation in Britain, *Phil. Trans. R. Soc. London.*, **B288**, 95–112.

Plant, J.A., Brown, G.C., Simpson, P.R. and Smith, R.T. (1980). Signatures of metalliferous granites in the Scottish Caledonides, *Inst. Min. Metall. Trans.*, **89**, B198–B210.

Plant, J.A., Simpson, P.R., Green, P.M., Watson, J.V. and Fowler, M.B. (1983). Metalliferous and mineralized Caledonian granites in relation to regional metamorphism and fracture systems in northern Scotland, *Instn. Min. Metall. Trans.*, **92**, B33–B42.

Plant, J.A., Watson, J.V. and Green, P.M. (1984). Moine/Dalradian relationships and their palaeotectonic significance, *Proc. R. Soc. Lond.*, **A395**, 185–202.

Plimer, R. (1978). Proximal and Distal Stratabound ore deposits, *Min. Dep.*, **13**, 345–353.

Plimer, R. and Finlow-Bates, T. (1978). Relationship between primary iron sulphide species, sulphur source, depth of formation and age of submarine exhalative sulphide deposits, *Min. Dep.*, **13**, 399–410.

Rankin, A.H. and Alderton, D.H.M. (1982). Fluid inclusion evidence for the evolution of hydrothermal fluids in SW England: abstract, *J. Geol. Soc. Lond.*, **139**, 96.

Read, H.H. (1961). Aspects of the Caledonian magmatism in Britain, *Liverpool and Manchester Geol. J.*, **2**, 653–683.

Roedder, E. (1977). Fluid inclusions as tools in mineral exploration, *Econ. Geol.*, **72(3)**, 503–525.

Sakai, H. and Matsubaya, O. (1974). Isotopic geochemistry of the thermal waters of Japan and its bearing on the Kuroko ore solutions, *Econ. Geol.*, **69**, 974–991.

Sheppard, S.M.F. and Taylor, H.P. (1974). Hydrogen and oxygen isotope evidence for the origins of water in the Boulder Batholith and the Butte ore deposits, Montana, *Econ. Geol.*, **69**, 926–946.

Simpson, P.R., Plant, J.A. and Cope, M.J. (1976). Uranium abundance and distribution in some granites from northern Scotland and southwest England as indicators of uranium provinces. In: *Geology, Mining and Extractive Processing of Uranium* (Ed. M.J. Jones), pp. 126–139; IMM Spec. Vol.

Simpson, P.R., Brown, G.C., Plant, J.A. and Ostle, D. (1979). Uranium mineralization and granite magmatism in the British Isles, *Phil. Trans. R. Soc. Lond.*, **A291**, 385–412.

Simpson, P.R., Plant, J.A., Watson, J.V., Green, P.M. and Fowler, M.B. (1982). The role of metalliferous and mineralized uranium granites in the formation of uranium provinces. In: *Uranium Exploration Techniques: Proceedings of the Symposium, Paris, June, 1982*, pp. 157–170; NEA, Paris.

Slack, J.F. (1982). Tourmaline in Appalachian-Caledonian massive sulphide deposits and its exploration significance, *Inst. Min. Metall. Trans.*, **91**, B81–B89.

Sleep, N. and Toksoz, M.N. (1971). Evolution of marginal basins, *Nature*, **233**, 548–550.

Smirnov, V.I. (1976). Geology of Mineral Deposits; MIR, Moscow.

Smith, C.G., McCourt, W.G., Fortey, N.J., Johnson, C.E., Parker, M.E., Coats, J.S. and Michie, U. McL. (1977). Investigation of stratiform sulphide mineralization at Meall Mor, South Knapdale, Argyll, *Min. Rec. Prog. Rep. Inst. Geol. Sci.*, **15**.

Smith, C.G., Pitfield, P.E.J., Burley, A.J., Howard, S.H.D., Parker, M.E., Michie, U. McL. and Fortey, N.J. (1977). Investigation of stratiform sulphide mineralization in parts of central Perthshire, *Min. Rec. Prog. Rep. Inst. Geol. Sci.*, **8**.

Smith, C.G., Pitfield, P.E.J., May, F., Burley, A.J., Howard, S.H.D. and Parker, M.E. (1977). Investigation of stratiform sulphide mineralization at McPhun's Cairn, Argyllshire, *Min. Rec. Prog. Rep. Inst. Geol. Sci.*, **13**.

Spooner, E.T.C. and Fyfe, W.S. (1973). Sub-sea-floor metamorphism, heat and mass transfer, *Contr. Mineral. Petrol.*, **42**, 287–304.

Tischendorf, G. (1973). The metallogenetic basis of tin exploration in the Erzgbirge, *Inst. Min. Metall. Trans.*, **82**, B9–B24.

Watson, J.V. and Plant, J.A. (1979). Regional geochemistry of uranium as a guide to deposit formation, *Phil. Trans. R. Soc. Lond.*, **A291**, 321–328.

Webb, J.S., Thornton, I., Thompson, M., Howarth, R.J. and Lowenstein, P.L. (1978). *The Wolfson Geochemical Atlas of England and Wales;* Oxford University Press, Oxford.

Westoll, T.S. (1977). Northern Britain. M.R. House, J.B. Richardson, C.H. Holland and T.S. Westoll. In: *A Correlation of Devonian Rocks of the British Isles.* Geol. Soc. Lond. Spec. Report No. 8.

Willan, R.C.R. (1980). Stratabound sulphide mineralization in the Dalradian Supergroup of the Grampian Highlands, Scotland, *Norges Geol. Unders.*, **300**, 241–258.

Willan, R.C.R. (1982). Isotopic evidence for the source of sulphur in the Aberfeldy baryte and sulphide deposits and the evolution of the hydrothermal system (abstract). Proc. Joint Meeting of Geol. Socs. of Great Britain: The origin of sediment-hosted stratiform metalliferous ores. Glasgow.

Willan, R.C.R. and Hall, A.J. (1980). Sphalerite geobarometry and trace-element studies on stratiform sulphide from McPhun's Cairn, Loch Fyne, Argyll, Scotland, *Inst. Min. Metall. Trans.*, **89**, B31–B40.

Wilson, G.V. and Flett, J.S. (1921). The lead zinc copper and nickel ores of Scotland, *Spec. Rep. Miner. Resources Gt. Britain*, **17**.

Winchester, J.A., Lambert R.St.J. and Holland, J.G. (1981). Geochemistry of the western part of the Moinian Assemblage, *Scott. J. Geol.*, **17**, 281–294.

Geochemical Patterns in the Granitic Terrain of Zimbabwe

K. A. VIEWING, N. J. TOPPING and R. S. HATHERLEY

Introduction

The Archaean granitic terrain of Zimbabwe underlies the eastern half of the country. It is characterized by thin residual soils and open savannah grassland with well-developed drainages to the Zambesi in the north and the Limpopo in the south. The mineral potential is poor, but it includes the Inyati copper mine, the Dorowa phosphate mine and a large number of small pegmatites, small gold-quartz vein deposits and small tungsten-skarn deposits. Most of these deposits occur in xenolithic remnants within the granitic rocks, in general close to the margins, but some are within the extensive areas of the granitic terrain. In the northern and eastern parts of the terrain there are extensive sheets of the Mashonaland dolerite (about 1.7 to 1.87 Ga). These dolerites have not been shown to be mineralized but the pyroxenite of the Madziwa nickel mine is of the same age.

Geochemical techniques have not been tested adequately in the granitic terrain, and the optimum sampling techniques are not established. A part of the problem, however, is to distinguish chemically between analytical data related to mafic and ultramafic xenolith remnants of Archaean age, and the geochemical character of the mafic rocks of Mashonaland age. The granitic rocks of the craton vary from tonalite, to granodiorite in the vicinity of Archaean schists, and to adamellite and granite in the central zones of some of the batholiths. The granitic rocks are of different ages, and the oldest are of about 3.5 to 3.6 Ga, and these were intruded at about 2.6 Ga (Nisbet *et al.*, 1981) by the Chilimanzi and Sessombe suites. The Chilimanzi suite is adamellitic and the Sessombe suite is tonalitic.

The oldest greenstone belt remnants constitute the Sebakwian Group and this includes metavolcanic rocks, ranging from komatiite to basalt; iron formations and some clastic and pelitic sediments. Chromite bearing ultramafic intrusions are a notable feature of the succession at Selukwe. The Bulawayan and Shamvaian Groups form the major greenstone belts. The upper part of the Bulawayan (Upper Greenstones) and the overlying, mainly sedimentary, Shamvaian Group are both about 2.7 Ga old. Together they constitute Zimbabwe's most extensive and best preserved greenstone belts.

In several areas, however, the lowermost part of the Bulawayan is distinguishable from the remainder, and from the Sebakwian. This lower sequence (Lower Greenstones) is associated with granites and gneisses dated at ca 2.9 Ga. The intrusion of the Great Dyke, at ca 2.5 Ga, was the last major event in the craton during the Archaean. In the Belingwe greenstone belt, the Lower Greenstones are divided into five formations. Major lithologies include mainly pyroclastic, felsic volcanics, with volcaniclastic sediments; komatiites and komatiitic basalts with interbedded banded iron-formation; and polymict clastic sediments. The Lower and Upper Greenstones of the Bulawayan Group are separated by an unconformity that is traced over most of the central craton. The Upper Greenstones are subdivided into western and eastern successions. The lithologies include massive and pillowed komatiitic and tholeiitic basalt flows, with pelites, iron formations and limestone. Calc-alkaline volcanics are limited to the western succession.

The Shamvaian Group unconformably overlies the Upper Greenstones in the west, east and southeast of the craton. The main rock types are arkose and sub-greywacke, derived from a mixed granitic-volcanic terrain. Most of the rocks of the schist belts are metamorphosed to the green-schist facies, but hornblende occurs in xenolithic fragments amongst the granitic rocks. In the north and south margins of the craton, bounded by the Zambesi and Limpopo metamorphic belts, the grade achieves sillimanite and granulite facies, respectively.

Rocks of the Sebakwian, Bulawayan and Shamvaian groups contain some 4000 deposits, and copper is associated in places with the auriferous hydrothermal lodes. By contrast, the Inyati copper deposit occurs in vein quartz within a schistose xenolith enclosed by the granitic-gneiss terrain, well away from the schist belts. Nickel sulphides occur in komatiitic rocks and in differentiated igneous intrusives within the Upper Greenstones. Archaean beryllium-lithium pegmatites, and tin and tantalum pegmatites are common and, together with emerald occurrences, appear to be associated with the Chilimanzi Suite of late granites. Asbestos occurs within the ultramafic complexes, particularly in the southern part of the craton. The Archaean schist belts contain extensive horizons of iron-formations, but iron ore is only mined at Bukwa, and at Kwekwe. Lenticular beds of limestone exist, often close to the iron-formations, and antimony, barytes, corundum, iron, pyrite and tungsten mineralization occur.

The granitic terrain thus offered an opportunity to test techniques of multi-element geochemistry on regional and local scales using rock and drainage material as the sampling media. Almost all of the granitic terrain is a part of the post-African land surface, of Miocene age (Lister, 1979). The older African surface (mid-Cretaceous to end of Oligocene) forms the watershed and also the western boundary to most of the granitic terrain. This younger, Pliocene surface, occupies the Zambesi valley to the north and, to an extent, the Limpopo valley to the south. In general, the soils are thin residual lithosols, and deep profiles of about 15 m occur only in the Eastern Highlands at elevations above 2000 m. Small patches of the Kalahari sand (Miocene) occur in parts of the Limpopo valley, and thin horizons of nodular calcrete are found adjacent to drainage systems amongst the basalts. There are no significant accumulations of silcrete or ferricrete.

The rainfall is seasonal and over most of the terrain it varies from 700 to

900 mm a^{-1}, rising to 1000 mm a^{-1} in the northern part of the granitic terrain. The granitic rocks are best exposed in the valleys of the drainage systems, and are uncommon in the watershed. The schists amongst the granitic rocks, and the dolerite sheets of Mashonaland age, form prominent ridges which rise from 20 m to about 150 m above the plain. The relief is therefore gently undulating from 1500 to 1650 m and is well drained by a network of small streams characterized by swift and seasonal flow. There are vlei drainage systems in the watershed areas, but the nick points are developed within a few kilometres from their sources. Most of the granitic terrain is used for subsistence farming in the Communal Lands, and the production of maize, sorgum and groundnuts. Tobacco and cotton may be produced and cattle, goats and fowl are kept. Commercial farms produce Virginia tobacco and this is fertilized heavily.

The application of lithogeochemistry

Geochemical patterns were first revealed from a study of chemical analyses in the records of the Geological Survey of Southern Rhodesia in 1967 (Viewing, 1968). The analyses of 88 samples of granitic rocks collected during early geological mapping of the country showed a marked regional distribution in which K is concentrated (in the range fiom 3.0 to 5.0%) in two broad geographic belts. These lie along the northern and southern margins of the Rhodesian craton adjacent to the Zambesi and Limpopo metamorphic zones. The patterns of Na : K ratios and the agpaitic ratio (NA + K)/Al have broadly similar trends. The K-rich granites occur in the vicinity of commercially important deposits of pegmatites, at the Karoi Ta–Sn deposits in the north; the Kamativi Sn deposits in the west and the Bikita Li, Sn, Ta and Be deposits, in the south.

Three other pegmatite belts: the Borrowdale pegmatites near Harare, the Shamva pegmatites which border the Chinamora batholith, and the Odzi pegmatites which extend across the Archaean granitic terrain of the Sabi Communal Lands, were not correlated with potassic phases for lack of analytical data. The scale of the compilation was too large and the data from the central region of the craton were inadequate to demonstrate an association within the craton. Accordingly, the pegmatite associations of the Chinamora batholith were investigated by lithogeochemistry, and a part of the Odzi belt by drainage sampling.

THE CHINAMORA BATHOLITH

The Chinamora batholith has an area of about 1500 km^2 and an almost circular form with a fringe of tonalite and granodiorite. The central zone varies from adamellite to granite. The age of the batholith is about 2650 Ga but (K/Ar) determinations indicate that the thermal activity in the central zone ceased at about 874 Ga (Viewing and Harrison, 1973). The Chinamora is surrounded by Archaean schists; in the south by the Harare Enterprise belt, extending into the Mazoe-Glendale belt in the west, and to the Bindura-Shamva belt along the northern boundary. The schists contain numerous deposits of gold–quartz, pegmatite bearing Li and Ta, sulphide facies iron-formations, and the Trojan nickel mine at Bindura.

The western part of the batholith, near Mazoe, includes 32 small gold mines and prospects, mainly as vein quartz in xenolithic remnants, and 8 pegmatites. These granitic rocks were sampled at random from 275 sites (Figure 1), a density of 1 sample per 5.5 km^2. An 0.5 g pulverized sub-sample was analysed from 1.0 kg of minus 4 mm rock. The analytical precision was about ±20% for all of the elements analysed, using a total attack and estimation by atomic absorption (Viewing *et al.*, 1979). The results indicated a series of geochemical patterns of different orientations. Those of Na and Ca are almost homogeneous, whereas those of K, Pb, Ba and Sr are concentric with K and Pb (Figure 2) concentrated in the central zone, and Ba and Sr at the periphery. The patterns of Ni (Figure 3) and Zn are in contrast, elongate and orientated east-west. The patterns of Cu, Mn, Li and F show a pronounced orientation to the north east with Li (Figure 4) concentrated in the vicinity of the pegmatite deposits at Borrowdale in the southwest, and near Tafuna Hill in the north east. Fluorine is associated with the Borrowdale deposits.

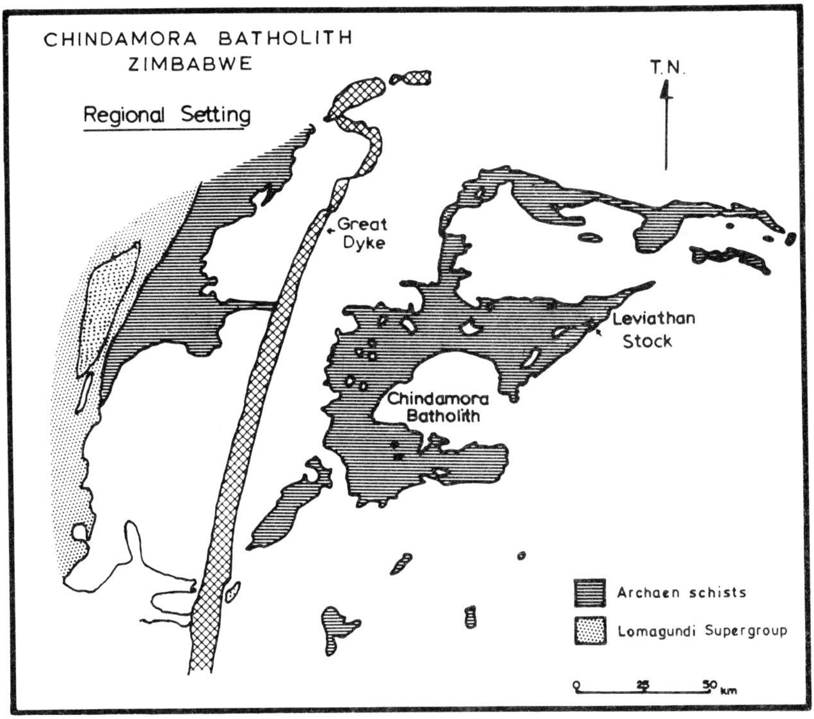

Figure 1 Chinamora batholith: regional setting and sample location

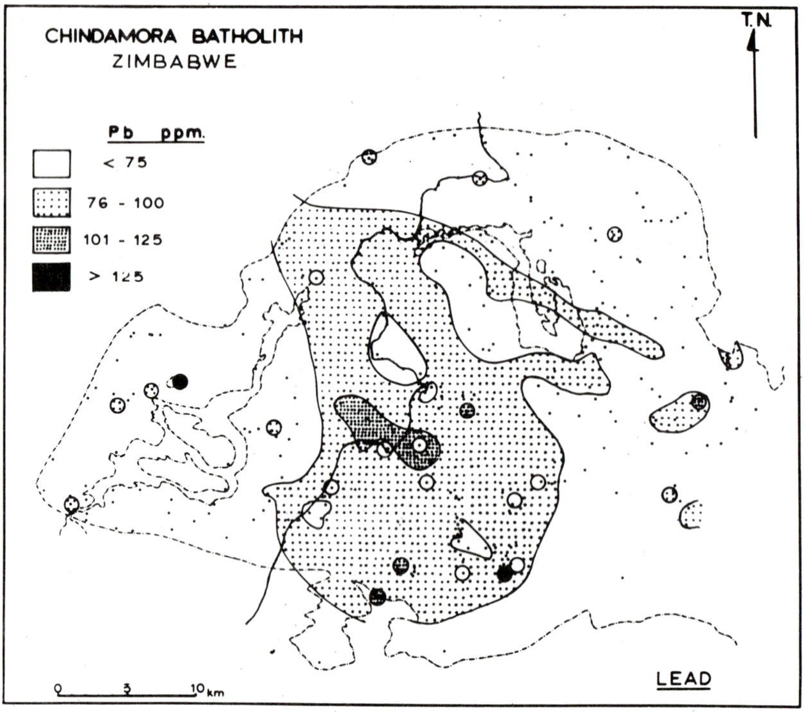

Figure 2 Geochemical patterns in the Chinamora batholith: lead

These geochemical patterns may be ascribed to the effects of K metasomatism, accompanied by intrusive phases of a younger granite in pre-existing granodiorite and tonalite. The patterns of those elements enhanced in mafic rocks indicate the presence of xenolithic material and it is believed the present level of erosion could be close to the top of the batholith. The Chinamora batholith is probably an ancient dome covered by mainly volcanic rocks of the schist belts. Metasomatic and, to an extent, intrusive phases relatively rich in K, Rb and Pb have subsequently modified the central part of the dome. The formation of the pegmatites appears, except for the intrusion of the dolerites which traverse the granitic rocks, to be the youngest phase of activity.

Certain elements (for example Mn, Li and F) relate to pegmatite deposits on a regional scale, but there is no evidence, at the density of sampling used or from the elements studied, of the distribution of many small gold-quartz deposits in the western part of the batholith. Lithogeochemical techniques were applied to the Leviathan gold deposit using a very dense sampling pattern.

THE LEVIATHAN TONALITE

The Leviathan tonalite lying to the north east and 25 km beyond the

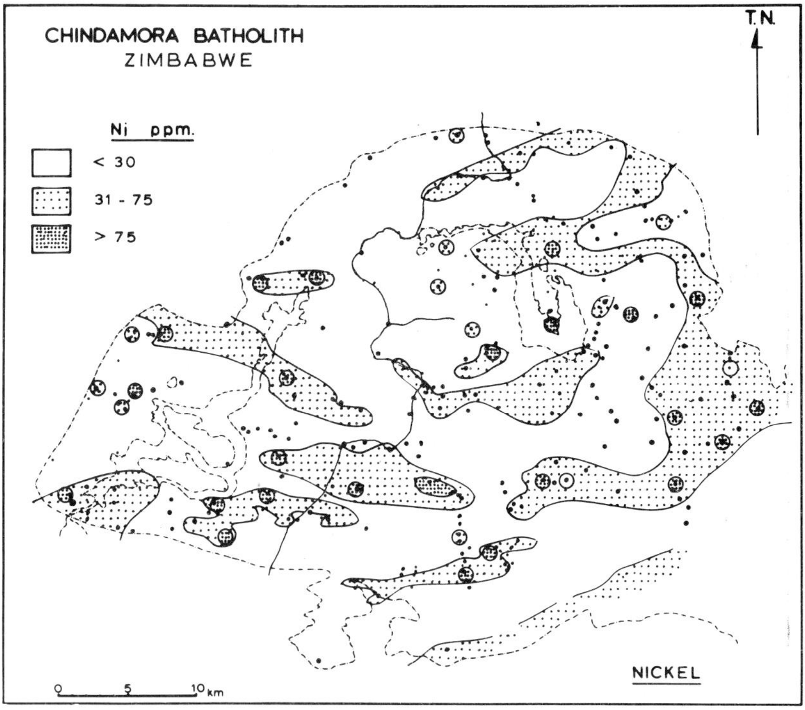

Figure 3 Geochemical patterns in the Chinamora batholith: nickel

Chinamora, is of about 0.7 km^2 in extent surrounded by Archaean schists of the Shamva belt. It contains a small gold–quartz prospect in which thin quartz veins with pyrite and gold, pervade the tonalite without any obvious control. The tonalite outcrops in drainage gullies and, in places, on the crests of ridges which extend to the east-northeast. Trenches were dug at intervals across the crests of the main ridges and an adit was excavated beneath (Figure 5).

Rock samples were collected, at a density of 65 per km^2, from 43 sites. Half a gram of pulverized material was analysed with a precision of better than 20%, using a total attack and estimation by atomic absorption (Viewing *et al.*, 1979). The results indicate coherent geochemical patterns in which the direction is significant. The elements Na, K, Ba and Pb (Figure 5b) are characterized by patterns which extend along the length of the tonalite. Patterns of Mg, P, Ti, Ni (Figure 5c), Cu and Zn all cut across that direction at a wide angle. Aluminium and Si (Figure 5d) follow this direction but their distributions are complementary. Thus evidence is obtained of silicification of the tonalite in a zone directed north-northeast across the strike of east-northeast assumed by the excavations.

The patterns of the alkali elements and Pb were taken to indicate the geochemical fabric of tonalite, whereas those of the transition elements reflect xenolithic material of mafic composition arranged parallel with the present axis of the schist belt. The distribution of the gold is consistent with the

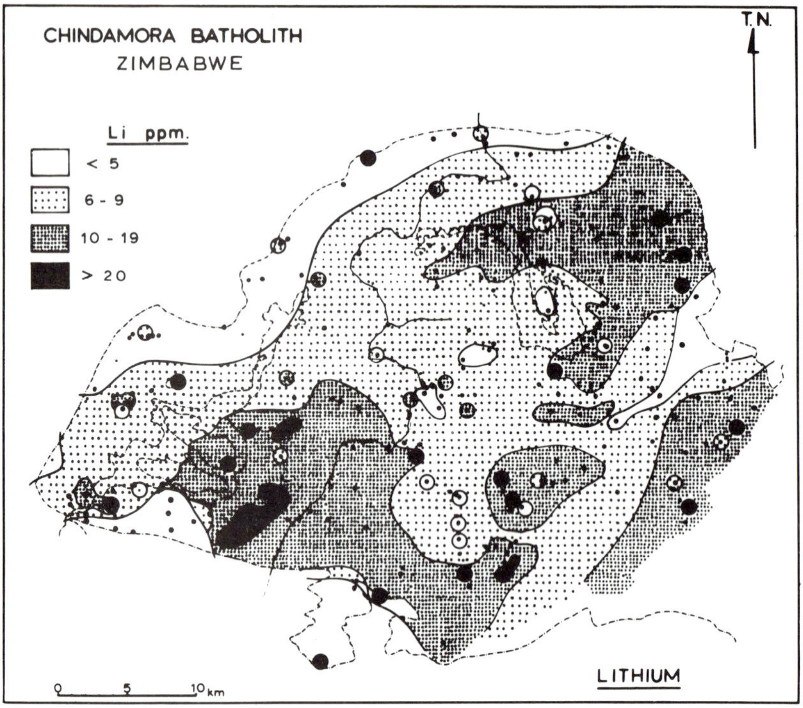

Figure 4 Geochemical patterns in the Chinamora batholith: lithium

direction of the zones of probable silicon enrichment, and of aluminium depletion. The pattern of Mn is distorted and could indicate the effect of mineralization at a more sensitive level than for the other elements studied.

The application of lithogeochemistry at different scales provides information of value to determine the nature and origin of granitic rocks and the mineral deposits which they may contain.

The application of multi-element drainage reconnaissance

The geochemical patterns obtained by rock sampling in the granitic terrain appeared to be so precise that it was believed that drainage sampling could be applied with a reasonable degree of confidence. The purpose was to evaluate the economic potential of the granitic terrain at a relatively low cost, and to interpret data following the classic objectives of J.S. Webb; in relation to lithology, the potential for mineral deposits, the secondary environment, application to agriculture, and relevance to urban and industrial pollution. The specific aim, however, was to produce a geochemical technique which was more precise than those applied previously in tropical and temperate terrains. Thus an analytical result from a sample set should be reliable and comparable with confidence to any other result obtained in a survey.

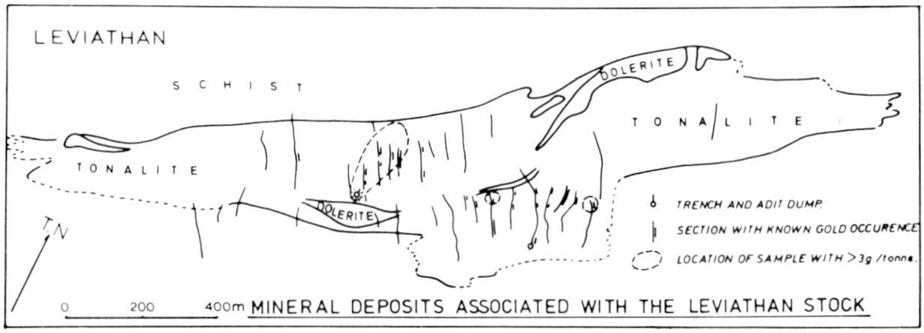

(a)

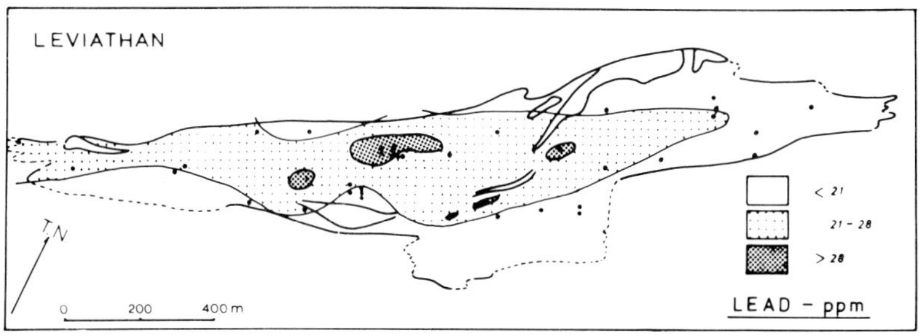

(b)

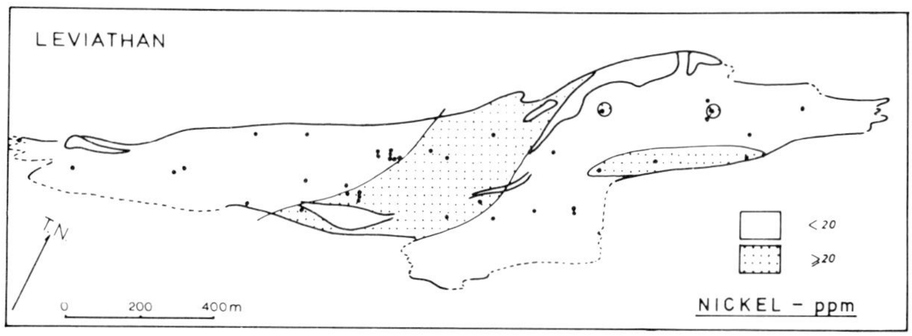

(c)

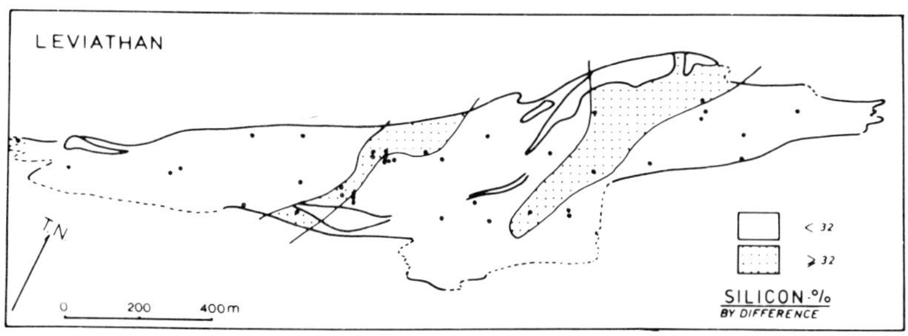

(d)

Figure 5 Geochemical patterns of the Leviathan tonalite: (a) mineral deposits; (b) lead; (c) nickel; (d) silicon

THE SABI TRIAL MULTI-ELEMENT GEOCHEMICAL DRAINAGE MAP

Nine orientation studies were conducted throughout the Archaean granitic terrain and the results of these were used in the preparation of a geochemical trial map of the Sabi Communal Lands. This map covered a rectangular area of 1350 km^2 granitic rocks. These include the south western extremity of the Odzi Archaean schist belt in the north east, extensive intrusives of Mashonaland dolerite, and the Shawa carbonatite in the north west.

Drainage samples were obtained from catchments of about 1 km^2, resulting in 1335 sample sites (Figure 6). Each sample was of about 1 kg and consisted of 3 or 4 sub-samples; these were dry sieved to minus 150μ and 1.0 g analysed by total extraction, with estimation for most of the elements by atomic absorption.

The geochemical maps were initially made by using concentric symbols to represent class intervals and their relevant populations. Subsequently R.J. Howarth produced 3 cell moving average maps (Howarth, 1971) but these were later replaced by 5-cell maps produced in Zimbabwe using an ICL 2946 computer (Cornu, 1982). The class intervals were selected from 15 cell histograms compared with cumulative probability distributions (Cornu, 1982).

The combined sampling and analytical error, and factors of depletion and enhancement in sediments as compared to rocks

The analytical error was determined by using a series of statistical control samples (Craven 1954) and the error was expressed as a precision over the range, despite its disadvantages. Bias was recognized from quality control charts. The orientation study included the collection of three sub-samples at each site to constitute bulk samples. The sub-samples from 10% of these sites were analysed separately but the combined sampling and analytical error (SAE) revealed no significant differences between the collection of 2 or 3 sub-samples. The analytical error, expressed as a precision at 95% confidence limits, varied between ±6% and ±21% of the reported value for all 19 elements determined. The precisions of 10 of the elements were better than ±11% and 14 elements were better than ±15%.

The error of the samples collected in the Sabi Trial Map was obtained from the analysis of duplicate composite samples from 4 sub-samples at 5% of the sites, or 61 pairs of samples from 1335 sites. These pairs of samples were analysed separately and the combined SAE was expressed as a precision at the 95% confidence limits. The combined SAE for the trial map varied from ±15% to ±59% of the reported value for all of the elements. However for Al and the alkali elements it was in the range of ±15% to ±33% and for the transition elements from ±34% to ±59% (Figure 7).

The relative sampling error for the elements K and Na is insignificant. For Al, Rb, Ba, Ca, Sr, Mg and Li, the relative error is of the same order as the analytical error, and is in the range of ±9% to ±17%. For the transition

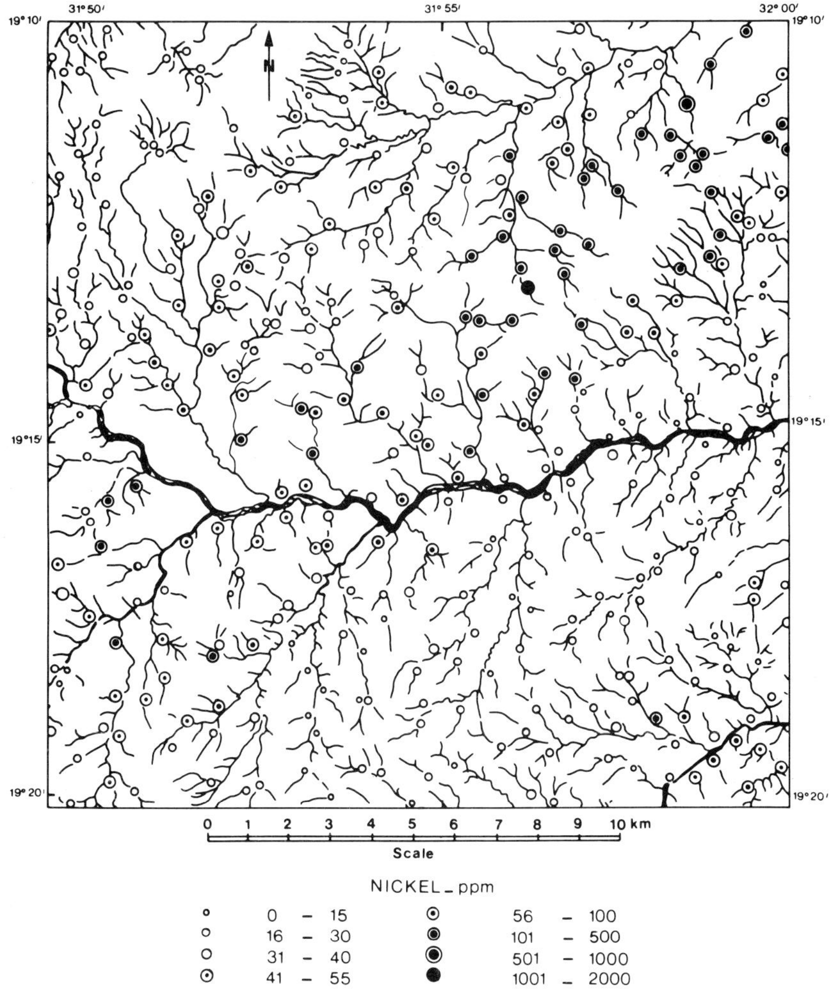

Figure 6 Sabi Trial Map: sample sites and results for nickel

elements the relative sampling error ranges from ±24% to ±46% and this is from 2 to 5 times greater than the analytical error.

Factors of depletion and enhancement of the elements were obtained from comparisons of mean values from drainage sediments and rocks. For the granitic rocks the elements Na, Zn, Pb, Ca, Ni, Sr, K, P and Fe were of the same level as in the sediments. Magnesium and Li were depleted and Co, Mn, Cu and Rb were enhanced by a factor of 2 or less. Titanium and Cr were enhanced by factors of 2.5 and 4.5 times, respectively. In the Mashonaland dolerites, the elements Co, Sr, Ni, Fe, Zn and Mn appeared unaffected by weathering processes, whereas Cu, Ca and Mn were depleted by factors of 1.9, 2.8 and 3.9. The elements Cr and Ti were enhanced by factors of 2.4 and 2.7.

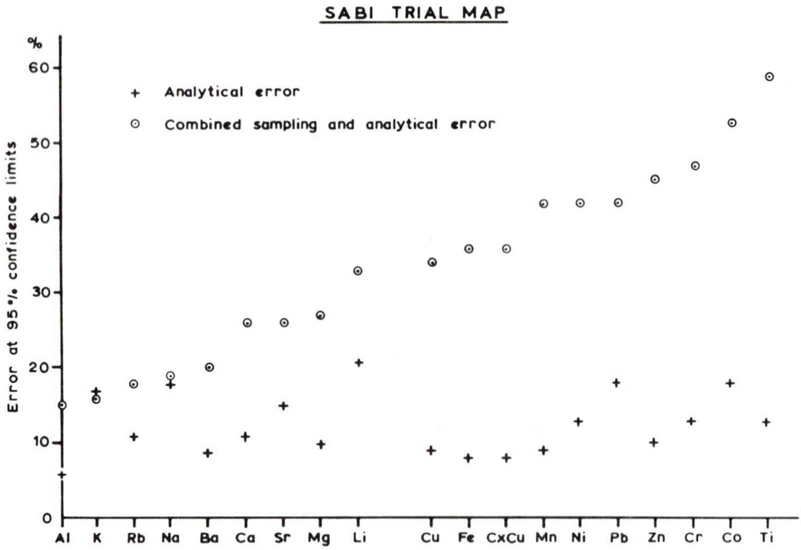

Figure 7 Combined sampling and analytical error as compared to the analytical error

Some elements of the Archaean ultramafics: Co, Cu, Zn, Cr, Sr and Mn, are little affected by dispersive processes, whereas Ca and Ni are depleted by 2.3 times and Mg by 4.3. In contrast Fe and Ti are enhanced by factors 2.3 and 3.5.

Correlation of the geochemical patterns with lithology

A knowledge of the combined sampling and analytical error and the degree of correlation with the bedrock provides confidence in the interpretation of geochemical patterns, particularly for those elements which vary in concentration within a particular lithological unit. Examples are taken to illustrate the high degree of fit of the 5 cell moving average geochemical patterns to the lithology, and also to illustrate certain variations amongst the granitic rocks.

Chromium (Figure 8) has a high combined sampling and analytical error of ±47%, with a high component of sampling error. However, the geochemical pattern outlines the ultramafic rocks of the Odzi Schist belt and of the Shawa complex. The drainage sediments accumulate Cr as compared to the bedrock, and so the Cr anomaly which extends beyond the schists to the southwest almost certainly reflects xenolithic material of ultramafic composition amongst the granitic rocks.

Copper (Figure 9) reveals the distribution of the Mashonaland dolerites. The fit of the patterns is close, and this reflects a combined sampling and analytical error of ±34% and the low analytical error of ±9%.

Rubidium has an analytical error of ±9% and a combined sampling and analytical error of ±18%. Its range and mean contents for granitic rocks are

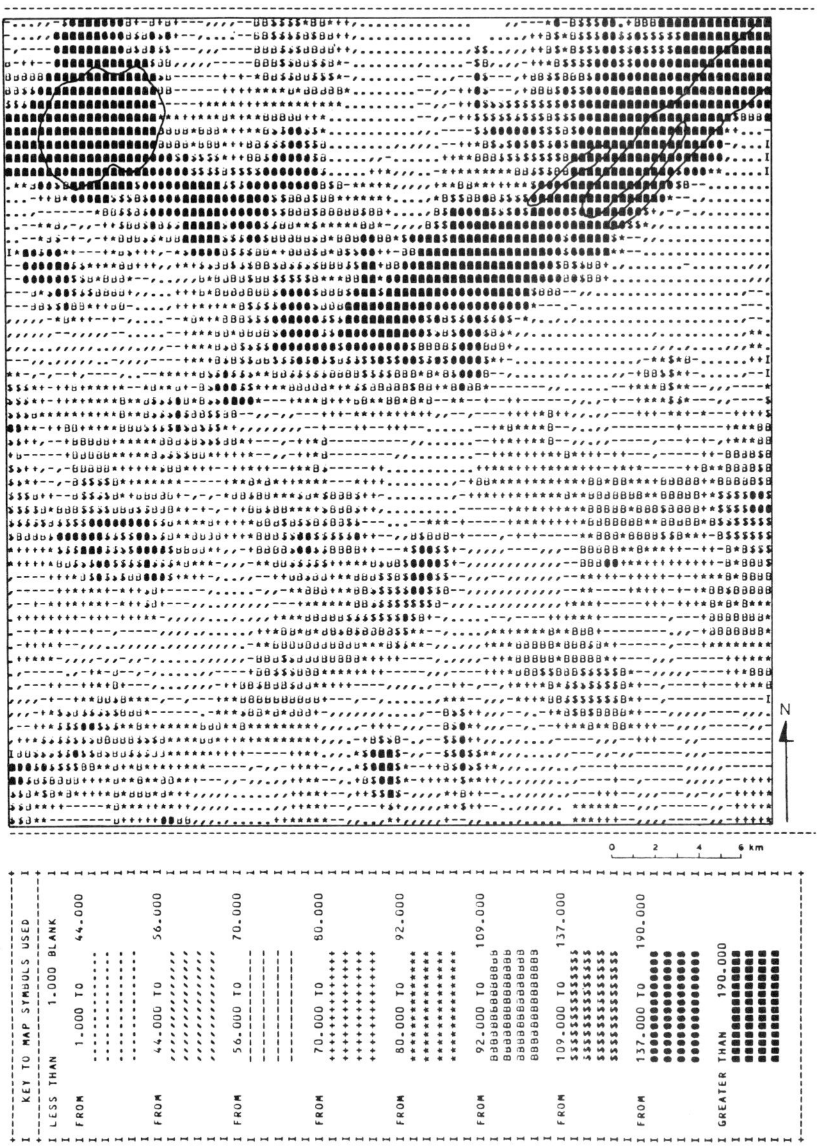

Figure 8 Sabi Trial Map: geochemical patterns of chromium

narrow, and it appears to be a reliable element in the granitic terrain. Like rubidium, strontium has a low combined sampling and analytical error, of ±26% including an analytical precision of ±15%. Strontium has a two-fold enhancement in drainage sediments as compared to granitic rocks and the ratios of these elements offer an advantage in mapping mafic rocks almost completely absorbed by the granites (Figure 10). The strontium levels are high

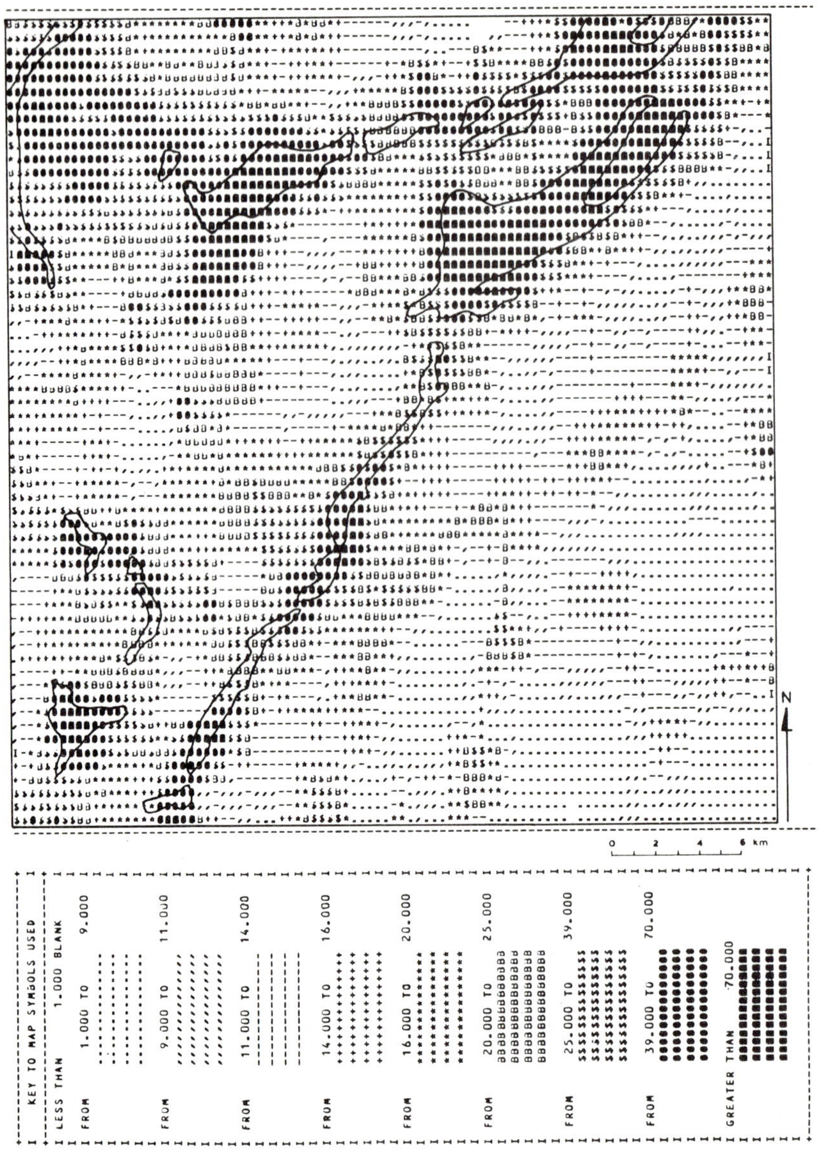

Figure 9 Sabi Trial Map: geochemical patterns of copper

and the copper levels are low as compared to those of the Mashonaland dolerites.

Varimax rotation was used subsequently to improve the resolution of the patterns for the ultramafic rocks and the Mashonaland dolerites (Kim, 1975). R-mode factor analysis with the factor Ni–Cr/Na–Al precisely fits the distributions of the ultramafic rocks, with the benefit of eliminating the

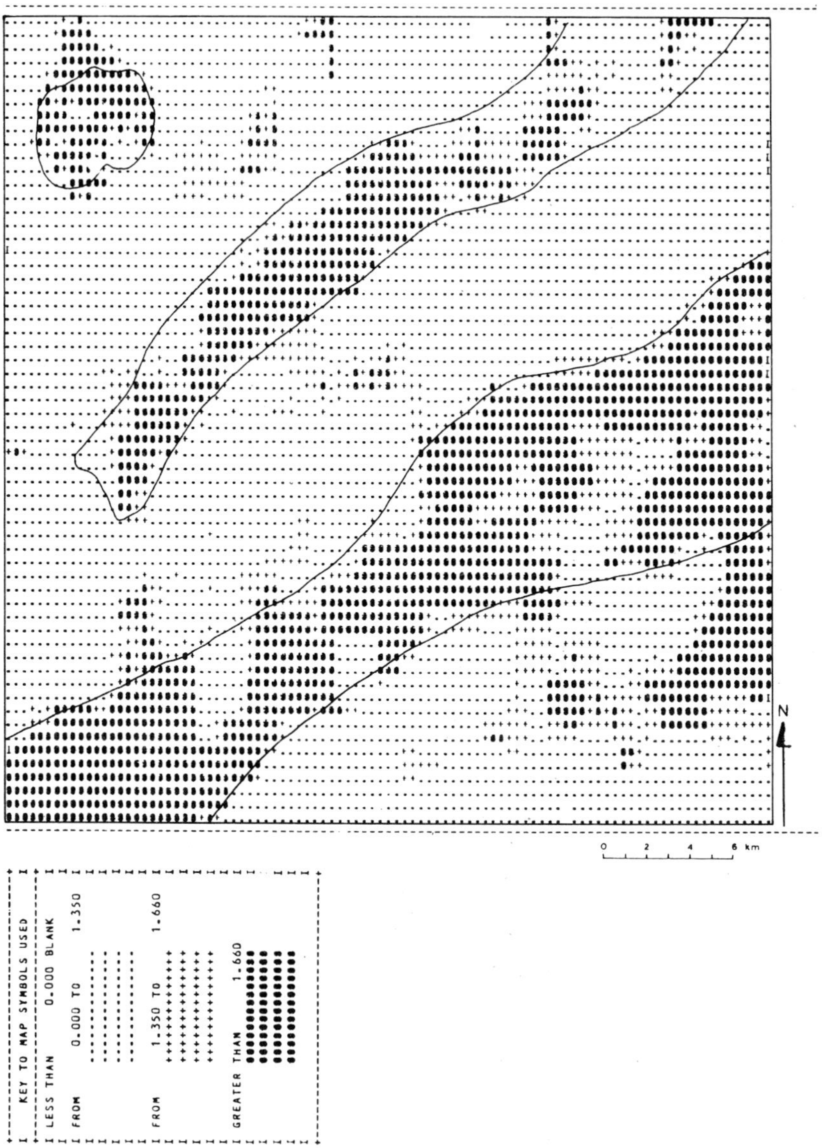

Figure 10 Sabi Trial Map: geochemical patterns of strontium: rubidium ratio

geochemical relief characteristics of the Mashonaland dolerites (Figure 11).

Conversely, Factor 1: Ti–Mn–Fe–Zn–Cu–cxCu characterizes the Mashonaland dolerites, but the resolution of the patterns is too precise and it offers no advantage as compared to the single element maps. The Sabi Trial Map demonstrated a relatively high degree of reliability in geochemical techniques. The patterns were found to be real, and the examination of heavy mineral

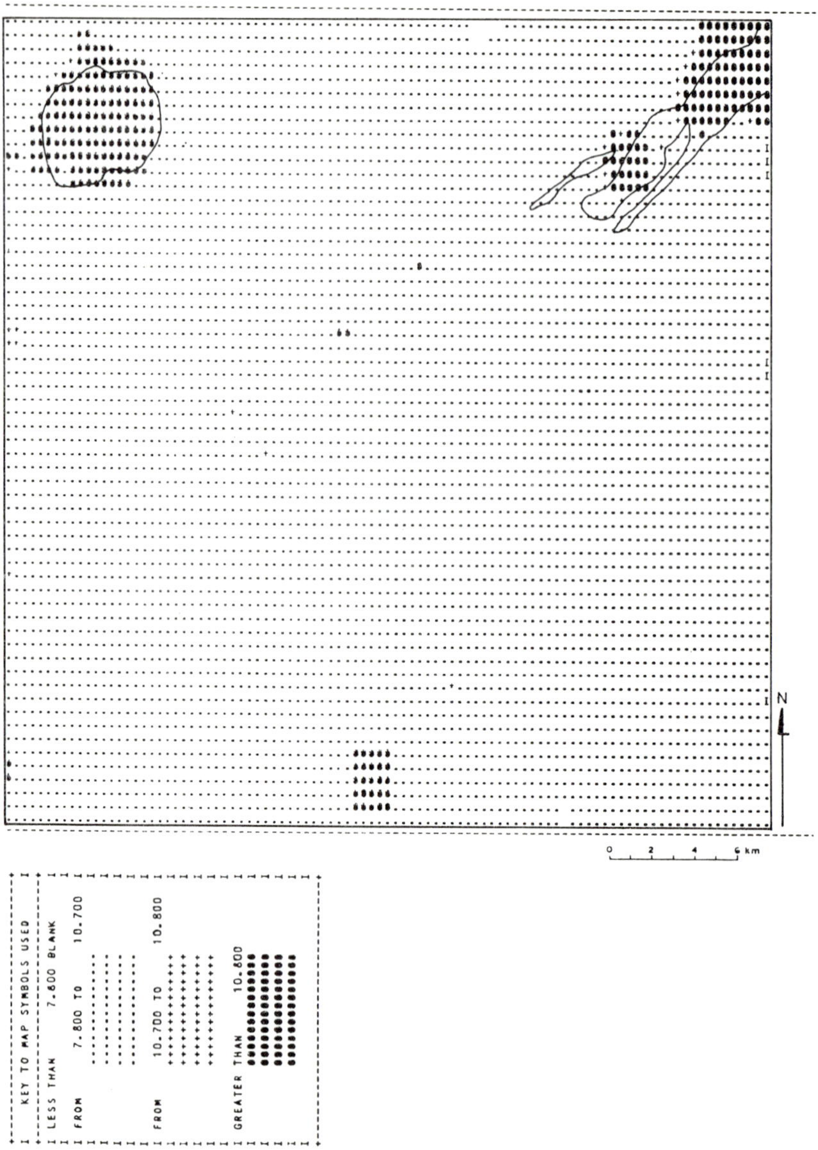

Figure 11 Sabi Trial Map: factor scores for N–Cr/Na–Al

concentrates subsequently confirmed the presence of a number of anomalies (Hatherly and Murahwi, 1980). Somewhat similar techniques were used to determine the geochemical background of a granitic terrain and to measure the effects of urban and industrial pollution.

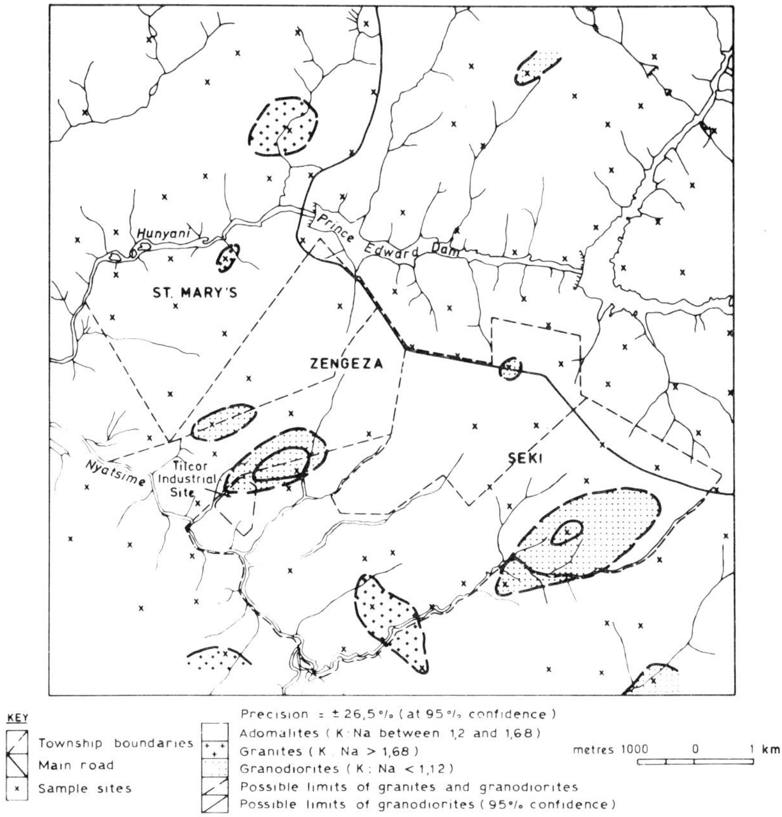

Figure 12 Geology of the Seki urban area based on the K : Na ratios in the bedrock

POLLUTION STUDIES AT SEKI URBAN DEVELOPMENT, NEAR HARARE

Seki Urban Development comprises a low-cost and high-density housing estate to the south of the capital city of Harare. The site is on the interfluve between the Hunyani and Nyatsime rivers and it includes the Henry Hallam and Prince Edward Dams. All of the drainage from the area is carried by these river systems to Lake McIllwaine and it is recirculated following filtration and chlorination. Seki consists of 4 principal settlements. St. Mary's was established on the south bank of the Hunyani prior to 1967 and an average of 25 houses per hectare has been established at an increasing pace since that time. New Seki and Zengeza were established in 1976 and 1978, respectively. The population of Seki is about 187 500 and a light industrial site is established (Hatherly and Viewing, 1982).

The topography at Seki forms gentle slopes toward the main drainage systems. These are characterized by glei soils and they form extensive

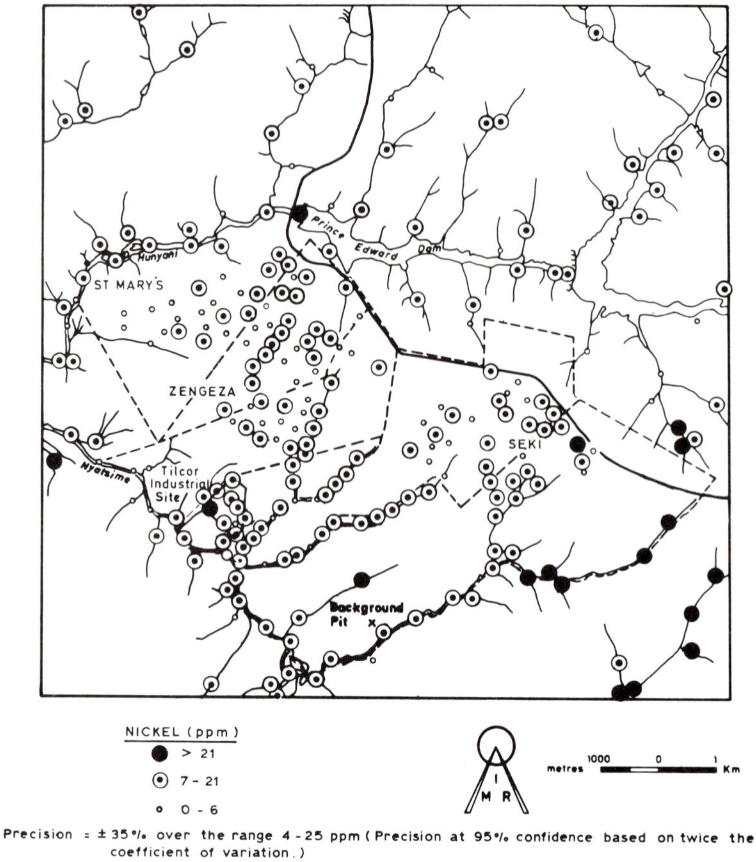

Figure 13 The distribution of lead in the Seki urban area

grasslands, which become wet and boggy during the rainy season. Outcrops of granitic rocks occur on the interfluves between them, and in places, within the drainages. The chemical analyses of outcrop samples revealed an adamellitic terrain within the area of 144 km^2 which includes two relatively small zones of granites in the north and south, and three zones of granodiorites beneath parts of Zengeza township, the light industrial site and a further extension of Seki (Figure 12).

The drainage sampling resulted in 13 geochemical anomalies which were ascribed to sources in the bedrock, to unusual effects of weathering and to pollution. The pollution anomalies were mainly due to discarded building materials and to metal scrap. However, three of them relate to urban development; motor exhaust, effluents and to waste disposal.

Lead (Figure 13) is concentrated in the urban areas, where Ni (Figure 14) indicates a homogeneous bedrock. The unpolluted natural drainage at Seki contains an average of 40 ppm Pb and it may rise to about 60 ppm depending

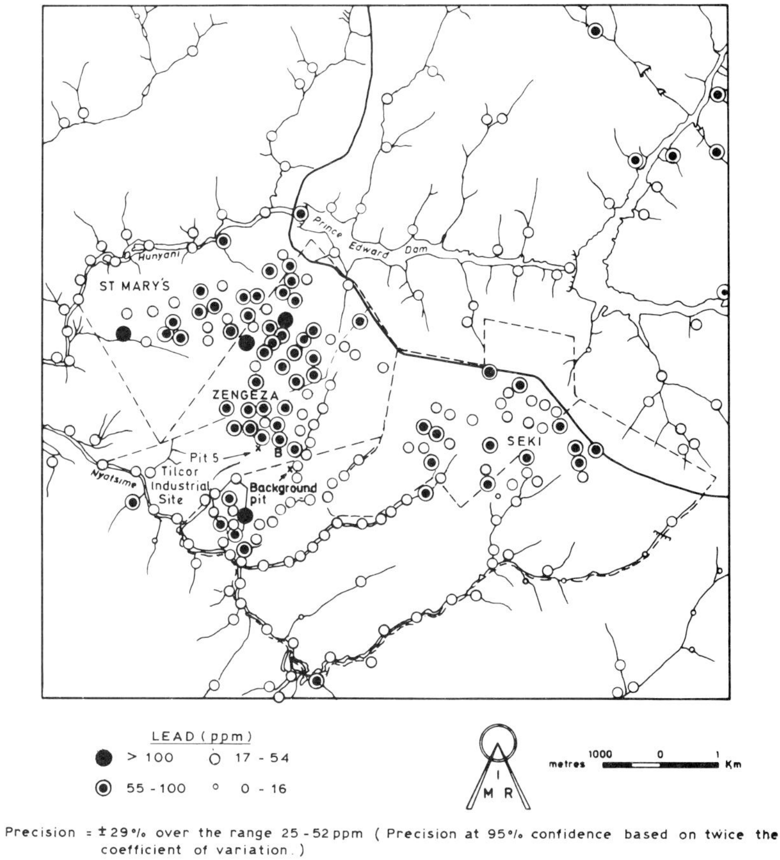

Figure 14 The distribution of nickel in the Seki urban area

upon its distribution in the adamellitic and granitic phases (Table 1 and Figure 13). The upper level of background for Pb in this granitic terrain is 40 ppm, and amounts of greater than 60 ppm are anomalous and related to urban pollution. The ratio of hxPb to cxPb reveals that the relatively high values of cxPb which occur in the older part of Zengeza, in the light industrial site and in New Seki (Figure 15) are probably caused by the density of traffic in these areas. The average increase of Pb is about 1 ppm a^{-1}, and a period of about 10 years is necessary for the Pb contents of artificial drainages to achieve a level of 60 ppm.

Geochemical backgrounds were established for the Seki area, (Table 2) but it is clear that these levels relate only to the area studied. The chemical variation amongst granitic rocks and their weathered products is so great that the critical levels for the measurement of pollution needs to be established, separately, for each area of interest (Siamsipa, 1978).

Table 1 **Lead content of sediments in artificial drainages**

Area	Approx. age (years)	No. of samples	Range in sedi-ments (ppm)	Mean (Pb) drainage sediments (ppm)		Ratio cx/Lx	Range in bedrock (ppm)
				Lx	cx		
Seki (older part)	15	5	32–78	61	14.8	0.24	54–67
St. Mary's	12	21	42–126	59	13.7	0.23	27–50
Zengeza (older part)	10	25	45–139	68	21.8	0.32	42–57
Tilcor Industrial Area	8	9	46–156	68	21.9	0.32	47–53
Zengeza (newer part)	5	15	32–81	57	13.1	0.23	50–69
Seki (newer part)	1	27	23–90	50	15.8	0.32	48–54
Seki background				40			50–97

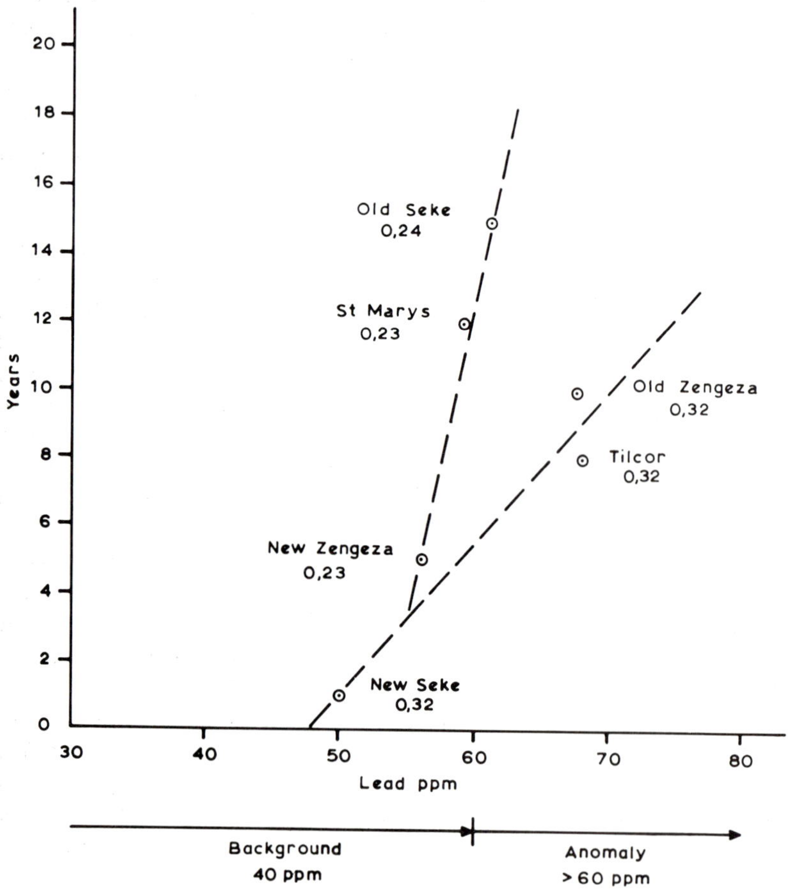

Figure 15 Identification of background levels of lead pollution in the Seki area

Table 2 **Mean threshold levels for heavy metals in the Seki urban area**

Element	Threshold	Element	Threshold
Copper	30 ppm	Chromium	30 ppm
ex Copper	12 ppm	Cobalt	9 ppm
Lead	60 ppm	Manganese	450 ppm
ex Lead	14 ppm	Nickel	12 ppm
Zinc	50 ppm	Titanium	3700 ppm
Iron	2.0%		

Discussion

These investigations of the chemical nature of granitic rocks reveal inhomogeneities upon different scales. Nevertheless, the patterns of distribution may be identified according to the density of sampling which is used, and their significance may be interpreted with confidence.

The compilation of analyses for the granitic rocks of the Rhodesian Craton revealed substantial gaps in the knowledge of the distributions of these phases. Younger granites are now distinguished from older in the geological map, but tonalites and granites of different ages are not separated.

The mineral potential of the granitic rocks of the Archaean remains low and economic deposits are unusual, whatever their character. The Dorowa and Shawa carbonatites, with their resources of apatite were obvious targets for further exploration and this was successful at Shawa. The significant vein quartz deposits of the Inyati Mine, mineralized with copper and silver, were recognized from outcrop and strong structural patterns. Geochemical methods of exploration for similar deposits need to distinguish copper liberated from the Mashonaland dolerites and mafic rocks of the schist belts. This is possible using multi-element techniques, but the low pH of the granitic environment and the leaching of copper from sulphide phases in vein quartz will lower the contrast.

Zones of silicification appear simple to detect, given a relatively high density of sampling and an analytical technique for the major elements in silicate rocks as for example X-ray fluorescence analysis. Similarly, the trends of pegmatite belts may be expected to be traced and perhaps predicted within a granitic terrain.

These techniques will be of value in the search for mineralized greissen zones in high alkali granites, but there is no history of such deposits from Zimbabwe. Rather, the application of the rock and drainage sampling techniques tested lies in the evaluation of granitic masses which are related, probably indirectly, to deposits of gold. For example, many gold deposits are spatially associated with the *ca.* 2.7 Ga intrusive soda-rich Sesombi Suite of tonalites and granodiorites, but very few occur within or near the predominantly adamellitic granites of the Chilimanzi suite which, from field evidence, post-date the Sesombi granites. This phenomenon may be related to the geochemistry and to the thermal history of the relevant granites which could be expected to control the nature of potentially mineralizing fluids and the degree of fracturing of the host rocks.

Studies of this nature should be extended to the quartz and feldspar porphyries, to the felsic extrusives of the Archaean (as in other countries), and to the cherty facies of the iron-formations where there is evidence of potential mineralizations.

The application of the drainage sampling method to the development and planning of agriculture, and in relation to environmental health, has been demonstrated by J.S. Webb, I. Thornton and others (Thornton and Webb, 1979). It has a direct application to the quality of life, particularly for developing countries where agriculture and health are the dominant influences upon development. The essential trace elements for the health of the higher plants and mammals are Fe, B, Cl, Cu, Co, I, Mn, Mo, Se and Zn (Underwood, 1962; Schutte, 1964). Of these elements, however, only Fe, Cu, Co, Mn, Mo and Zn were in the analytical programme for the Sabi Trial Map. Phosphorus was included, for this element is commonly deficient in the ranching areas of Zimbabwe.

The interpretation of the analytical date from the Sabi Trial Map is incomplete but the results of EDTA-extractable Cu and Zn in the soils of certain test areas, with mean values of 0.38 ppm and 0.58 ppm respectively, indicate what deficiencies are likely to occur in the Communal Lands. The mean values of EDTA-extractable Co and Mn are 0.9 ppm and 114 ppm, respectively, but no critical deficiencies for these elements have been established in Zimbabwe.

Conclusions

1. The results of multi-element lithogeochemical and stream drainage surveys confirm that these techniques are directly applicable to the evaluation of the Archaean granitic terrain of Zimbabwe. The data obtained reveal coherent patterns of distribution of the chemical elements which yield information of value to the history and formation of the rocks, and depending upon the sample density, to the mineral deposits which are contained.
2. Drainage samples at 1 per km^2 of catchment with large samples taken at the stages of collection and analysis result in low values for the combined sampling and analytical error; so that the geochemical maps have a high degree of resolution. The factors of depletion and enhancement for the chemical elements in drainage sediments are not significant except for Mg, Cr and Ti.
3. Geochemical drainage maps are not applied to land use planning and agricultural extension studies in Zimbabwe. Relatively low levels of Co, Cu, Zn and P indicated in the Sabi Trial Map are revealed and warrant

attention. These potential deficiencies are yet another factor in the difficulty of raising livestock and crops in the Communal Lands.

4. A combination of lithogeochemical surveys and drainage surveys is of value in establishing geochemical backgrounds to identify the level of urban and industrial pollution. The techniques are sensitive and precise, and leave no room for doubt concerning the chemical composition of the bedrock and the products of weathering.

ACKNOWLEDGEMENTS

This contribution summarizes the results of investigations in applied geochemistry carried out at the Institute of Mining Research, University of Zimbabwe. Many members of the staff contributed to this work, particularly the former Chief Chemists, D.A. Wenden, H.A. Stephens and B. Witkin. Gold Fields Prospecting Company of Zimbabwe provided the financial support for the work at Leviathan, and the Tribal Trust Land Development Corporation of Zimbabwe (TILCOR) for the Sabi project. The authors wish to acknowledge the generous advice given in the field by Dr. I. Thornton of the Applied Geochemistry Research Group. Dr R.J. Howarth encouraged the use of computer-produced 'grey scale' maps and moving average maps early in the Sabi project, and he is thanked for his assistance. Assistance with the development of computer programs for the ICL 2946 machine was given particularly by J.G. Cornu and J-L. Baron under the scheme of French Technical Cooperation.

References

Cornu, G. (1982). Fortran IV: Moving average program for grey level plotting of geochemical data. In: Inst. Mining Res., Univ. of Zimbabwe, Report No. C 292.

Craven, C.A.U. (1954). Statistical estimation of the accuracy of assaying, *Inst. Min. Metall. Trans.*, **63**, **12**, 551–563.

Hatherly, R.S. and Murahwi, C. (1980). Investigation of the heavy mineral fraction of stream sediments from the Sabi Trial Trust Land—a preliminary report. In: Inst. Min. Res., Univ. of Zimbabwe, Report No. C 196.

Hatherly, R.S. and Viewing, K.A. (1982). The solid phase: a study of pollution benchmarks on a granitic terrain. In: *Lake McIllwaine—The Eutrophication and Recovery of a Tropical African Lake* (Ed. J.A. Thornton), pp. 77–79.

Howarth, R.J. (1971). Fortran IV programme for grey-level mapping of spatial data, *Math. Geol.*, **3**, 95–121.

Kim, J.O. (1975). Factor analysis. In: *Statistical Package for the Social Sciences* (SPSS) (Ed. N.H. Nie *et al.*).

Lister, L.A. (1979). The geomorphic evolution of Zimbabwe Rhodesia, *Trans. Geol. Soc. S. Afr.*, **82**, 363–370.

Nisbet, E.G., Wilson, J.F. and Bickle, M.J. (1981). The evolution of the Rhodesian Craton and adjacent Archaean terrain: Tectonic models. In: *Precambrian Plate Tectonics* (Ed. A Kroner), pp. 161–183; McGraw-Hill, New York.

Schutte, K.H. (1964). *The Biology of Trace Elements*; Crosby Lockwood and Sons, London.

Siamsipa, F. (1978). The dispersion of heavy metal pollution around Salisbury. In: Inst. Min. Res., Univ. of Zimbabwe, Ninth Annual Report, pp. 59–61.

Thornton, I. and Webb, J.S. (1979). Geochemistry and Health in the United Kingdom, *Phil. Trans. R. Soc. Lond.*, **B288**, 151–168.

Underwood, E.J. (1962). *Trace Elements in Human and Animal Nutrition*; Academic Press Inc., New York and London.

Viewing, K.A. (1968). Regional geochemistry and the Rhodesian Basement Complex: Proc. Symposium on the Rhodesian Basement Complex., *Geol. Soc. S. Afr., Rhod. Branch*, 21–32.

Viewing, K.A. and Harrison, N.M. (1973). A geological reconnaissance of the Chinamora Batholith near Salisbury, Rhodesia, *Spec. Publ. Geol. Soc. S. Afr.*, **3**, 419–432.

Viewing, K.A., Hodgson, W.A. and Wenden, D.A. (1979). The geochemistry of the two granites associated with mineral deposits, near Salisbury, Rhodesia, *Spec. Publ. Geol. Soc. S. Afr.*, **5**, 253–261.

The Role of Computing in Applied Geochemistry

R. J. HOWARTH and R. G. GARRETT

Introduction

The last decade has seen a steady increase in the number of papers in geochemistry with a significant use of numerical techniques, either for data display or analysis (Figure 1). These applications may be broadly divided into categories, which include quality control at the field or laboratory level, database management, univariate and multivariate statistical applications and computer mapping (Figure 2). In this paper, we do not attempt an in-depth assessment and explanation of the methods used in all these areas, and

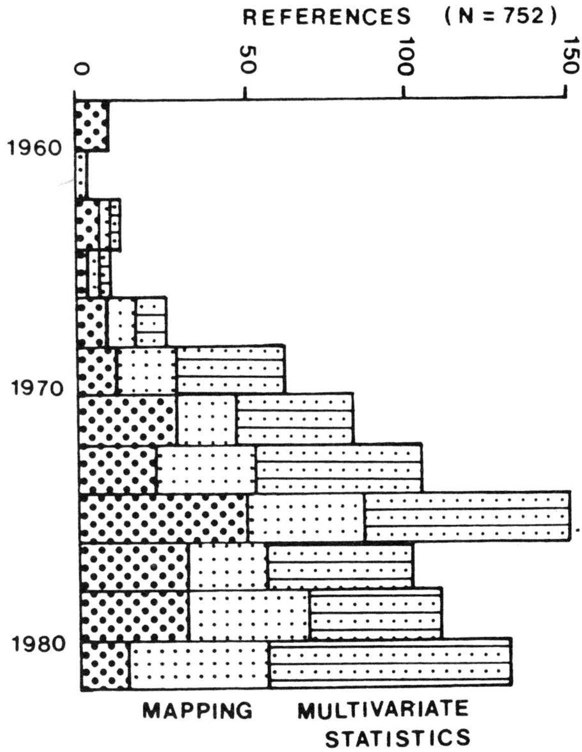

Figure 1 Growth in multivariate statistics, mapping and other computer applications in geochemistry since 1960.

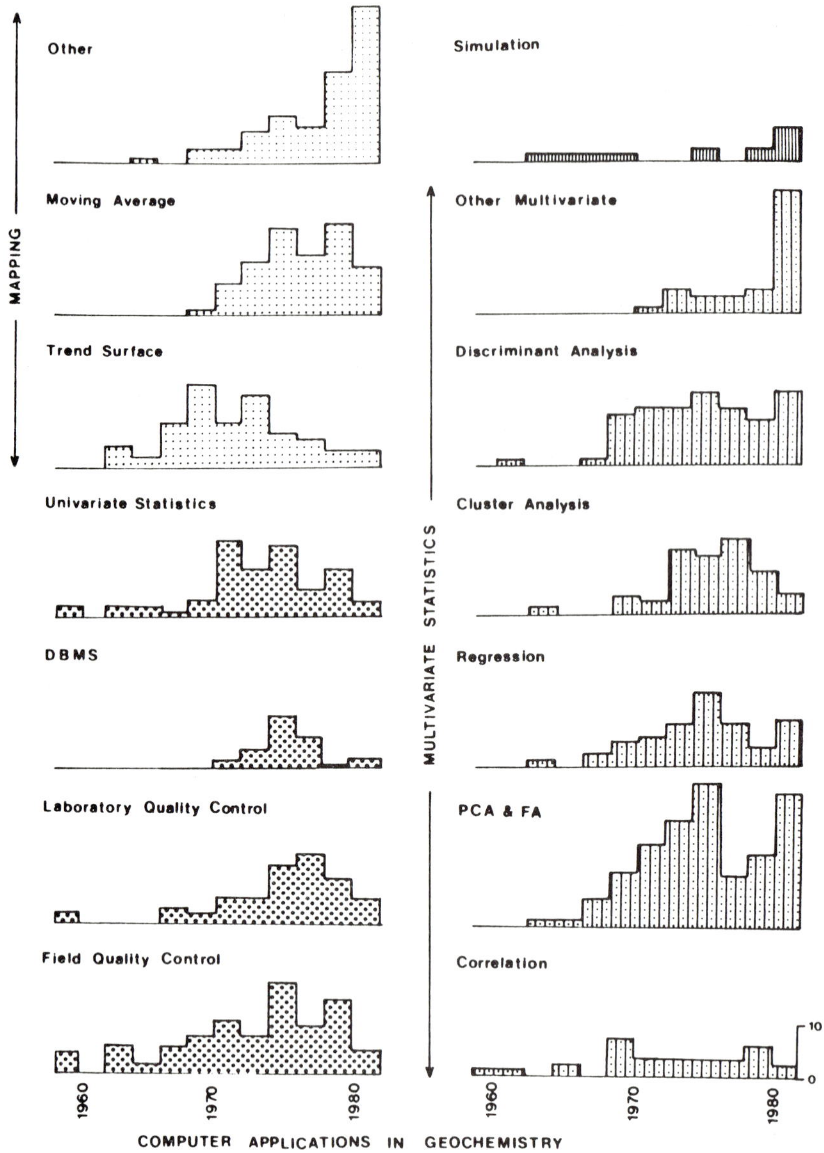

Figure 2 Growth in computer methods in geochemistry since 1960 by principal application area.

refer the reader to the more detailed material in Howarth (1983a) for geochemical prospecting applications and LeMaitre (1982) for more petrological applications of computers to geochemical data analysis.

What has motivated geochemists to become involved in data processing and analysis? We believe there are three main reasons:

1. The rigour of preparing data for analysis and presentation encourages

better organization, and the careful consideration of data requirements, from both geochemical and data handling viewpoints, for a project before sampling and geochemical analysis commences. This organization leads to consistency in execution and preserves the long term value of the data, as a correctly documented project file may be archived for future use by others than the originator.

2. The use of data processing allows a standardization of approach. If all files are generally similar in content, i.e. types of observations, similar products may be easily prepared for each data set processed. This facilitates the computer assisted interpretation learning process through comparison, and the development of new procedures.
3. The cost of fieldwork is increasing more rapidly than that for geochemical analysis. Both of these are increasing faster than data processing costs. The major motivation in using data processing is to produce, rapidly and reliably, data summaries, analyses and graphical plots cost effectively. In fact, one of the few things which is decreasing in cost today is the unit cost of data processing.

Advances in computer applications in exploration geochemistry have to a large extent paralleled the growth in memory capacity and processing speed of mainframe computers (e.g. the IBM 1620 (1960), 7094 (1962), 360 (1965) and 370 (1970)); the introduction of large-scale integrated circuits in 1971 led directly to the new generation of 8-bit microcomputers in 1974, 16-bit (1981) and 32-bit (1983), as well as super-computers such as the CDC 7600 and Cray in the mid- to late-1970s. Parallel developments in storage media allow modern desk-top microcomputers, with perhaps 256 Kbytes of RAM and 1 to 10 Mbytes of floppy and/or hard (Winchester) disc storage, to out-perform the mainframe computers of the early 1960s, giving the exploration geochemist the opportunity to perform statistical analyses and plot survey results in the field. By analogy, this capability may in the long term become as significant as the availability of rapid analytical field techniques was to geochemistry in the 1950s.

Until 1965, the use of statistics to aid laboratory quality control or instrumental calibration, and in the interpretation of geochemical results, had largely been through graphical procedures. The major impediment to progress was the sheer volume of data generated in geochemical surveys. In the early 1960s, digital computers became available to geochemists, firstly in North America, notably at the Canadian and United States Geological Surveys, and slightly later in Europe. It was the installation by IBM of a 7090/1401 system in Imperial College in 1964 that finally allowed members of the Applied Geochemistry Research Group to investigate the role of data processing in applied geochemistry. Since that time, the spread of computer usage among exploration geochemists has been steady, particularly in North America (Garrett, 1983a), Western Europe and Scandinavia. An interesting insight into computer applications in Eastern Europe and the Soviet Union was given at the 1981 Irkutsk Symposium on Methods of Applied

Geochemistry (Kolosovsky *et al.*, 1981). Computer analysis of survey data is also being undertaken in South Africa, Australia and China amongst other countries, and in the course of aid programmes in many of the developing countries. Without doubt, the power of the modern micro- and mini-computers, combined with their low cost compared to large mainframe computers, will have a significant impact on mineral exploration in the next decade, particularly in the poorer countries of the world if rugged equipment is backed up with reliable maintenance networks.

In the data acquisition phase of a project, the computer can significantly reduce human error in recording sample site coordinates by their direct input from a stable base map using a digitizing table. The digitized points may be subsequently replotted for comparison as an overlay to the original field map for a final check of the recorded position and site (sample) number. Field observations may also be directly entered into the computer data base for later comparison with analytical data, often using a microcomputer as an intelligent terminal to a remote mainframe computer. This has the advantage that time-consuming data validation can be carried out by computer, and the data edited and checked by the field geologist, prior to any further processing. Digitization of geology, drainage patterns, etc. may also be carried out at this stage for use in later data analysis and/or displays.

Microprocessors are rapidly becoming an integral part of modern analytical laboratory instrumentation (e.g. atomic absorption spectrophotometers, inductively coupled plasma spectrometers and gas chromatographs) for data acquisition and, often automatic, calibration and correction for inter-element interferences. A number of maufacturers are now developing Laboratory Information Management Systems to aid the tracking of samples through large multi-instrument laboratories as an aid to overall quality control. Such systems may be linked to a central computer system (sometimes remote from the analytical laboratory), so that the analytical data can be merged with the field data in a project-oriented database for further statistical analysis and mapping.

Sampling and search

Following the wartime development of search theory (Koopman, 1980), a few major organizations have used computers to study theoretical sampling schemes by computer simulation in order to arrive at optimal search strategies, e.g. Savinskii (1965), Singer (1972, 1975) and Drew (1979). These have now been generalized to the use of regular square, triangular and hexagonal sampling patterns and stratified random sampling designs (Garrett and Goss, 1980b) over elliptical targets. However, it is also highly desirable to be able to assess the relative proportions of natural sampling and sub-sampling variability between samples from different geological units, and field versus laboratory sources of error. Following the pioneering work in this field by Miesch (1964, 1967, 1976a) who initially used balanced sampling designs and analysis of variance (ANOVA) to assess the partitioning of variability between the various levels in the design, Garrett and Goss (1979, 1980a, 1980b) have extended applications to hierarchical unbalanced sampling designs. This allows much more efficient sampling designs to be achieved, which may represent savings of as much as 50% or more in the number of samples collected and analysed where four or more levels are studied (e.g. between analyses, samples at a site, sites in a subarea, between subareas and regions) over completely balanced equivalent designs. Such methods afford a valuable guide to the sources of maximum variability in the data, and if carried out in a reconnaissance or orientation survey indicate the points in survey (and laboratory) strategy at which most attention to replication (or reduction of error by bulking samples) should be paid. However, recent work in the Geological Survey of Norway (Ottesen *et al.*, 1983) demonstrates that for at least some elements in the stream sediment environment both total and sampling error increase at high concentrations in addition to the well-known increase of analytical error towards the detection limit. Similar results have been obtained in the Applied Geochemistry Research Group with stream sediment data from the West of England. This non-constancy of variance with concentration is more complex than the usual models assumed for heteroscedastic geochemical data, and may require a different approach to the usual transformation to normality or homogeneity of variance using log or Box-Cox power transforms (Howarth and Earle, 1979) in connection with the Burr-Foster Q Test (Anderson and MacLean, 1974).

Laboratory quality control

One of the problems associated with chemical analysis of geochemical exploration data is that of maintaining adequate quality control over a wide range of concentrations in samples which may often have very differing matrices. The Applied Geochemistry Research Group has been associated with work in this field since Craven (1954) described a method based on a 'statistical series' of control samples, made by mixing a pair of high- and low-concentration standards in known proportions. Although many did not realize the fact, this method was one of the earliest uses of a least-squares statistical procedure by exploration geochemists. The method combined monitoring of accuracy and precision and could be carried out by hand calculation, and thus saw wide usage (James, 1970). Stern (1959) criticized the method on statistical grounds, but his proposed improvements yielded such a complex procedure that it was hardly used in practice until it was implemented in 1965 on a digital computer (Garrett, 1966). Thompson (1983) points out that the main drawbacks to the 'statistical series' approach are that it yields a mean precision for the concentration range, thus underestimating the true precision at background levels. Also, there are practical difficulties in preparing series of standards in such a way that they will be well mixed and at the same time have a grain-size representative of the actual samples being analysed.

Garrett (1969) proposed a method based on ANOVA techniques, and subsequent workers have also followed this approach. Thompson and Howarth (1973) suggested a new method based on estimation of the variation of standard deviation with concentration from duplicate analyses of run-of-the-mill samples. The approach can either be used with large numbers of duplicates to derive precision as a function of analyte concentration, or (when the number of duplicated samples falls below 50) to check that a preset precision limit is maintained.

There is a need for substantial work to be done in the area of multi-element quality control (as many modern laboratory instruments produce such data), and determination of optimum strategies for re-analysis in such situations. The effect of lack of analytical precision on the probability of misrecognition of exploration targets (both false alarm and false dismissal) has been discussed by Howarth and Martin (1979).

Database management

One of the most useful aspects of modern computers is their ability to store enormous quantities of data for subsequent retrieval and data analysis or display. There is now a wide range of commercially available Database Management Systems (DBMS) on micro-, mini- and mainframe computers developed for the storage and retrieval of data. The commercial systems are entirely general in nature. However, some DBMS have been specially developed for geological users and may incorporate relevant graphical and statistical procedures. These include: GRASP (Bowen and Botbol, 1975), G-EXEC (Jeffery and Gill, 1976; NERC, 1977), SIGMI (Kremer *et al.*, 1976), RASS (van Trump and Miesch, 1977) and UREP-DPS (Kane, 1980). Principles of database design generally follow those described in Date (1975). A significant trend at the present time is the spread of DBMS usage in major organizations concerned with long-term acquisition and usage of regional and exploration geochemical data.

Statistical analysis

The first introduction many geochemists had to statistics was in the use of histograms and cumulative frequency plots, meticulously prepared by hand, and then the computation of means and standard deviations with slide rules, mechanical calculators or log tables from the grouped data. The papers of Tenant and White (1959) and Lepeltier (1969) described early work in the use of statistics to aid in threshold selection. By the early 1960s the use of the 'mean plus two standard deviations' procedure to attempt to identify the top two per cent of the data was well entrenched.

Geochemists' earliest experiences with computers were in undertaking routine tasks rapidly and accurately that previously had been error prone and tedious in the extreme, e.g. generating histograms, computing means and standard deviations, etc. However, it did not take long before two other areas were investigated. Firstly, map plotting and contouring, and secondly, multivariate statistics.

Most geochemists wish to inspect summary statistics and histograms or cumulative frequency plots in order to gain an insight into their data, and help select clearly 'anomalous' samples and thresholds. A great variety of software is now available to help in these tasks, varying from desktop calculator programs to package (e.g. Minitab, BMDP, SPSS, etc.) and custom-written programs for large mainframe computers. A major advantage of data processing is that many empirical element combinations, ratios, etc. that have proven useful in the past, or should theoretically prove useful, may be rapidly computed. Without the aid of a computer such tasks are so onerous they are usually not attempted, and the opportunity of considering the data from alternative perspectives is lost.

Two topics have been of continuing recent interest in univariate statistical analysis. Firstly, the transformation of data using optimization methods to achieve 'optimum' transforms in the sense that the transformed distribution is made symmetric and near-Normal. A variety of alternatives have been proposed such as the Box-Cox (Howarth and Earle, 1979) and three-parameter $\log(x + c)$ transforms (Kane, 1979); both require computer optimization methods for fast, routine, solution. Secondly, continuing interest has focussed on the problems of splitting frequency distributions into their component parts, and the definition of threshold. For example, Sinclair (1974, 1976) furthered the work based on new models, and Bridges and McCammon (1980) are typical of those who have implemented the graphical approach of Sinclair using interactive computer graphics. Miesch (1981) has proposed the use of a new 'gap statistic' test for identification of threshold separating 'background' and 'anomalous' populations. In this type of work it is often assumed that background and threshold distributions are both normal, either in their original or transformed form. However, problems in recognition of sub-populations may arise if one has an exponential or linear decay away from a source with superimposed random variation, as in till or drainage sampling (Bjorklund, 1983). The bias which can be caused in statistics such as means and standard deviation by censoring of data below a detection limit or by an outlier, has long been known. The computer has made it much easier to use the cumulative probability plot to routinely assist in the recognition of anomalous (outlying) observations (Figure 3).

This last point focusses on differences between the data and objectives of the classical statistician and the exploration or applied geochemist. The traditional approach has been interest in statistical summaries of population parameters and hypothesis testing (e.g. could the data have come from a Normal population, could two or more data sets have been drawn from the same population, or is there evidence for several populations?). The exploration geochemist may well have questions such as these but, in contrast,

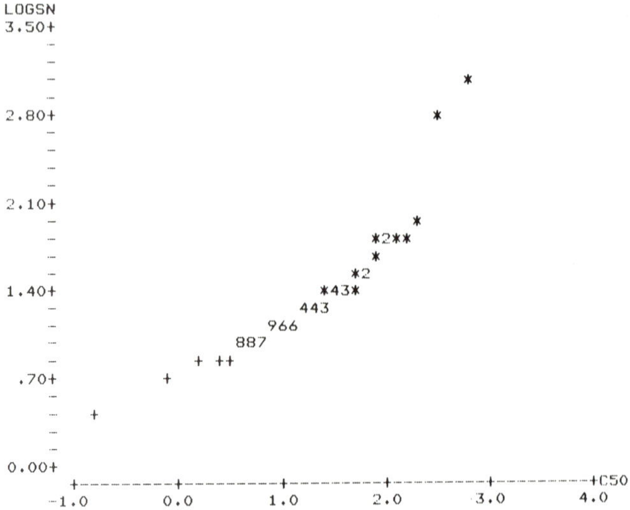

Figure 3 Cumulative probability distribution for $\log_{10}Sn$ in 269 rocks, versus quantiles of the Normal distribution. Two outliers (5880 and 12420 ppm) are apparent. (Individual data points are shown as an asterisk; 2-a coincident points (at the scale of the plot) by a figure; 10 or more by a plus sign)

will be particularly interested in identifying the outliers, 'anomalies' of geochemical jargon, as it is often these that lead to the recognition of mineral deposits. In the last decade there has been increasing interest by some statisticians in what Tukey (1977) called 'exploratory data analysis' (EDA), in contrast to the hypothesis testing of classical statistics which is often referred to as confirmatory data analysis (CDA). A major concern in EDA is in identifying outliers which, apart from any importance they have in their own right, can deleteriously affect the results and integrity of the CDA most statisticians wish to employ at some later stage of their work.

Many of the EDA techniques employ graphics of one kind or another. The human eye is a superb instrument for detecting irregularities or nonlinearity on appropriately prepared plots. As a result, a major thrust in EDA, and of geochemists involved in data analysis and display, has been in the development of graphics which facilitate visualizing various properties and irregularities of data.

For example, Tukey suggested two useful univariate displays of data, in which percentile points are used in determining the characteristics of the data display. These have the advantage (particularly in large data sets) that they are 'robust', i.e. not influenced by the effects of outliers in the data.

The first is the histogram-like stem-and-leaf display (Figure 4) in which data values are sorted into ascending order and summarized graphically in a histogram-like way to display the individual values. The median is first determined, and the so called upper and lower hinges are defined as the median of the lower and upper halves of the data respectively, i.e.

```
STEM-AND-LEAF DISPLAY OF ZN
LEAF DIGIT UNIT =    1.0000
1 2 REPRESENTS 12.


   4     0   5899
  10     1   001445
  18     2   01133478
  28     3   2355667777
  40     4   000134567789
  60     5   01122333445556788999
  83     6   01122223445555667788899
 101     7   001112223355566678
 129     8   0001222333333445566677889999
(38)     9   00000111112222223333444455566777788899
 102    10   0000000112233344456666667799
  76    11   001223334445556677777788899
  51    12   000011339
  42    13   1124557788
  32    14   0029
  28    15   33588
  23    16   19
  21    17   4889
  17    18   123

        HI   208, 218, 223, 224, 243, 279, 286, 286, 296, 299, 301
             350, 460, 672
```

Figure 4 Stem-and-leaf plot for Zn in 269 rock samples

approximately the lower and upper quartiles or 25th and 75th percentiles. The hinge-width is the difference between the hinges. The leading digits of the values form the common stem, and the trailing digits the leaves. Hence in Figure 4 we see values of the sorted data in the pseudo-histogram run from 5 (min) to 183 (max), values in excess of the upper and lower hinges plus-or-minus 1.5 times the hinge width respectively are printed out individually, these are values of 208, 218 . . . , 672 ppm. The leftmost column of the display (Figure 4) shows the cumulative number of values away from the tails towards the median. For example, the first two rows of the stem-and-leaf plot correspond to the 10 lowest values, 5, 8, 9, 9, 10, 10, 11, 14, 14 and 15 ppm Zn. The class containing the median has 38 values in it, ranging from 90 to 99 ppm Zn. The display clearly has the advantage over a conventional histogram that it conveys more information about the actual data values, while still indicating the shape of frequency distribution.

The second type of display, the boxplot, is extremely useful for the comparison of data subsets. Figure 5 shows a boxplot for the data of Figure 4 corresponding to each of 10 lithological types. In each case, the central box encloses the range between the lower and upper hinge, the median being indicated by a cross. Whiskers extend out from this to the values of the extreme samples still lying within 1.5 hinge widths of the hinges, samples outside these limits are shown individually, by the letter O in the figure, and could be regarded as extremes for their subset. Subset 3 in Figure 5 clearly tends to have enhanced concentrations of Zn associated with it.

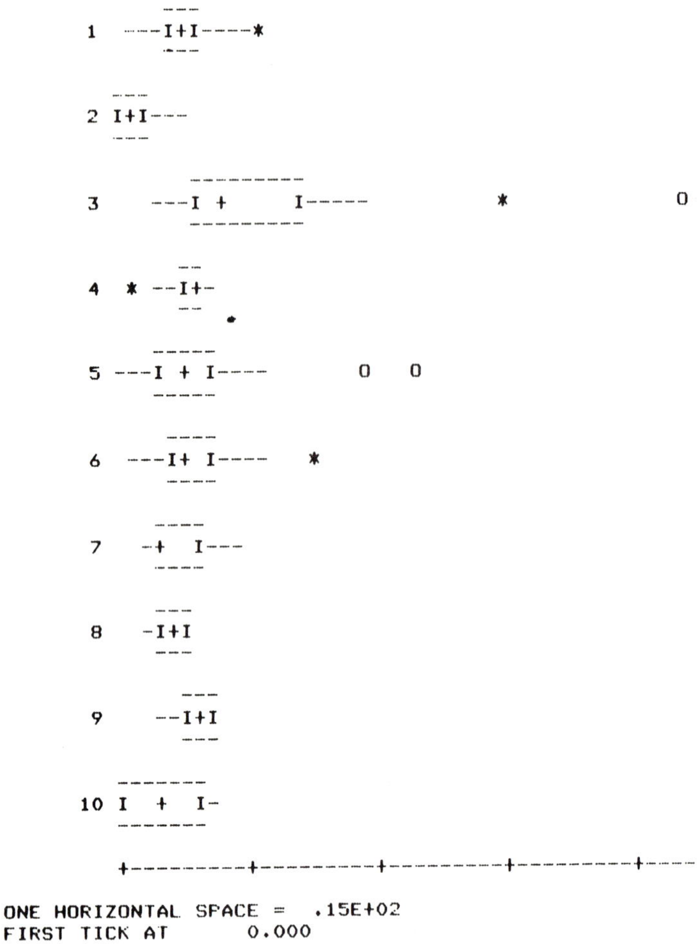

Figure 5 Boxplot for Zn in 269 rock samples divided into 10 lithological subsets

Probability plots of the ordered values for each element against the quantiles of the Normal distribution, can also be used as EDA tools for recognition of outliers. Many more graphical techniques exist for non-parametric statistical analysis, and have been summarized by Fisher (1983).

Graphical displays are also particularly helpful for understanding the differences between samples in terms of their multivariate composition. Figure 6 shows the composition of 25 stream sediments from a drainage basin in Norway (Howarth and Sinding-Larsen, 1983) in the 'tree' representation devised by Kleiner and Hartigan (1981). Samples down drainage from known mineralization (13, 15, 16, 12 and 9) and with enhanced base-metal values caused by Fe-Mn scavenging (22, 23) show up very clearly. Figure 7 shows the same data plotted as Chernoff faces (Chernoff, 1973), in which the size of particular facial features are linked to selected elements.

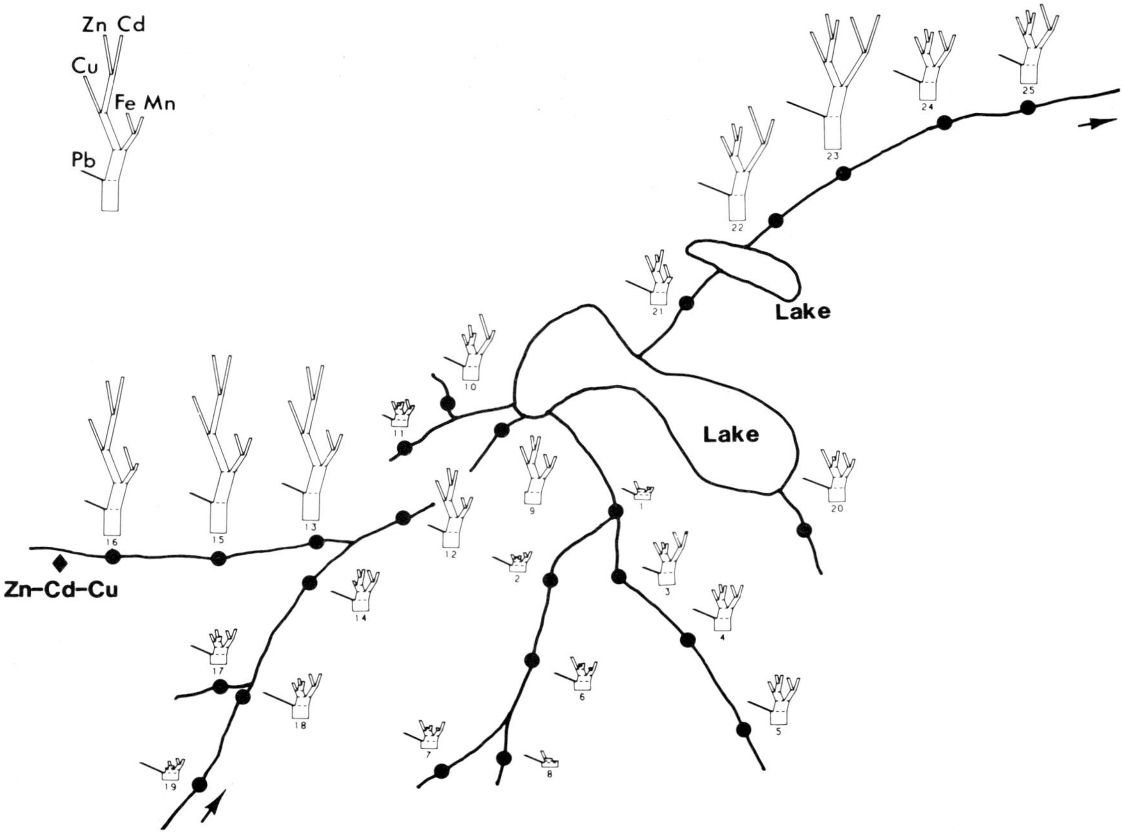

Figure 6 Kleiner-Hartigan tree representation of the chemistry of 25 stream sediment samples from a drainage basin in Norway (Garrett, 1983b; data of Howarth and Sinding-Larsen, 1983)

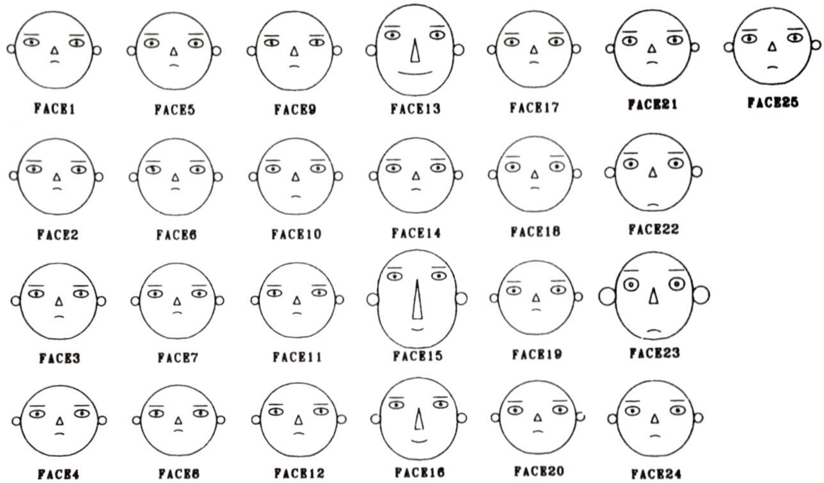

Figure 7 Chernoff face representation of stream sediment data of Figure 6

In some senses this has brought us full circle, in the pre-computer days there was a heavy reliance on graphics by geochemists. However, the practical problems of preparing these graphics for large data sets militated against extensive application. Even when computers became available it took some time for graphic output devices of adequate quality to become widespread. Now suitable equipment is at hand and a great variety of plots can be prepared which allow visual inspection and display of data in differing ways. At last the human interpreter can take full advantage of graphics machines, for we know that most people process images far more reliably than they do numbers and tables.

By their very nature geochemical data are multivariate, rarely do we have only one geochemical measure *and* no field parameters. Thus there is a great interest in multivariate statistical methods which help the geochemist move from numerical responses in the data to numerical measures of the geological and geochemical processes active in an area.

The earliest computer work in geochemical exploration applied Q-mode (between samples) factor analytic techniques, based on the Manson and Imbrie (1964) program, to the identification of end-member geochemical compositions in stream sediment data (Garrett, 1966). At about the same time Chayes (1960, 1962, 1964) identified a number of problems concerned with assessing the statistical significance of induced correlations in a correlation matrix based on 'closed' data, i.e. data whose constituents add to an essentially constant sum. Petrologists have continued to address these problems and have suggested a variety of solutions, but none really seemed satisfactory until the recent work of Aitchison (1982, 1983). Principal components analysis (Howarth and Sinding-Larsen, 1983) has provided an extremely useful tool for assessing multi-element associations (provided sensible sets of relatively homogeneous data, in a geological sense, have been used). More recent variants of this approach have been the use of the eigenvector bi-plot (Howarth and Sinding-Larsen, 1983) and, particularly in the French school of geochemistry, correspondence analysis (David and Dagbert, 1975). A more recent variation of the analysis of 'closed' data matrices has been the development of a Q-mode factor analytic solution, which allows resolution into petrological end-members, by Miesch (1976b, c).

Regression analysis has also been frequently used, particularly for correction of drainage geochemical data for the effects of Fe and Mn scavenging. The majority of applications have been based on stepwise addition to, or deletion of, variables from the regression equation. Typical of these are the work of Austria and Chork (1976) and Selinus (1983). However, it has been recognized in recent years that the effect of outliers on regression equations can be extremely deleterious (Chapman, 1976). A variety of so-called 'robust' regression methods have been proposed to overcome this problem (Tukey, 1977; Huber, 1981) in which the influence of samples lying far from the multivariate mean is down-weighted in computing the least-squares regression solution. Such methods have met with encouraging results when applied to test geochemical data (Garrett, 1983b). Associated with this new revival of interest in regression, is a greater interest among statisticians in good methods to detect influential data points among the predictor data (as opposed to simple analysis of residuals for the element being predicted) using criteria such as those discussed by Cook and Weisburg (1982). A related development is the current investigation into the utility of the 'ridge regression' (Marquardt and Snee, 1975) approach for correct estimation of regression coefficients when the predictor variables, which should in theory be uncorrelated, are in reality correlated to some degree. Results so far (Turner, 1980; Howarth, 1984) indicate that the method can be extremely useful if it is important that correct estimates of the regression coefficients are obtained for interpretational purposes. Coupled with this is an increased awareness of the importance of analysis of regression residuals, and the iterative approach to regression using cyclic identification of outliers, followed by their omission and re-analysis (Malmqvist, 1978; Howarth and Sinding-Larsen, 1983).

The assignment of samples to groups of similar composition has been examined using a wide variety of cluster-analysis techniques (Everitt, 1974; Howarth and Sinding-Larsen, 1983). Until recently, the number of samples which could be analysed at one time was often limited, by the necessity to store the inter-sample similarity matrix, to about 500 samples at maximum. The tree-diagram (dendogram) has always been a popular tool for this (e.g. Obial, 1970). Alternative presentation methods include non-linear mapping, similar to multi-dimensional scaling as used in the Social Sciences (Howarth, 1973), and related techniques (Friedman and Rafsky, 1981), offer improved scope for representing the affinities of samples in high-dimensional data with a minimum of distortion in their two-dimensional representation. Figure 8 illustrates the use of the Friedman and Rafsky non-parametric planing technique to the Howarth and Sinding-Larsen (1983) stream sediment test data set (Garrett, 1983b).

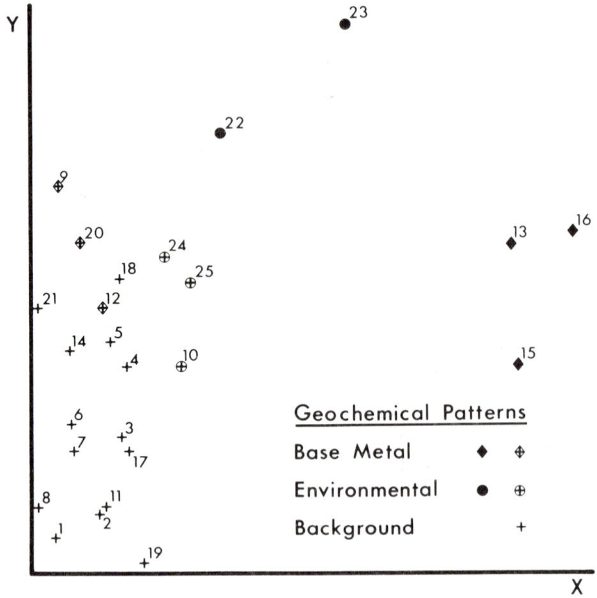

Figure 8 Cluster display of the stream sediment data of Figure 6 using the Friedman-Rafsky planing technique

Discriminant analysis has long been used for the classification of samples of unknown affinity into groups, based on training sets of known samples, using either linear separating boundaries (e.g. Cameron *et al.*, 1971; Clausen and Harpoth, 1983) or non-parametric methods such as the empirical discriminant function (Howarth, 1971). Other developments include the gradually increasing use of so-called 'jack-knife' methods for estimating the reliability of the classifier using the training set as an independent test set by successively eliminating one sample at a time and treating it as an unknown test sample (Howarth, 1973), the use of logistic models (Chung, 1978; Chung and Agterberg, 1980; Simard, 1980) and classification using Characteristic Function analysis (Botbol *et al.*, 1978). These methods are now being successfully applied in industrial or government exploration work.

Mapping

Machine plotting of data is routine for any geochemical group using data processing. Whereas many early applications used a line printer to output symbol maps, nowadays high quality drum, flat bed, or direct photographic plotters are more widely available. As a result, good quality maps and overlays, in a wide variety of sizes, may now be rapidly produced, saving many hours of handwork. Large drum plotters may be used to prepare maps and profiles many metres long and 0.8 m wide; flat bed and other drum plotters plot on a variety of sizes of material, however the largest standard size (ISO A0 or ANSI E) is about 0.84×1.19m.

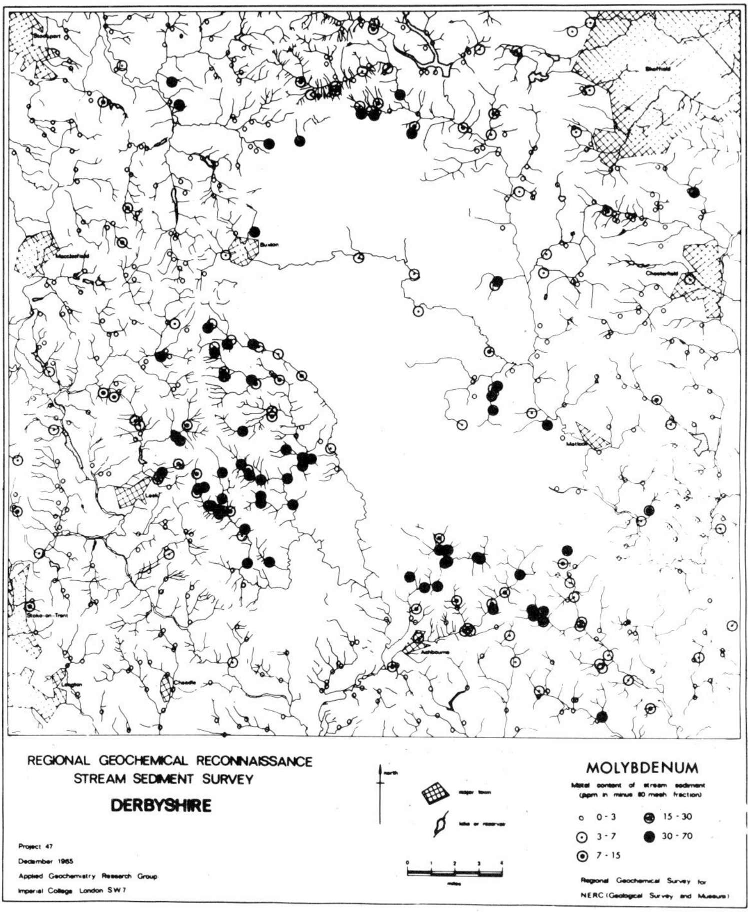

Figure 9 Example of an early computer-plotted geochemical atlas, Mo in stream sediments Derbyshire; Applied Geochemistry Research Group, December, 1965. Plotted symbols superimposed on a hand-drafted basemap

In the early days access to large plotting devices was limited and many maps were drafted by hand on the basis of computer output from line printers. One early attempt in machine plotting was, in retrospect, markedly ahead of its time. In 1965, the Imperial College group, with the aid of Dr J.G.T. Jones of the Aldermaston Atomic Weapons Research Establishment, prepared geochemical maps on film directly from data (Figure 9) using a combination of the computer facilities at Imperial College and a Stromberg-Carlson 4020 photo-plotter at Aldermaston (Nichol *et al.,* 1966). These maps consisted of symbols, similar to those developed for use in the Namwala project (Webb *et al.,* 1964) in Zambia, which have subsequently been extensively used elsewhere in the world.

Following this early start, geochemists used polynomial trend surfaces as a method of obtaining regional geochemical surfaces. However, this soon gave way to the moving average (low-pass filtering) technique which has become standard in regional geochemical mapping. With the availability of modern laser, microfilm, ink-jet plotters and colour television monitors, regional maps are increasingly plotted in colour. Examples elsewhere in this volume from the *Wolfson Geochemical Atlas of England and Wales* attest to how far the techniques have developed with increasing sophistication of computer peripherals. A detailed survey of current exploration geochemical mapping practice is given by Howarth (1983b).

The future

Geochemists passed from the pre-computer era to having investigated a wide range of mathematical, statistical and data processing tools in a remarkably short time. The subsequent 15 years have been a consolidation of the initial work, investigation of techniques borrowed from other sciences and the development of new techniques necessary due to unique aspects of geochemistry and its problems. Concomitant with all this work there has been the growth in experience, which now helps in the selection of the most appropriate tools to help solve the geochemical problem at hand.

This new ability to rapidly manipulate and display data will, we hope, help bridge the gap between the geochemist and his data on one hand, and the adept geochemical data analyst and the main mass of geochemists on the other. Clearly, the development and testing of new techniques to aid data analysis and interpretation has run ahead of their widespread application by applied and exploration geochemists. However, it is the wider availability of computing power, through time sharing services and personal micro-computers, and the greater use and availability of graphics, which presents the opportunity to close the development-implementation gap that has formed.

What does the future hold for us in the next 20 year period, a doubling of our activity time?

Whatever others may think, it is not the objective of the geochemical data processor to replace the geochemist with an interpreting machine. It is unrealistic to think that this can be done in the foreseeable future. Even if it could, it would require supercomputers of the type which are only just now beginning to be designed; and an enormous (perhaps impossible) programming effort to even begin to simulate the kind of associative thinking that goes on in a geochemist's brain during an interpretation. It is far more efficient, portable, 'reprogrammable' and cost-effective to train a geochemist! If this is the case, what will be the role of data processing: several well-defined objectives need to be met, and the systems will be developed to meet these.

Firstly, data are becoming more expensive to acquire. Notwithstanding the advances in multi-element analysis, which have done much to keep analytical costs very acceptably low, the major cost-component of any geochemical survey is still that of sampling and fieldwork. Therefore, there is an increasing incentive to organize and archive data from surveys so that they may be reviewed at a later date in terms of new genetic models for mineral deposits, or different commodities from those that originally motivated a survey. The only truly effective way to index and manage the maps, reports and data from a multiplicity of surveys is with the aid of data processing techniques. This is not to say that the files will be only digital, material such as maps and interpretive reports are often best left in written form. These materials should be properly indexed and a computer-based information management system used to identify their physical location. The hard numeric survey data should be maintained in file, or data base management, systems so that new studies can be readily undertaken. A number of major international companies are now following procedures such as these and their use will become more widespread as the cost of people-related tasks (e.g. sampling, draughting, etc.) continues to rise.

In a similar fashion, there will be an increased demand for data from geochemical laboratories in machine-readable form. This demand will provide the impetus for greater laboratory automation and the next decade will see the introduction of complete computer-linked laboratories. One of the side benefits of such systems may well be more formal approaches to quality control for monitoring short term precision and longer term analytical drift. In general, competent laboratories have little problem with short term precision, and in most cases the variability due to field sampling is significantly greater than that for analysis. The major concern has to be long term drift; if data are to be archived and compiled over many projects and years one has to know the extent of long term drift, and if at all possible it must be kept within geochemically acceptable limits.

The major challenge in the next period of our history will be to bring data processing and statistics to the aid of geochemists in ways that are effective and that they can use themselves. This task may be taken in two parts. Firstly, what are the best tools to give the geochemist, and secondly, how should those tools be presented and used.

Computer graphics is fast coming of age, and at prices that are no longer unreasonable. As is well known, it is far easier to assimilate the important features of a data set from looking at graphs and plots than it is from tables. Perhaps even more importantly, explorationists are critically interested in the relatively few, geochemically unusual, 'anomalous' cases. These cases are often far easier to pick off some form of plot, than recognize through statistical summaries that do not adequately reflect their presence. In the past decade, statisticians have become more concerned with situations where outliers may effect the outcome of a data analysis, or where the distributional assumptions may not be valid. As a result, there has been a very considerable development of robust statistical methods, where statistical outliers are identified and eliminated or downweighted, and non-parametric techniques (which make no assumptions as to the underlying data distributions). The robust methods are particularly interesting as the statisticians' outliers are often the anomalous samples we seek as geochemists. The estimates obtained with these methods are also superior measures of the properties and parameters of geochemical backgrounds when outliers are present. In many surveys where orientation data are sparse or missing, it is very difficult to determine the parameters of the 'anomalous' distributions. The only easy way to fulfil the task of recognizing outliers or anomalies is to do it in reverse, by recognizing the cases which do not fit the background model. The superior estimates of 'background' population parameters available through robust methods have genuine advantages to offer.

The first task is therefore to review the interpretational problems of geochemists in the light of recent advances in statistical data analysis. From our experience, there are clearly a variety of recently developed tools that have potential. These must be introduced to geochemists via case histories, and eventually hands-on experience. Some problems may remain unique to the analysis of exploration geochemical data, and in such cases the analysis techniques must be developed by geochemists and others from standard statistical building blocks.

Secondly, the mode of presentation of the selected techniques. The microcomputer revolution that is now gaining real momentum is capable of providing a user-friendly environment with the fast response time required for interactive graphics. However, the task of making the program packages user-friendly is a major one, the objective being to achieve a dialogue between the system and the user to be carried out in the user's own terminology through the medium of prompts, question and answer sessions, and item selection from menus. In addition, on the basis of the designer's experience, the system should carry out some work ahead of the user so that intelligent prompts, and therefore guidance, can be provided. Such 'expert' systems are now being developed in statistical research laboratories for use by engineers who are not well versed with the finer points of data analysis. There is no reason why such systems cannot be developed for exploration geochemists; in fact, we would be surprised if one or more did not exist ten years from now.

One of the advantages of expert, or expert-like, systems is that they allow guidance to be given, and therefore instruction. Whereas today's computer packages transfer the mechanics of statistical procedures, the expert systems will allow us to transfer at least part of the interpretational experience base. As we have pointed out there are a number of new approaches and techniques for geochemical data analysis which we bélieve to be useful. However, these are as yet unfamiliar to most geochemists. Their incorporation into user-friendly systems will allow them to become familiar to geochemists, and perhaps more importantly, be evaluated as tools in a broad range of problems and surveys.

The combination of graphics and user-friendly individual computing, where there is no forbidding computer centre to deal with on a daily basis, is possibly an ideal way of closing the development-implementation gap that has arisen in the recent past. The objective, as always, is to help the geochemist fulfil the interpretational task, the computer will be used to do what it is best at: manipulate data, and present thought-provoking graphics and tables. In the final analysis, it is up to the geochemist to take the decisions, and we hope that they will wish to take part in the development of these sophisticated new systems which will help bring the real power of data processing to geochemists who feel they need it, and can use it to a project's advantage.

ACKNOWLEDGEMENTS

The authors wish to express their sincere thanks to Dr Carolyn Morgan of General Electric, Schnectady, for her help in preparing the Chernoff face plots used in this paper. The order of authorship of the paper has to be credited to nature and the computer industry, it was decided by the flip of a digital coin; and both authors wish to thank those staff at their institutes who assisted with the preparation of typescripts and figures. This paper is published with the permission of the Director-General, Geological Survey of Canada.

References

Aitchison, J. (1982). The statistical analysis of compositional data. *J.R. Stat. Soc. Ser. B.* **44**, 139–177.

Aitchison, J. (1983). Principal component analysis of compositional data, *Biometrika*, 70, 57–67.

Anderson, V.L. and MacLean, R.A. (1974). *Design of Experiments–a Realistic Approach;* Marcel Dekker, New York.

Austria, V. and Chork, C.Y. (1976). A study of the application of regression analysis for trace element data from stream sediment in New Brunswick, *J. Geochem. Explor.*, **6**, 211–232.

Bjorklund, A. (1983). Effects of exponentially decaying spatial patterns on the probability distribution of anomalous values, *J. Geochem. Explor.*, **19**, 349–359.

Botbol, J.M., Sinding-Larsen, R., McCammon, R.B. and Gott, G.B. (1978). A regionalized multi-variable approach to target selection in geochemical exploration, *Econ. Geol.,* **73,** 534–546.

Bowen, R. and Botbol, J.M. (1975). The Geological Retrieval and Synopsis Program (GRASP), *US Geol. Surv., Prof. Pap. 966.*

Bridges, N.J. and McCammon, R.B. (1980). DISCRIM: a computer program using an interactive approach to dissect a mixture of normal or lognormal distributions, *Comput. Geosci.,* **6,** 361–396.

Cameron, E.M., Siddeley, G. and Durham, C.G. (1971). Distribution of ore elements in rocks for evaluating ore potential. Nickel, copper, cobalt and sulphur in ultramafic rocks of the Canadian Shield. In: *Geochemical Exploration,* (Eds. R.W. Boyle and J.I. McGerrigle) pp. 298–314; Canadian Institute of Mining and Metallurgy, Spec. Vol. 11.

Chapman, R.P. (1976). Limitations on correlation and regression analysis in geochemical exploration, *Inst. Min. Metall. Trans., Ser. B,* **85,** 279–283.

Chayes, F. (1960). On correlation between variables of constant sum, *J. Geophys. Res.,* **65,** 4185–4193.

Chayes, F. (1962). Numerical correlation and petrographic variation, *J. Geol.*, **70**, 440–452.

Chayes, F. (1964). Variance-covariance relations in some published Harker diagrams of volcanic suites, *J. Petrol.,* **5,** 219–237.

Chernoff, H. (1973). Using faces to represent points in k-dimensional space graphically, *J. Am. Stat. Ass.,* **68,** 361–368.

Chung, C.F. (1978). Computer program for the logistic model to estimate the probability of occurrence of discrete events, *Geol. Surv. Can. Pap.*, 78–11.

Chung, C.F. and Agterberg, F.P. (1980). Regression models for estimating mineral resources from geological map data, *Math. Geol.,* **12,** 473–488.

Clausen, F.L. and Harpoth, O. (1983). On the use of discriminant analysis for classifying chemical data from panned heavy-mineral concentrates—central East Greenland, *J. Geochem. Explor.,* **18,** 1–24.

Cook, R.D. and Weisberg, S. (1982). *Residuals and Influence in Regression;* Chapman and Hall. London.

Craven, C.A.U. (1954). Statistical estimation of the accuracy of assaying, *Trans. Inst. Min. Metall.,Lond., Sect. B,* **63,** 551–563.

Date, C.J. (1975). *An Introduction to Database Systems;* Addison-Wesley, New York.

David, M. and Dagbert, M. (1975). Lakeview revisited: variograms and correspondance analysis —new tools for the understanding of geochemical data. In: *Geochemical Exploration 1974* (Eds. I.L. Elliott and W.K. Fletcher), pp. 163–181; Elsevier, Amsterdam.

Drew, L.J. (1979). Pattern drilling exploration: Optimum pattern types and hole spacings when searching for elliptical shaped targets, *Math. Geol.*, **11**, 223–254.

Everitt, B. (1974). *Cluster Analysis;* Heinemann, London.

Fisher, N.I. (1983). Graphical methods in nonparametric statistics: a review and annotated bibliography, *Internat. Stat. Rev.*, **51**, 25–58.

Friedman, J.H. and Rafsky, L.C. (1981). Graphics for the multivariate two-sample problem, *J. Am. Stat. Ass.*, **76**, 277–291.

Garrett, R.G. (1966). Regional Geochemical Reconnaissance of Eastern Sierra Leone, *PhD Thesis, University of London* (unpubl.).

Garrett, R.G. (1969). The determination of sampling and analytical errors in exploration geochemistry, *Econ. Geol.*, **64**, 568–571; see also **68**, 282–283.

Garrett, R.G. (1983a). Mathematical and Statistical Activity in North America. In: *Statistics and Data Analysis in Geochemical Prospecting* (Ed. R.J. Howarth), pp. 341-360; Elsevier, Amsterdam.

Garrett, R.G. (1983b). Opportunities for the 80s, *Math. Geol.*, **15**, 385–398.

Garrett, R.G. and Goss, T.I. (1980a). UANOVA—A FORTRAN IV program for unbalanced regional geochemical surveys. In: *Geochemical Exploration 1978*. (Eds. J.R. Watterson and P.K. Theobald), pp. 371–383; Association of Exploration Geochemists, Rexdale Ontario.

Garrett, R.G. and Goss, T.I. (1980a). UANOA—A FORTRAN IV program for unbalanced nested analysis of variance, *Comput. Geosci.*, **6**, 35–60.

Garrett, R.G. and Goss, T.I. (1980b). The appraisal of survey effectiveness in regional geochemical surveys of Canada's Uranium Reconnaissance Program, *Math. Geol.*, **12**, 443–458.

Howarth, R.J. (1971). Empirical discriminant classification of regional stream-sediment geochemistry in Devon and East Cornwall, *Trans. Inst. Min. Metall., London, Sect. B*, **80**, 142–149.

Howarth, R.J. (1973). The pattern recognition problem in applied geochemistry. In: *Geochemical Exploration 1972* (Ed. M.J. Jones), pp. 259–273; Institution of Mining and Metallurgy, London.

Howarth, R.J. (Ed.) (1983a). *Statistics and Data Analysis in Geochemical Prospecting. Handbook of Exploration Geochemistry. Vol. 2;* Elsevier, Amsterdam.

Howarth, R.J., (1983b). Mapping. In: *Statistics and Data Analysis in Geochemical Prospecting* (Ed. R.J. Howarth), pp. 111–205; Elsevier, Amsterdam.

Howarth, R.J. (1984). Statistical applications in geochemical prospecting: a survey of recent developments, *J. Geochem. Explor.*, **21**, 41–61.

Howarth, R.J. and Earle, S.A.M., (1979). Application of a generalised power transform to geochemical data, *Math. Geol.*, **11**, 45–62.

Howarth, R.J. and Martin, L. (1979). Computer-based techniques in the compilation, mapping and interpretation of exploration geochemical data. In: *Geophysics and Geochemistry in the Search for Metallic Ores* (Ed. P.J. Hood), pp. 545–574; Geol. Surv. Can., Econ. Geol. Rept.

Howarth, R.J. and Sinding-Larsen, R. (1983). Multivariate statistics. In: *Statistics and Data Analysis in Geochemical Prospecting* (Ed. R.J. Howarth), pp. 207–289; Elsevier, Amsterdam.

Huber, P.J. (1981). *Robust Statistics*; Wiley, New York.

James, C.J. (1970). A rapid method for calculating the statistical precision of geochemical prospecting analysis, *Trans. Inst. Min. Metall., Lond. Sect. B*, **79**, 88–89.

Jeffery, K.G. and Gill, E.M. (1976). The design philosophy of the G-EXEC system, *Comput. Geosci.*, **2**, 345–349.

Kane, V.E. (1979). Fitting a three-parameter lognormal distribution with applciations to hydro geochemical data from the National Uranium Resource Evaluation program, Tech. Rept. K/UR-27, Oak Ridge Gaseous Diffusion Plant, Oak Ridge, Tenn.

Kane, V.E. (1979). Fitting a three-parameter lognormal distribution with applications to hydro *Proc. 1980 Statistical Computing Section, American Statistical Association, Washington,* 38–41.

Kleiner, B. and Hartigan, J.A. (1981). Representing points in many dimensions by trees and castles, *J. Am. Stat. Ass.*, **76**, 260–269.

Kolosovsky, J., Lomonosov, D.E.S., Cherkashina, G.A., Pomorova, L.V. and Volkova, D.M. (1981). *Methods of Applied Geochemistry, Abstracts of Second International Symposium, Vols. 1 and 2;* Vinogradov Geochemistry Institute, Irkutsk.

Koopman, B.O. (1980). *Search and Screening—General Principals with Historical Applications;* Pergamon, Oxford.

Kremer, M., Lenci, M. and Lesage, M.T. (1976). SIGMI—a user-orientated file-processing system, *Comput. Geosci.*, **1**, 187–193.

Le Maitre, R. (1982). *Numerical Petrology. Developments in Petrology, Vol. 8;* Elsevier, Amsterdam.

Lepeltier, C. (1969). A simplified statistical treatment of geochemical data by graphical representation, *Econ. Geol.*, **64**, 538–550.

Malmqvist, L. (1978). An iterative regression analysis procedure for numerical interpretation of regional exploration geochemistry data, *Math. Geol.*, **10**, 23–41.

Manson, V. and Imbrie, J. (1964). Fortran program for factor and vector analysis of geologic data using an IBM 7040 and 7094/1401 computer system, *Kansas Geol. Surv., Spec. Distribution Publ. 13.*

Marquardt, D.W. and Snee, R.D. (1975). Ridge regression in practice, *Am. Stat.*, **29**, 3–20.

Miesch, A.T. (1964). Effects of sampling and analytical error in geochemical prospecting. In: *Computers in the Mineral Industries* (Ed. G.A. Parks), Vol. 1, Stanford Univ. Publ. Geol. Sci., **9**, 156–170.

Miesch, A.T. (1967). Theory of error in geochemical data, *US Geol. Surv. Prof. Paper 574-A.*

Miesch, A.T. (1976a). Geochemical survey of Missouri—methods of sampling, laboratory analysis and statistical reduction of data, *US Geol. Surv., Prof. Paper 954-A.*

Miesch, A.T. (1976b). Q-mode factor analysis of compositional data, *Comput. Geosci.*, **1**, 147–159.

Miesch, A.T. (1976c). Q-mode factor analysis of geochemical and petrologic data matrices with constant row-sums, *US Geol. Surv., Prof. Paper 574-G.*

Miesch, A.T. (1981). Estimation of the geochemical threshold and its statistical significance, *J. Geochem. Explor.*, **16**, 49–76.

NERC (1977). *The Generalised Data Management System* G-EXEC (Ed. J.A. McGinnety); NERC, London.

Nichol, I., Garrett, R.G.,and Webb, J.S. (1966). Automatic data plotting and mathematical and statistical interpretation of geochemical data, *Geol. Surv. Can., Paper 66-54*, 195–210.

Obial, R.C. (1970). Cluster analysis as an aid in the interpretation of multi-element geochemical data, *Trans. Inst. Min. Metall., Sect. B,* **79**, 175–80.

Ottesen, R.T., Bolviken, B., Ekremsaeter, J. and Volden, T. (1983). Stream sediment geochemistry: Error versus concentration, *J. Geochem. Explor.*, **19**, 403.

Savinskii, I.D. (1965). *Probability Tables for Locating Elliptical Underground Masses with a Rectangular Grid;* Consultants Bureau, New York.

Selinus, O. (1983). Regression analysis applied to interpretation of geochemical data at the Geological Survey of Sweden. In: *Statistics and Data Analysis in Geochemical Prospecting* (Ed. R.J. Howarth), pp. 293–301; Elsevier, Amsterdam.

Simard, R. (1980). Logistic model applied to geophysical data for sulphide nickel exploration, *CIMM Bull.*,**73**, 96–100.

Sinclair, A.J. (1974). Selection threshold values in geochemical data using probability graphs, *J. Geochem. Explor.*, **3**, 129–149.

Sinclair, A.J. (1976). *Applications of Probability Graphs in Mineral Exploration;* Association of Exploration Geochemists, Rexdale, Ontario, Spec. Vol. 4.

Singer, D.A. (1972). ELIPGRID, a FORTRAN IV program for calculating the probability of success in locating elliptical targets with square, retangular and hexagonal grids, *Geochem. Programs, London, No. 4.*

Singer, D.A. (1975). Relative efficiencies of square and triangular grids in the search for elliptically shaped targets, *US Geol. Surv.,J. Research,* **3**, 163–167.

Stern, J.E. (1975). A statistical Problem in Geochemical Prospecting, *MSc Thesis, London*

University, (unpubl.).

Tenant, C.B. and White, M.L. (1959). Study of the distribution of some geochemical data, *Econ. Geol.,* **54,** 1281–1290.

Thompson, M. (1978). Interpretation of duplicated data in geochemical analysis, *Comput. Geosci.,* **4,** 333–340.

Thompson, M. (1983). Control procedures in geochemical analysis. In: *Statistics and Data Analysis in Geochemical Prospecting* (Ed. R.J. Howarth), pp. 39–58; Elsevier, Amsterdam.

Thompson, M. and Howarth, R.J. (1973). The rapid estimation and control of precision by duplicate determinations, *Analyst,* **98,** 153–160.

Tukey, J.W. (1977). *Exploratory Data Analysis;* Addison-Wesley, Reading, Mass.

Turner, M.St.J. (1980). A comparative study of multiple regression techniques in geochemistry, *MSc Thesis, London University* (unpubl.).

van Trump, G. and Miesch, A.T. (1977). The US Geological Survey RASS-STATPAC system for management and statistical reduction of geochemical data, *Comput. Geosci.,* **3,** 475–488.

Webb, J.S., Fortescue, J., Nichol, I. and Tooms, J.S. (1964). Regional Geochemical Maps of the Namwala Concession areas, Zambia, *Appl. Geochem. Res. Gp., Imperial Coll.Sci.Technol., Tech. Comm. 47.*

Geochemical Analysis: The Role of the Consulting Laboratory

P. E. CROFT

Introduction

The role of independent private laboratories in providing geochemical and chemical services to mining and exploration companies and to industry at large puts them, at times, in direct competition with the universities and other publicly funded bodies. Nevertheless, there is a high degree of liaison between these various bodies, and it is pertinent at the start of this paper to record the cooperation and help received from Professor Webb and his colleagues at Imperial College over a number of years. His pragmatic technical approach and vast experience make him a giant among geochemists, and we at Robertson Research value greatly the contact that we have had with him.

The consulting laboratory tries to offer as complete a service as possible and the various techniques available for geochemical exploration at Robertson Research will be described, although at the present time this plays a relatively small part in the varied work of the laboratory.

General aspects of analysis for mineral exploration

Sample preparation is arguably the most important but often neglected aspect of geochemical analysis. For any method of analysis it is necessary to have a finely-powdered sample of the soil or rock specimen, and it is critical not to contaminate the sample with metallic elements during the preparation. Tungsten carbide, although very hard, is to be avoided as a grinding medium, since it contaminates not only with tungsten, but also with cobalt, chormium and nickel as well. We prefer to use plain carbon steel or agate for grinding as usually the addition of a small quantity of iron or silicon can be tolerated. For soil analysis, a preliminary sieving of the soil through an 80-mesh sieve is still taken as standard, although it is many years since this was chosen by the Applied Geochemistry Research Group at Imperial College. Only nylon and plastic containers are used for any sieving operations.

Ideally, on any particular sample, it would be desirable to have figures for the concentration of the whole periodic table of elements. This is possible by mass spectrometry, but in practice the technique is hardly ever used for geochemical exploration on account of its high cost and low sample throughput. Also, the consulting laboratory tends to analyse only samples connected with mineral exploration projects and not with pure geochemistry. Thus, the requirement is usually to determine specific groups of elements only.

In the early days of geochemical exploration, rapid colorimetric methods were used. These were designed to be carried out in test-tubes and were specific for particular elements (e.g. Cu, Pb, Zn, Sn, W). Many of the methods could be used either in the field or in the laboratory. The contribution of the Applied Geochemistry Research Group, under the direction of Professor Webb, was invaluable in providing a series of publications which contained what became virtually standard methods. Modern instrumental methods have largely replaced these colorimetric methods, but in our laboratory a few still remain in use, e.g. the determination of tungsten with the organic reagent dithiol.

Instrumental methods

The laboratory of Robertson Research has been carrying out geochemical analysis for over twenty years and during this time a number of instrumental methods have been used for exploration programmes. These include optical emission spectrography (OES), atomic absorption spectrophotometry (AAS), inductively coupled plasma spectrometry (ICP) and X-ray fluorescence spectrometry (XRFS). In order to obtain maximum element coverage to low limits of detection it is necessary to use a combination of these techniques. Their relative merits will be discussed with particular regard to cost effectiveness as this is an important consideration when investing considerable capital sums in equipment.

OPTICAL EMISSION SPECTROGRAPHY

This is the oldest established instrumental method of analysis. It has the advantage of using a small sample of powdered rock or soil without the need to produce a solution. A 20 mg sample contained in a cup in a graphite electrode is subjected to heating, atomization and excitation by means of a constant current DC arc running at 12A. A photograph of the spectrum of the sample is obtained by means of a spectrograph. This is a particularly useful method for the rapid identification of the major constituents of a sample as well as obtaining semiquantitative trace element information on a wide range of elements. The limits of detection vary upwards from a lower level of about 1 $\mu g\ g^{-1}$. The elements Be, B, Co, Cr, Cu, Ga, Pb, Mo, Ni, Ag, Sn all have a limit of detection of 5 $\mu g\ g^{-1}$ or better thus making OES singularly useful when these elements are to be included in a multi-element scan. High sample productivity can be achieved since it is possible to record the spectra of seventy samples on one photographic plate. Although the capital cost of equipment is relatively low the method has the disadvantage of being very dependent on operator skill and experience. Without the dedication of the older school of spectroscopist this technique could fall into disuse.

ATOMIC ABSORPTION SPECTROPHOTOMETRY

It was the introduction of AAS that revolutionized geochemical analysis. It provided a simple, rapid method for analysing large numbers of samples for a limited, but important range of elements (e.g. Cu, Pb, Zn, Cd, Ag, Ni, Co). The method determines the elements in solutions and existing techniques of sample dissolution from the days of colorimetric procedures were easily adapted for AAS determination. The usual procedure consists of treating a 0.2 g sample in a test-tube with a perchloric-nitric acid mixture, the solution being evaporated to dryness overnight on a hot air-bath. Special racks which hold one hundred test-tubes are used for the evaporation. The residue is then taken up with hydrochloric acid and diluted with water. The solution is nebulized into an air-acetylene flame through which a beam of light from a hollow cathode lamp is passed. The cathode is made from the element to be determined. The atomic absorption phenomenon results from the element in its atomic ground state in the flame absorbing the light from the hollow cathode lamp in proportion to its concentration, and obeys Beer's law at low levels. With present day instruments a digital readout of the concentration calculated back to the sample can be obtained. The method is relatively free from interference (Pb and Ag usually require background correction which is easily achieved automatically on modern instruments) and requires only a modest capital outlay. Inexperienced operators can quickly learn to produce reliable results for large geochemical exploration programmes. The consulting laboratories seized on this method, as did numerous small operators, many of whom made fortunes working from sheds and garages just on the geochemical determination of the standard three elements Cu, Pb and Zn. The main disadvantage of AAS is that it is not a simultaneous multi-element technique in that each element has to be measured sequentially using an individual lamp.

INDUCTIVELY COUPLED PLASMA SPECTROMETRY

As the theoretical aspects of ICP are discussed elsewhere in this volume only our experiences with geological trace element applications will be considered here.

The first reports on the use of ICP as an excitation source for trace element analysis were published in 1964. Many analysts purchasing an ICP spectrometer hoped for an application where the solution produced for AAS could be used for simultaneous multi-element determinations. This could not be easily achieved for reasons outlined below. Also, the sensitivity of the important element Pb is poor. When assessing the claims for the high sensitivity of the technique, analysts should be wary of the published data since much of the work was based on pure single element solutions. When solutions produced from real geological samples are analysed various limitations on producing reliable results become apparent. These are generally associated with spectral interferences from overlapping and coincident emission lines or from background. Careful selection of analytical lines can generally overcome spectral interference, or correction factors can be applied. The most serious effects are changes in background intensity caused by the emission of molecular species or from stray light scattering in the spectrometer. The elements Ca, Mg and Al are the most troublesome in this respect since they are usually quite variable but abundant in rocks and soil samples. It is possible to measure the background intensity adjacent to the analytical line by moving the entrance slit of the spectrometer. To carry out all the necessary corrections sophisticated computer control and software is required for the spectrometer. It is not therefore as simple a technique as AAS to apply to geochemical trace analysis and the high capital cost precludes its use when less than about six elements are to be determined on a sample. In fact for the group of elements Cu, Pb, Zn, Cd and Ag, AAS may be preferable since the sensitivity is better. However, the multi-element capability of ICP is very attractive and it definitely scores over AAS for the refractory elements Be, Ba, Sr, Cr, V, B, Sn and Zr, but the old analytical problem of finding an attack that will bring all elements into solution still remains.

X-RAY FLUORESCENCE SPECTROSCOPY

If our geochemical laboratory were confined only to one instrumental technique probably XRFS would be the choice. It does not require solution preparation and has limits of detection of a few $\mu g\,g^{-1}$ for most elements down to F in the periodic table. The instrument would need to be a fully computer-controlled wavelength dispersive spectrometer fitted with automatic sample changer. Most consulting laboratories could not justify the high capital outlay (£150,000) for this type of installation. We use XRFS but as our instrument is a manual spectrometer we tend to use it for a limited number of elements rather than multi-element scans. We find XRFS to be very satisfactory for the elements U, Th, Sn, W, Nb, Ta and Rb, particularly as their determination is difficult by other techniques. Soil and rock samples need to be finely ground (<250 μm) and about 7 g is required for analysis. The sample is pressed into a pellet and subjected to X-ray analysis using a rhodium X-ray tube. For accurate trace analysis it is necessary to carry out mass absorption corrections and the most versatile approach is provided by the use of an incoherently scattered (Compton) X-ray tube line where intensity is inversely proportional to mass absorption at the wavelength of the element to be determined. The data from the spectrometer is fed directly into a minicomputer and the necessary calculations carried out to produce quantitative results. The accuracy of XRFS is dependent upon standard samples of known composition. Although a fairly large number of certified standards are now available, the range and quality of some of the reference data is very variable. Often, reliance must be placed on a technique such as the use of spiked additions in order to create viable calibrations. With the continued improvements in X-ray spectrometers and computing facilities and also the development of energy dispersive instruments, XRFS must be the first technique to be considered for large exploration programmes when its cost effectiveness could be proved.

Methods for specific elements

In spite of the various multi-element techniques, certain elements must be determined by specific methods to obtain low enough limits of detection for geochemical applications.

MERCURY

This important pathfinder element needs to be determined to a limit of detection of 0.01 $g\,g^{-1}$ or less. Numerous procedures have been used over the years, but they are all based on measuring evolved mercury vapour. We find a flameless atomic absorption method following acid digestion to be the most reliable for a wide range of sample types.

ARSENIC, ANTIMONY, SELENIUM, TELLURIUM AND BISMUTH

This interesting group of pathfinder elements all have the property of forming gaseous hydrides when a solution is reduced with sodium tetrahydroborate. This is the basis of a very sensitive method of analysis since the gases formed can be swept into an inductively coupled plasma by a stream of Ar and the elements determined by normal spectroscopic means. Limits of detection of 0.1 $\mu g\,g^{-1}$, when calculated back to soil samples, are easily achieved.

URANIUM

Although interest in this element has waned in the last few years, there are a number of interesting techniques used in exploration programmes. Total uranium can easily be determined by X-ray fluorescence, but it is usually of more interest to know soluble (mobile) uranium. This is determined on a nitric acid extract of the sample with determination being carried out fluorimetrically in a cast disc after an initial extraction by ethyl acetate. Various techniques for measuring Rn have been used for detecting U anomalies, but being a gas the near surface concentration at any point is determined not only by the rate of supply from the disintegrations of U and its immediate daughter products at depth, but also on such effects as variation in atmospheric pressure and diurnal changes in temperature, surface wind velocity, precipitation etc. We have developed a geochemical procedure for determining ^{210}Po in the soil; this element is towards the end of the ^{238}U decay chain and it is equivalent to a Rn measurement integrated over many years, which minimizes the influence of short term fluctuations.

GOLD

This element captures the attention as much now as it did a few thousand years ago. Considerable efforts are made to find workable deposits and geochemistry is very important. There are various theories concerning pathfinder elements, but it is often felt that the only way to find Au is to determine Au itself in the soil, rocks and stream sediments. We have developed methods to determine Au to very low levels using an *aqua regia* attack and an organic solvent extraction with AAS measurement. Due to the sampling problems associated with Au, we prefer to take a large sample (up to 500 g) and achieve a limit of detection of better than 0.01 $\mu g\,g^{-1}$. Another method that is still widely used is the fire-assay determination of Au and the noble metals. It is probably the oldest assay method. The Chinese about 500 BC were known to have used it, and there are descriptions in the Bible that refer to a knowledge of fire assay procedures. It is basically a smelting method that recovers the Au from the sample by Pb collection and cupellation. Although it is not as accurate as AAS or neutron activation, especially at low levels, it has the great attraction of producing an Au bead rather than a number from an instrument.

PLATINUM METALS

The fire assay procedure is also very useful for the determination of Pt, Pd and Rh. There are no simple geochemical methods for these elements, but good sensitivity can be achieved by carrying out a fire assay separation method with Pb collection, using Ag as a carrier and then subjecting the Ag cupellation bead to emission spectrographic analysis for the determination of the platinum metals.

Conclusions

An important area that the consulting laboratory becomes frequently involved in is contributing figures to the certification of standard samples. In the early days there were virtually no rock or soil samples with agreed figures for major and trace elements. Now, however, there are a considerable number of internationally recognized reference materials. These materials are generally analysed by classical chemical methods which are time consuming but produce reliable results. It is important to use methods that are independent of possible instrumental bias. The inclusion of reference materials, as well as hidden duplicate samples, forms a significant part of quality control procedures in the laboratory.

With the development of simultaneous multi-element techniques (e.g. ICP, XRFS) there is now the possibility of including major elements as well as trace elements in comprehensive schemes of analysis. The data produced can be stored on computer discs and even incorporated directly into geological mapping programmes. The consulting laboratory is thus still endeavouring to provide industry with a cost effective analytical service which is in keeping with the demands of the 1980s.

The Future Role of Inductively-Coupled Plasma Atomic Emission Spectrometry in Applied Geochemistry

M. THOMPSON

Multi-element methods in exploration geochemistry

The value of multi-element analysis in exploration geochemistry has long been appreciated. However, as always in this field, there is a lag between appreciation of the technical merit of an exploration strategy and the capability for executing it cost-effectively. The lag is due soley to the shortfall in analytical technology. Indeed there is little incentive to develop an analytical method unless there is a clearly recognized use to which the new type of data can be put. On the other hand no exploration manager would even consider multi-element analysis unless a cheap reliable method were available.

Historically, the first analytical method used in systematic exploration geochemistry was the spectrograph, itself a multi-element method. Although undoubtedly cheap, spectrography was not especially reliable, the main shortfalls being in the limited accuracy and precision which could be obtained, and the fact that many elements of interest to the geochemist could not be determined at background concentrations under the compromise conditions used for multi-element analysis. Consequently, as more reliable methods became available (firstly colorimetric, then atomic absorption methods), the popularity of spectrography began to wane, despite the fact that elements could not be determined simultaneously by the newer methods.

In the 1980s, there are essentially three multi-element methods with capabilities approaching the needs of the exploration geochemist, viz. X-ray fluorescence (XRF), instrumental neutron activation analysis (NAA) and inductively-coupled plasma atomic emission spectrometry (ICPAES). It is not my purpose to argue in detail the relative merits of these three methods, but simply to give an account of the potential of one of them, ICPAES. Indeed such is the pace at which new developments appear, it would be quite risky to try to guess which method would be the cheapest and most reliable in 1990. However to put the discussions into the geochemical context the following points should be born in mind:

1. Each one of the methods has a range of elements which cannot easily be

determined in a multi-element context, although these elements are different in each case.

2. Low detection limits cannot easily be obtained by XRF, but can usually by ICPAES and NAA.
3. XRF is popular among non-chemists because no chemical dissolution is required, in contrast to some other methods. (Actually, for the most precise work, dissolution methods *are* used in XRF, because the sample is presented as a *solid solution* in the borate bead.)
4. At present an ICPAES instrument can produce more element determinations in unit time than can the other techniques, but the instrument needs more attention than in XRF.
5. ICPAES is at an earlier stage of development than either XRF or NAA, consequently there is more scope for improvement.
6. As will come out in the subsequent discussions, ICPAES is a very adaptable and versatile method, and therefore radically new methods of applying it to samples may be expected with a probability bordering on certainty.
7. The real cost of NAA is much higher than the apparent cost, because of the enormous cost of building a nuclear reactor in which the samples are activated.

However, despite potential advances in these techniques, it seems unlikely that any single method will ever satisfy all of the analytical requirements for exploration or environmental geochemistry. However it is quite likely that several methods, including ICPAES, will achieve a close approximation to this. The use of ICPAES for geochemical analysis is already widely practiced and has been discussed by a number of authors, including Church (1981) and Thompson and Walsh (1983).

Analytical characteristics of ICPAES

ICPAES is a spectrometric method with a uniquely effective emission source. The source is in fact a plasma, induced in a stream of (usually) argon gas which flows in a specially constructed torch (Figure 1). The ionization of the gas renders it a good conductor, and therefore power can be transferred into it by means of the two-turn induction coil. This keeps the temperature of the plasma at about 10 000 K and maintains the ionization. The orifice of the injector tube and the flow rate of the injector gas are balanced to produce a toroidal (doughnut-shaped) plasma, through which the sample material passes. This special geometry, combined with the high temperature (8000 K) to which the sample is subjected, confers on the ICP its near-ideal characteristics. Sample material passing through the plasma is completely atomized, the atoms are excited with high efficiency, and emission takes place from an 'optically thin' source. Emission spectra are analysed with a high resolution spectrometer within the wavelength range of 170–780 nm. The advantages of ICPAES are summarized in the following paragraphs.

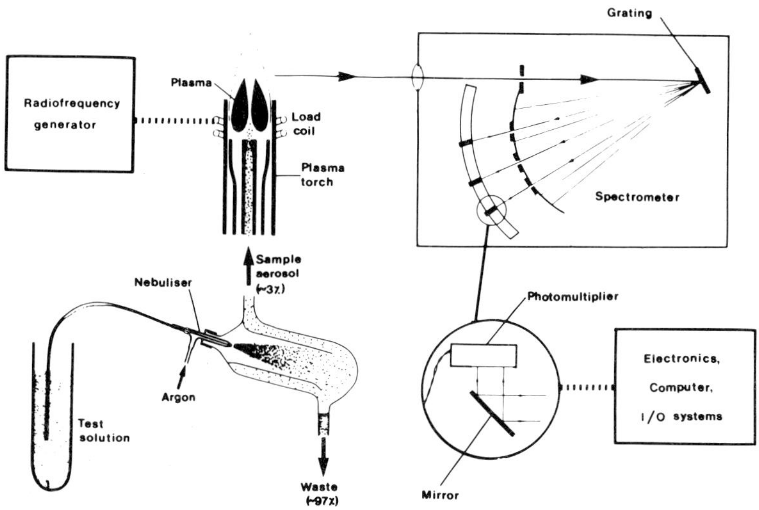

Figure 1 Schematic diagram of an ICPAES system for simultaneous multi-element analysis. An aerosol of droplets of test solution is produced by the nebulizer/spray-chamber assembly and injected into the plasma by a stream of argon. Power is supplied to the plasma by the RF generator. The sample is largely atomized by the high temperature of the plasma. Light emitted just above the bright zone by the excited atoms and ions is analysed by the grating spectrometer, which would normally have 30–50 separate channels

SIMULTANEOUS MULTI-ELEMENT ANALYSIS

In principle all of the elements can be determined, bar argon which is the normal plasma gas. In practice there are some limitations, for example radioactive elements would cause a problem of fume disposal, and some elements (F, Cl, Br) would need special optics to allow access to emission lines below 170 nm. Simultaneous analysis requires a polychromator spectrometer and it is unusual to find such instruments which are capable of accommodating more than 50 lines. Scanning monochromators are more flexible in this respect, but can provide only a sequential determination of the elements. This seems unlikely to prove attractive in exploration work, as the work output of sequential instruments is substantially less than that of simultaneous instruments.

LOW DETECTION LIMITS

Robust detection limits (i.e. capable of being regularly obtained in real analysis) are given in Table 1. For many metallic elements detection limits are comparable with those of conventional flame AAS. For non-metals and 'refractory oxide' metals ICPAES detection limits are markedly superior. For a few elements (e.g. Zn, Na, K, Rb) AAS or flame emission give lower

Table 1 **Robust detection limits for elements simultaneously determined by ICPAES with the conventional pneumatic nebulization of aqueous solutions**

Element (nm)		*DL ($\mu g\,l^{-1}$)	Element (nm)		*DL ($\mu g\,l^{-1}$)	Element (nm)		*DL ($\mu g\,l^{-1}$)
Ag	(328.1)	10	Ho	(345.6)	10	Ru	(240.2)	50
Al	(309.3)	10	I	(178.3)	20	S	(180.7)	100
As	(193.7)	50	In	(230.6)	100	Sb	(206.8)	100
Au	(242.8)	20	Ir	(224.2)	50	Sc	(361.3)	2
B	(249.8)	10	K	(766.5)	100	Se	(196.1)	100
Ba	(455.4)	1	La	(398.8)	20	Si	(288.2)	50
Be	(313.0)	0.2	Li	(670.8)	5	Sm	(359.2)	50
Bi	(223.1)	50	Lu	(261.5)	2	Sn	(189.9)	50
C	(193.1)	50	Mg	(279.5)	0.2	Sr	(407.8)	0.5
Ca	(393.4)	0.2	Mn	(257.6)	5	Ta	(226.2)	50
Cd	(214.4)	5	Mo	(202.0)	20	Tb	(350.9)	50
Ce	(418.6)	50	Na	(589.0)	50	Te	(214.3)	50
Co	(228.6)	10	Nb	(309.4)	50	Th	(283.7)	100
Cr	(267.7)	5	Nd	(430.3)	100	Ti	(337.2)	10
Cu	(324.8)	5	Ni	(231.6)	20	Tl	(190.8)	50
Dy	(353.1)	10	Os	(225.5)	1	Tm	(313.1)	10
Er	(390.6)	20	P	(178.3)	10	U	(385.9)	100
Eu	(381.9)	5	Pb	(220.3)	50	V	(311.1)	5
Fe	(259.9)	10	Pd	(340.4)	50	W	(207.9)	50
Ga	(294.4)	50	Pr	(422.3)	50	Y	(371.0)	5
Gd	(335.0)	20	Pt	(214.4)	50	Yb	(328.9)	2
Ge	(209.4)	50	Rb	(780.0)	20	Zn	(202.5)	5
Hf	(277.3)	20	Re	(197.3)	10	Zr	(349.6)	20
Hg	(194.2)	10	Rh	(233.4)	50			

*DL—detection limit. Some detection limits may be degraded by interference.

detection limits but among this group the performance of rubidium alone falls short of geochemical requirements in ICPAES.

LONG LINEAR CALIBRATIONS

ICPAES calibrations are linear over at least 4 orders of magnitude of concentration and, for the most part, usable for 5 orders of magnitude. This greatly simplifies calibration, and is *essential* for genuinely simultaneous analysis, to avoid dilutions for some analytes out of range.

GOOD PRECISION AND ACCURACY

Under conditions of stringent care, a working precision of 1% RSD is achieved with ICPAES. With conditions prevailing in routine analysis for geochemistry, analytical precisions of 5% RSD should always be bettered where the analyte is well above its detection limit.

ICPAES is not completely free from interference effects, contrary to the original contention, but chemical interference effects which plague AAS are absent. Translational interference due to spectral overlap and background enhancements are common, but they are well understood and can easily be corrected. (All ICPAES instruments have computer facilities which carry out this correction automatically.) Matrix effects are noticeable but normally they are easily contained within the limits set by the routine precision.

ADDITIONAL FEATURES

Other features of merit for the ICPAES include: (a) good reliability; (b) use of a small sample volume (0.5–2.0 ml typical range); (c) a fast analysis time; and (d) a great flexibility for adaptation to different methods of sample introduction. The conventional technique (nebulization of an aqueous solution) can be readily replaced by the novel methods discussed below, and the ICP can easily be adapted as a sensitive element-specific detector for various chromatographic separations.

Current limitations of ICPAES in applied geochemistry

Much less widely appreciated than the advantageous aspects of ICPAES discussed above are a number of limitations which currently affect its applicability to exploration technology and environmental studies. These limitations are first summarized in this section, and methods of alleviating them are subsequently suggested. Some of these suggestions are existing realities, whereas others are rather more speculative.

SAMPLE DISSOLUTION PROBLEMS

Geochemical samples are usually multiminerallic and some elements of very considerable importance occur in resistate phases. These minerals (e.g. rutile, zircon, barite, chromite, cassiterite, wolframite) do not dissolve in acid mixtures under conditions appropriate for geochemical analysis. Consequently the elements cannot be included in a conventional multi-element analysis, a noteworthy deficiency in the ICPAES method, as compared with (say) XRF. Modified decomposition methods could be envisaged (e.g. sealed bomb attacks) but these would be too cumbersome for routine application. Fusion methods, especially those based on lithium metaborate, would overcome the insolubility problem. However the minimum dilution factor usable with lithium metaborate is 500, so that the most vital elements would normally occur below their detection limits in the resultant solutions. Some critics have argued that the necessity for dissolving the sample at all is a severe disadvantage of ICPAES. While this is not the view of the author, it should be noted that some progress towards the injection of solid samples has been made, which technique may have a role in the future of geochemical analysis.

LOW ABUNDANCE ELEMENTS

Many important elements occur in typical geochemical samples at such low concentrations that they fall below the detection limits of ICPAES after the dilution due to dissolution. The minimum practical dilution factor for a simple acid digestion is about 50 and under these conditions a number of ore elements (U, Mo, W, Bi, Cd, Ag) are not detectable, and lead is only just quantifiable, at crustal abundances (Figure 2). A number of other elements, interesting for pathfinder applications and their environmental effects (As, Sb, Se, Hg), are also not included among the elements which can be directly determined.

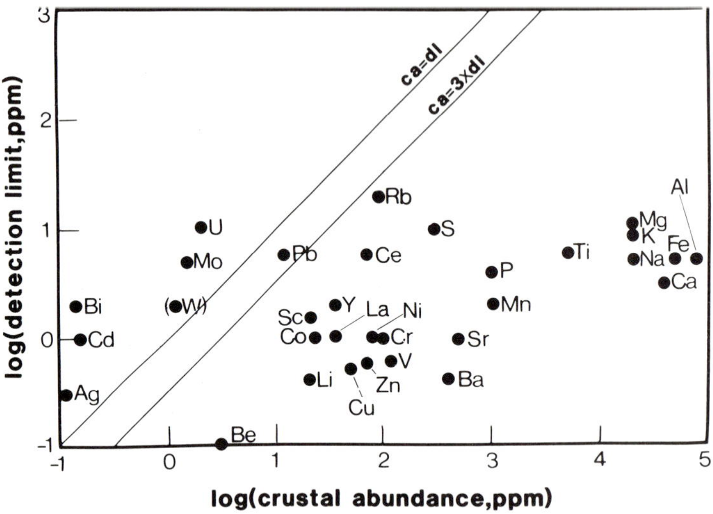

Figure 2 A comparison between ICPAES detection limits and crustal abundances for a number of elements. The digestion envisaged is an $HF/HNO_3/HCLO_4$ procedure with a final dilution factor of 40. It can be seen that a number of important elements are not quantifiable at crustal abundance (i.e. the crustal abundance is less than 3× the detection limit), although they may be detectable at anomalous levels

SPECIATION

The ICPAES cannot distinguish between different species of the same element. Consequently analysts have to rely on selective dissolution procedures, chromatography, etc. if specific forms of an analyte are required. Selective leachants often contain a high dissolved salt-content which are difficult to handle by ICPAES because of nebulizer problems.

INTERFERENCE EFFECTS

As previously mentioned, interference effects are not negligible, although usually they can be easily managed by computer correction. However, the correction terms are subject to some uncertainty, amounting to about 5% of the term. This has the effect of adding about 10% of the term to the detection limit, in some cases substantially degrading the detection limit. Aluminium is the most consistent interferer, for instance changing the detection limit of molybdenum from 0.6 $\mu g\, g^{-1}$ in aluminium-free samples to about 9 $\mu g\, g^{-1}$ at typical levels of aluminium (5% m/m) (Thompson and Walsh, 1983, p. 124ff). This can be very significant for elements with background concentrations close to the detection limit.

COST-EFFECTIVENESS

The capital cost of a simultaneous ICPAES is rather high (about £80 000), and the productivity normally obtained is within the range 100–200 samples for an 8 hour day's work. Very little work has been carried out with a view to optimizing the geochemical use of ICPAES for cost-effectiveness.

HARDWARE PROBLEMS

The useful operation of the ICP instrument in the conventional mode relies on the consistent behaviour of the nebulizer. Unfortunately this is the part of the instrument most prone to problems. The most common difficulties stem from (a) nebulizer blockage by foreign particles and (b) nebulizer salting-up. The former problem can be minimized by careful working procedures. The latter occurs when solutions containing more than about 1% of dissolved solids are nebulized. Although a variety of nebulizers have been tried, this is still a problem area in ICPAES.

The potential of sensitivity-enhancing techniques

Sensitivity-enhancing techniques can overcome many of the above mentioned limitations of ICPAES as applied to geochemistry. Two methods are considered here, namely gas-phase sample injection after hydride generation, and preconcentration.

HYDRIDE GENERATION

A group of metallic and non-metallic elements (Ge, Sn, Pb, As, Sb, Bi, Se, Te) are readily reduced in aqueous solution to hydrides which at ambient temperatures are both volatile and not very soluble in water. Thus it is relatively simple to inject these elements into the ICP in the gas phase. This style of gas-phase injection confers two benefits on the analysis. Firstly the analytes are almost completely separated from the matrix and many possible interference effects are thus eliminated. This tends to decrease the background noise and improve detection limits. Secondly, the rate at which the analytes are injected is increased by a factor of 200 or more because (a) the uptake rate of solution is greater and (b) the efficiency of hydride generators approaches 100% for most of the elements. The efficiency of normal nebulizer systems rarely exceeds a few per cent. These two factors (high injection rate and low matrix analysis) combine to provide a method with very low detection limits and good accuracy. Robust instrumental detection limits are given in Table 2.

Table 2 **Instrumental detection limits for elements determined by ICPAES after hydride generation**

Analyte (nm)	Detection limit ($\mu g\ l^{-1}$)
Ge (303.9)	0.03
Sn (189.9)	0.03
Pb (220.3)	4
As (193.7)	0.1
Sb (206.8)	0.2
Bi (223.0)	0.2
Se (196.1)	0.2
Te (214.3)	0.2
*Hg (184.9)	0.05

*Determined after reduction to element.

Hydride generation linked to ICPAES has been a possibility since Thompson *et al.* (1978a) designed a continuous hydride generator. (Discrete-addition generators for AAS are not compatible with the ICP.) A schematic hydride generator is shown in Figure 3. Currently several manufacturers are offering hydride generators based on continuous mixing. The hydride/ICPAES method has been applied to rock, soil, sediments, dust, water and plant material. Because of the diverse chemistry of these elements, determination has to be preceded by special chemical digestion procedures. (The general methods of geochemistry are not suitable.) Indeed, few decomposition methods are capable of solubilizing more than a few of the hydride elements simultaneously, although most of the elements (except tin and germanium) can be reduced simultaneously. Practical detection limits for solid samples are in the range 0.02–0.5$\mu g\ g^{-1}$ depending on the preliminary procedures. Even when preceded by lithium metaborate fusions (i.e. with a large dilution factor), hydride generation-ICPAES gives detection limits of about 0.3 $\mu g\ g^{-1}$ for Ge and Sn. The process of hydride generation is subject to some interference effects especially from copper and nickel, but these can

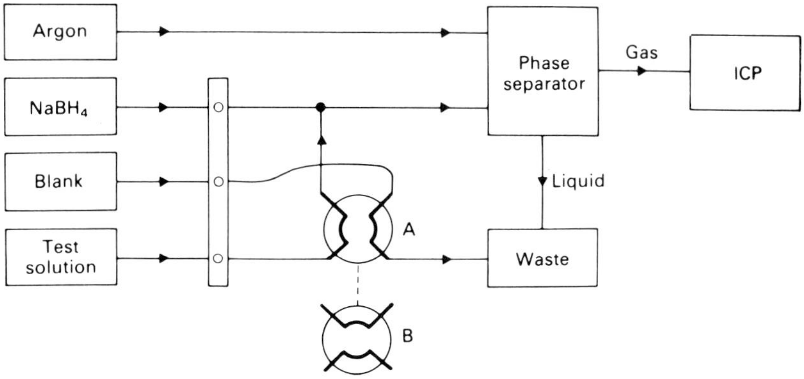

Figure 3 A schematic diagram of a continuous hydride generator for ICPAES showing the valve in position (A) for running the test solution and in position (B) for running blanks or changing test solutions. (Reproduced with permission from Pahlavanpour *et al.* (1981). *Analyst*, **106**, 467–471. Copyright Royal Society of Chemistry)

normally be circumvented by suitably modified procedures. Mercury can be reduced to the metal vapour simultaneously with hydride formation and determined with very low detection limits by ICPAES.

PRECONCENTRATION

Preconcentration can be used to enhance sensitivity in ICPAES just as it can in AAS. Two essential differences separate the fields, however. A multi-element determination by ICPAES requires only a small volume of test solution, so that the final concentration factor can be made higher. Secondly *multi-element* sample dissolutions and extractions are required to make the best use of ICPAES. As usual, an additional virtue of preconcentration is that it can often be used to separate trace analytes from the sample matrix and so provide an interference-free analysis. This is essential in the fairly common situation when the detection limit is determined by the interferent levels.

Several applications of preconcentration to geochemical analysis by ICPAES have been reported, the materials being water, rocks, soils and sediments (Table 3). However it is clear that a good, general purpose, rapid, multi-element preconcentration method which is selective against interfering matrix elements (Al, Ca, Mg, Fe, Na, K) is still lacking, and would easily repay the considerable thought and investment in development time.

The ICP can be used with organic solvents, so it is possible to envisage the solvent extraction preconcentration and separation illustrated in Figure 4, where a large volume of aqueous solution is equilibrated with a few millilitres of organic liquid which is subsequently analysed directly. If such a technique were applied after a fusion with lithium metaborate, then an overall dilution factor of only 8 could be achieved. Most of the elements occurring in minerals which are resistant to acid attacks would be completely solubilized by the

Table 3 **Some examples of preconcentration applied to geochemical samples for analysis by ICPAES**

Material	Analytes	*Preconcentration Method	Factor	Authors
Rock, soil etc.	Ag, Au, Bi, Cd, Cu, Pb, Zn	SE	0.2	Motooka *et al.* (1979)
Rocks	La, Ce, Pr, Nd, Sm, Eu, Gd, Dy, Ho, Er, Yb, Lu	IE	0.1	Walsh *et al.* (1981)
Seawater	Fe, Cu, Pb, Cd, Zn, Ni, Co	IE	25	Sturgeon *et al.* (1980) Berman *et al.* (1980)
Seawater	Cd, Cu, Fe, Mo, Ni, V, Zn	SE	250–500	McLeod *et al.* (1981)
Freshwater/ Seawater	Cr, Mn, Co, Ni, Cu, Cd, Pb	C	240	Hiraide *et al.* (1980)
Freshwater/ Seawater	As, Sb, Bi, Se, Te	C	20	Thompson *et al.* (1981b)
Freshwater	Al, Ba, Bi, Ca, Co, Cd, Cr, Cu, Fe, Hg, K, Li, Mg, Mn, Na, Ni, P, Pb, S, Se, Sr, V, Zn	E	10	Thompson *et al.* (1982)

*The preconcentration factor is the ratio of the analyte concentration in the test solution to that in the original sample. Preconcentration methods are: SE—solvent extraction; IE—ion exchange; C—coprecipitation; E—evaporation.

fusion (e.g. W, Sn, Nb, Cr, Be, etc.) and the attractive range of detection limits shown in Table 4 could be achieved. Whether this exact projection is capable of fulfillment is not yet clear, but there is certainly a great need for such a scheme.

Prospects for direct injection of solid samples and allied techniques

The direct injection of solids, (which would avoid the problems of total solubilization of samples) has been an alluring prospect from the earliest days of ICPAES. It was thought that the high temperature of the ICP might suffice to atomize the solid sample completely and avoid the problems of selective volatility which are familiar in arc spectrography. Early attempts at continuous powder sample introduction utilized spark-elutriation and

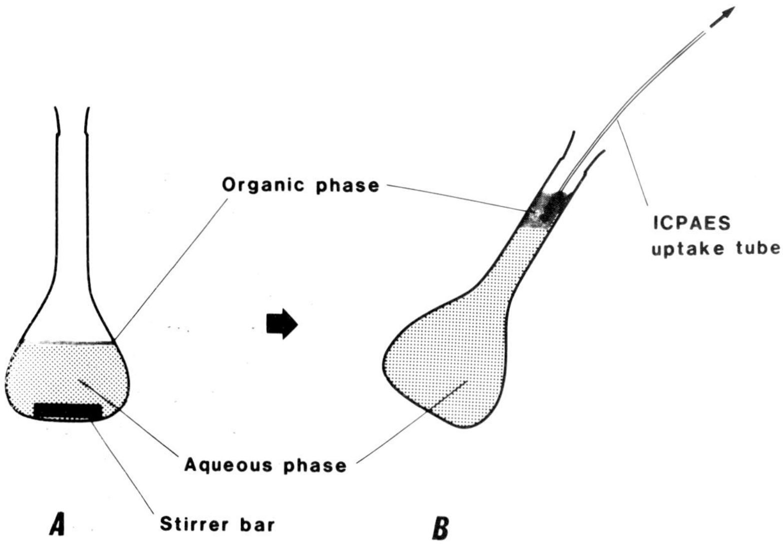

Figure 4 A projected scheme for ICPAES determination of resistate ore metals after a lithium metaborate fusion and extraction of the aqueous leachate into a small volume of organic reagent solution, showing (A) the extraction and (B) the determination. Such a method could give good detection limits for a number of otherwise difficult elements

Table 4 **Projected detection limits (DL) for a possible multi-element extraction following a lithium metaborate fusion and ICPAES determination from the organic phase**

Element	DL ($\mu g\,g^{-1}$)	Element	DL ($\mu g\,g^{-1}$)
W	0.3	Co	0.04
Mo	0.06	Ni	0.1
U	2	Cu	0.04
Sn	0.2	Zn	0.01
Nb	0.3	Cd	0.02
V	0.1	Pb	0.3
Cr	0.04		

Note: The overall dilution factor would be 8.

insufflation, but a number of problems, including particle segregation, were encountered. More recently, introduction of slurries by means of special nebulizers has been shown to have some merit. Discrete sample introduction by means of electrothermal vaporization, direct insertion, decrepitation and laser ablation have all recently achieved a measure of success and may prove to be a valuable addition to the ICP repertoire. Before some of these methods are discussed in more detail, it should be noted that the introduction of samples in solution does have special advantages of its own, relating to the homogeneity of solutions and the ease of making up matrix-matched standard solutions for calibration. Therefore whatever the future success of solid sample introduction, it is unlikely to become the predominant method.

NEBULIZATION OF SLURRIES

Several authors have claimed a reasonable degree of success in nebulizing aqueous slurries of solid geological materials (Thelin, 1981; Fuller *et al.*, 1981). The sample has to be very finely ground and suspended in a mixture containing thixotropic additives to prevent premature separation. Concentrations of about 1% of solids were used. Nebulizers used for this work were of the Babington-type, where the liquid for nebulizing is pumped over a small (200 μm) orifice through which gas is passing at high linear velocity. The film of slurry or solution is disrupted as it interacts with the gas stream, and produces an aerosol of fine droplets. As the liquid does not have to pass through a fine orifice (in contrast to other pneumatic nebulizers) there is no possibility of blockage due to the suspended particles. Fuller and co-workers were able to obtain good accuracy and precision for major and minor elements after calibration with certified reference rocks etc. under carefully standardized conditions. However it seems that low detection limits were not obtained.

Slurry nebulization seems to be capable of some improvement, but is likely always to require very careful physical pretreatment of the sample, which may turn out to be as time consuming as a chemical digestion. It is likely to be used, therefore, only where chemical digestion is a real problem.

LASER ABLATION

The energy of a laser beam can be used to vaporize small amounts of material which are then subjected to a spectroscopic analysis of some kind. Such a combination was almost the first geological application of ICPAES in the 'Lasertrace' system developed by Barringer Research (Abercrombie *et al.*, 1978). In this system the sample material (usually airborne particulate matter) is supported on transparent adhesive tape which is fed automatically into the path of the laser beam. When the sample is stationary in a specially designed cell, the pulse of infrared laser radiation volatilizes the sample without affecting the transparent substrate. The vaporized sample condenses rapidly and is swept as an aerosol into the plasma. The transient signals obtained from the ICP (Figure 5) are quantified by integration in the normal way during the passage of material through the plasma. As the sample-feed and data-logging are automatic in this system, a cycle time of 10 s per sample for a multi-element analysis can be maintained, which may constitute the fastest analysis ever achieved! Calibration is achieved by the volatilization of reference materials finely ground and spread on to the tape.

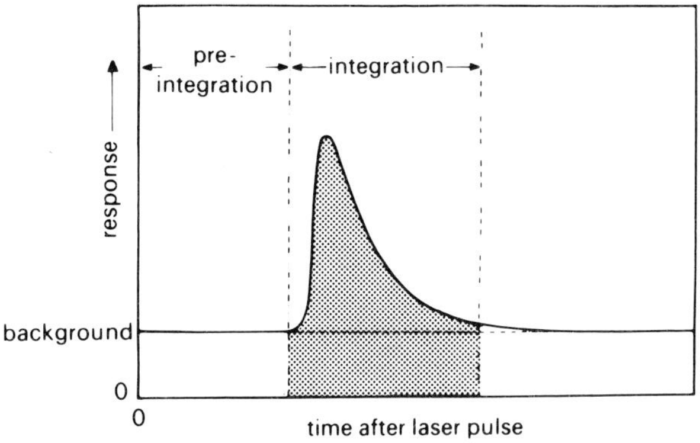

Figure 5 The time-resolved response of a single channel of an ICPAES after a discrete sample injection. The integration period shown (shaded area) includes most of the peak and as little of the background as possible. (Reproduced with permission from Thompson and Walsh (1983). Copyright Blackie)

A simpler system described by Thompson *et al.* (1981a) (Figure 6) employs a commercially-available laser microprobe. The laser is mounted in an otherwise-conventional high-power binocular microscope, which is used to locate the small objects (particles, inclusions, small crystals, etc.) for analysis. When the object is in focus at the cross-wires, the laser is fired through the microscope optics and volatilizes 1–5 μg of material from the surface, leaving a crater of about 200 μm diameter. The sample is enclosed in a small cell with an optical silica window and the volatilized material is swept into the plasma as a condensed aerosol. Signal quantification is carried out as previously described.

With fairly homogeneous samples (such as metallic alloys) consistent results are obtained and when allowance is made for the weight of sample ablated (by internal standardization) excellent calibrations could be obtained by means of reference materials. With heterogeneous materials such as soils (finely ground and pelletized) results are not completely satisfactory (Hale and Thompson, 1983) because of variations in composition of the ablated sub-samples. Nevertheless useful geochemical information can be obtained (op. cit.) and it seems likely that improvement could be obtained by the use of homogeneous fused beads of some kind. Other geochemical samples analysed by this system include manganese dioxide coatings on small pebbles, and individual grains from pan concentrates produced from stream sediments.

Calibration of this system has been facilitated by the discovery that, under controlled conditions, element sensitivity ratios established by nebulization of aqueous solutions are remarkably accurate for laser ablation also. Thus the need to prepare solid standards which closely simulate the physical properties of the samples is largely circumvented. Only the need for reasonable sample homogeneity remains, and this is readily tested by making replicate ablations on fresh portions of sample surface. Concentration detection limits for laser

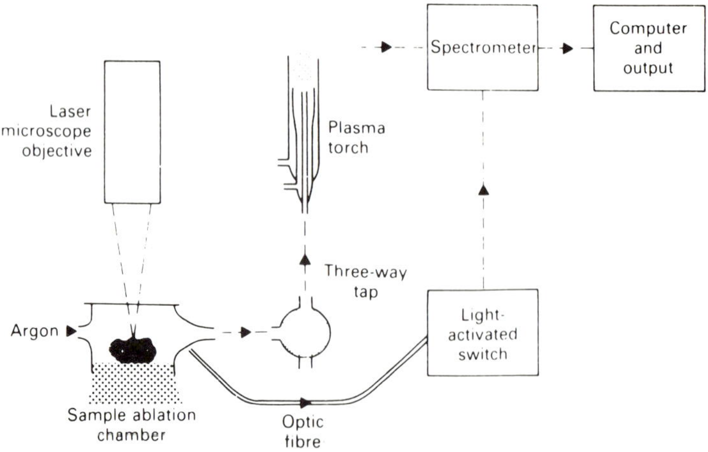

Figure 6 Schematic diagram of the laser/ICP microprobe assembly. Material volatilized from the surface of the sample rapidly condenses into a fine aerosol and is transported into the ICP. The measuring cycle of the ICPAES instrument is initiated by a light sensitive switch which is activated by scattered light from the laser pulse (Reproduced with permission from Thompson *et al.* (1981a). Copyright Royal Society of Chemistry)

ablation are not as low as those obtained by nebulization, because the total amount of sample injected is smaller. (Absolute detection limits are about the same for a particular element, regardless of the mode of injection.)

RAMP-HEATING AND DECREPITATION

These two methods of sample introduction are both being considered for use in geochemical exploration. Although they are conceptually very different, they share a common technique and illustrate very effectively the versatility of the ICPAES as a multi-element detector for a variety of physical processes. The simple equipment illustrated in Figure 7 is used to heat the sample up to 500–1000°C over the course of about two minutes, simply by inserting the cold sample tube into the furnace. Techniques for a linear rate of heating would undoubtedly improve the results, but are more difficult to achieve for a fast sample cycle time.

Ramp-heating

Many elements which fall in the right-hand side of the periodic table are sufficiently volatile (either as the element or as a compound) to be detected in the gas phase by ICPAES when samples containing them are heated at temperatures of up to 1000°C. Elements falling into this group include Zn, Cd, Hg, C, Pb, N, As, Sb, Bi, S, Se, Te and I, i.e. a number of ore-elements and pathfinders. If the ICPAES response for these elements is monitored continuously, elements liberated at different temperatures will appear during different stages of the heating, presumably during the decomposition or

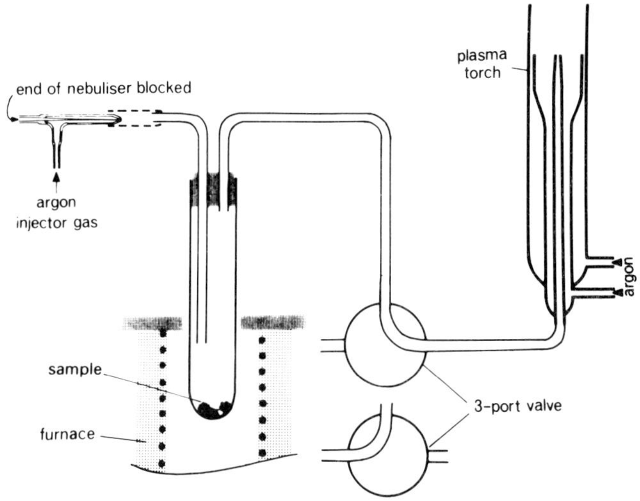

Figure 7 Equipment (schematic) for the ramp-heating or decrepitation method of introducing sample material into the ICP. The flow rate of injector gas is determined by the blocked nebulizer. The same nebulizer (and flow rate) is subsequently used to establish sensitivity ratios by nebulizing aqueous standards. (Reproduced with permission from Thompson and Walsh (1983). Copyright Blackie)

recrystallization of particular phases or species. Thus a degree of specificity is introduced into the analysis, and this may provide a more useful diagnostic tool in exploration than the total content of the element. In addition to this the method is very sensitive, because of the large sample weight (e.g. 0.1–0.5 g) and the high efficiency. Thus detection limits are below 1 $ng\,g^{-1}$ (ppb). The method has been relatively little explored, but may provide a useful extention to the vapour geochemistry methods for exploring in transported overburden.

Decrepitation

When crystals containing fluid inclusions are heated, the internal pressure generated is often sufficient to burst the crystal and eject the contents of the inclusion. If the heating is carried out in a stream of argon the fine ejecta can be swept into an ICP for analysis. By this method a number of elements not previously detected in the inclusions have been identified and approximately determined (Thompson *et al.*, 1980; Rankin *et al.*, 1982; Alderton *et al.*, 1982). It is thought that the passage of metalliferous fluids through granite rocks may result in fluid inclusions in the quartz which contain appreciable concentrations of ore metals. They can be identified by the decrepitation technique, and the rock thus differentiated from similar barren material. At this time, however, the technique is still relatively undeveloped, and the results correspondingly difficult to interpret.

Possible improvements in cost-effectiveness of ICPAES

Cost-effectiveness can be improved by two methods: decreasing the cost or increasing the effectiveness. Methods by which costs of ICPAES, both capital and recurrent, can be reduced are in the hands of instrument manufacturers and will be considered in the final section. Increasing the effectiveness of existing instruments, however, is a matter of increasing the number of samples analysed in a unit time, and is quite within the scope of the determined user.

UNIT TIMES FOR ANALYSIS

Figure 8 shows a breakdown of the time spent in analysing a single sample solution by ICPAES (conventional nebulization technique). The timings shown for the different stages vary widely from one laboratory to another, but are approximately average values at the time of writing. Each of these timings can be substantially reduced by the relatively easy means detailed below, providing a three-fold improvement in sample throughput.

1. Cleanout and signal stabilization time is the time elapsed from the withdrawal of one sample to achieving a stable signal from the next. If insufficient time is allowed for this, then there will be a degree of 'memory-effect' between samples which will be worse when a low-analyte sample follows one with much higher levels. This period can be minimized by attention to the design of the nebulizer and spray-chamber system. It is particularly important to avoid dead space where eddy currents can retain material from a previous sample. It is also important to minimize the volume of connecting tubing etc. by using the smallest possible diameter. By these means this period can be reduced to 15 s with no danger of memory effects.

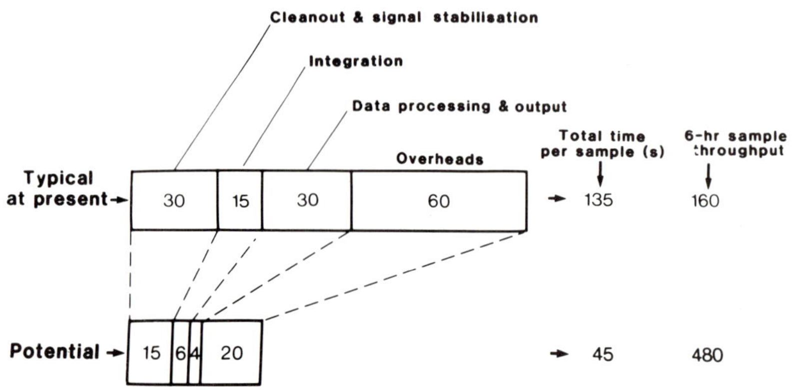

Figure 8 Diagram showing how in ICPAES the time required for the analysis of a single sample solution can be reduced substantially by attention to instrumental and operational details

2. Integration times. According to an analyst's elementary rule-of-thumb standard deviation of the measurement will be decreased in proportion to the square root of the integration time, i.e. longer integrations give better precision. This is *not* true in most practical analysis by ICPAES where most of the variation is attributable to medium-term drifts rather than short term noise. Reduction of the integration period to 2 s from 5 s is unlikely to produce a discernable worsening in overall precision.
3. Data processing and output. The computations needed for a multi-element analysis are accomplished within a few seconds, but usually the computation is followed by the production of a hard-copy of the results (often including intermediate stages in the calculation, relative standard deviations, etc.). Much time can be saved if the data is not presented in hard copy, but on a VDU terminal, with simultaneous storage on a disc system. The only data that the analyst really needs during the batch run are the various checks on accuracy, etc. The sample data can be printed or otherwise processed at the end of the run, at a slack time in the use of the instrument, or even on a different computer.
4. 'Overheads' consists of time when the instrument is necessarily running but not analysing samples. This includes periods for 'warming up' the instrument, system checks, calibration, accuracy checks, etc. This period cannot be substantially reduced, but if the remaining part of the shift can be used more effectively, the overhead time can be distributed over a larger number of samples.

AUTOMATION AND ARTIFICIAL INTELLIGENCE

The automation currently available on analytical instruments amounts to little more than a robot system feeding the test solutions into an instrument. There is no element of feedback where the system monitors its own performance and makes the necessary adjustments. Thus in ICPAES frequent (perhaps hourly) tests for drift need to be carried out by the operator and, if it is judged to be necessary, a secondary calibration is made to adjust the baseline levels and sensitivities (normalization). However there is nothing in this procedure which cannot be carried out by computer, and the amount of instrumentation and software development which would be required is relatively trivial. This would enable the ICP to run for long periods (e.g. overnight) without attention. Any rare problems encountered outside the scope of the software could be made to halt the analysis and activate an alarm for operator intervention. Clearly such a system could be extended to include other features, such as warning for abnormal conditions or samples, checks and adjustments for correction factors and a large part of routine data quality control. This type of 'expert system' would leave the chemist free to do what he is best employed for, i.e. to control the chemistry involved in preparing the test solutions from the raw samples.

Fundamental developments in ICP instrumentation

The above discussion centres on the possibilities of extending the use of existing ICPAES instrumentation to increase its efficiency and scope in relation to applied geochemistry. In concluding, this paper summarizes the possibilities for significant new developments in ICP hardware.

THE MINI-PLASMA

One possible method of producing a cheaper ICPAES instrument lies in the development of a mini-plasma. It has been demonstrated that plasmas which are 30–50% smaller than the conventional size can be equally effective in terms of analytical characteristics. The potential advantages of a smaller plasma are two-fold. Firstly, it will run at a lower RF power consumption, which may allow the use of a smaller, solid-state power supply, which will be cheaper. Secondly, it will require much less argon gas for running and hence will be noticeably cheaper to run. This is especially important in non-industrialized countries where argon is very expensive.

MORE STABLE HARDWARE

The precision and detection limits obtainable by ICPAES are currently limited by the stability of certain components, essentially those parts dealing with sample injection and the introduction of power to the sample. The main source of the trouble encountered in practice is the nebulizer. Much effort is being devoted to the production of a stable nebulizer. Another parameter affecting stability performance is the power delivered. This is currently stabilized to levels of about 0.2% relative (short term stability), but a further improvement would be welcome. A different approach to the improvement of precision consists of producing a stability enhancement by computer methods, by making use of a number of reference channels ('data massaging'). This corresponds roughly to the old idea of using internal standards, although the mathematical procedures used would be much more sophisticated. By means of such a combination of hardware and software developments it could be anticipated that ICPAES precisions will be reduced to about 0.1% relative within this decade, with corresponding reductions in detection limits and extentions of linear calibration range.

SPECTROMETER IMPROVEMENTS

Two improvements in spectrometers which would be welcomed by ICPAES users are (a) in resolution and (b) in cost. Enhanced resolution can be obtained in the echelle spectrometers, but the rectangular spectrum obtained is not very convenient for simultaneous analysis with photomultipliers. Image-dissecting vidicon tube detection systems have been tried but do not seem to be sensitive enough at present. An improvement in the latter method of detection would provide a remarkable enhancement of the capabilities of ICPAES, as it could allow, in principal, simultaneous access to all of the lines in the spectrum. This, in combination with other possible improvements outlined above (especially with real-time 'data massaging') would be even more effective.

A second possible substantial improvement may stem from the extension of Fourier transform spectrometry (FTS) into the visible and ultraviolet regions of the spectrum. FTS is well established as an infrared technique, but mechanical and computing difficulties have prevented its use at shorter wavelengths. This situation may now be changing, mainly because of the falling costs of computing power. Several groups are now working on such a spectrometer. Whether a workable system can be devised remains to be seen, but an FTS would substantially improve upon either light-gathering power or spectral resolution. In theory, it would also make every line in the spectrum available, which would greatly increase the power of ICPAES.

APPLICATIONS OF THE ICP OTHER THAN IN ATOMIC EMISSION

Two currently available techniques in an early stage of development that are based on the ICP deserve a mention here, viz.: (a) atomic fluorescence in the ICP (ICPAF); and (b) the use of the ICP as an ion source for mass spectrometry (ICPMS).

ICPAF

This simple technique does not require a spectrometer at all. Sample material is injected into the ICP in the normal fashion, and the plasma is illuminated by the light of a pulsed hollow-cathode lamp. Light from the plasma is sampled through a narrow bandpass filter by a photomultiplier tube, and the fluorescence light is separated from other emissions by electronic filtering. The plasma can be surrounded by a number of element-specific modules carrying the appropriate lamps, filters and photomultipliers. Advantages claimed for this device include freedom from spectral overlaps and a very stable baseline. Detection limits are better than those obtained by ICPAES for a few elements and somewhat worse for most others. However detection limits could be improved simply by increasing the intensity of illumination (Demers and Allemand, 1981).

ICPMS

A plasma as an ion source for mass spectrometry was pioneered by Gray (1975) and has recently been brought to a reliable operating state (Date and Gray, 1983). The technique of sampling ions from the ICP involves the use of a cooled probe with a pinhole (70 μm or 400 μm). The sampled ions pass into the mass spectrometer through several pressure reduction stages. The main advantages of the ICP over the conventional spark source for mass spectrometry are two-fold. Firstly, the spectra obtained are fairly simple, as relatively few elements give rise to a high proportion of doubly charged ions in the ICP. Secondly, sample changeover is as quick as in ICPAES, viz., about 1 sample per minute. Detection limits obtained are lower than ICPAES (usually by one or two orders of magnitude), and isotope ratios obtained compare very closely with spark source results. This exciting technique will clearly have a major impact on the use of isotope geochemistry in applied fields.

Conclusions

We have discussed here the capabilities of ICPAES, both as presently practiced and in the light of developments which can be foreseen occurring in the rest of the decade. Clearly, many improvements can be expected which are entirely related to progress in electronics and computing. This is largely in the hands of manufacturers of the instrumentation. However, there is a large amount of work that needs to be accomplished in respect of the chemical pretreatment of samples, and this is clearly a task to be accomplished by users of ICPAES equipment. One such enterprise is the development of multi-element preconcentration techniques referred to previously.

Despite the work which remains to be done, ICPAES emerges as a technique with extraordinary power and versatility which I believe will become a major part of geochemical practice. Rather ironically, but also excitingly, the development of sophisticated instrumentation for multi-element analysis has posed a greater challenge to the ingenuity of analytical chemists than ever before.

References

Abercrombie, F.N., Silvester, M.D., Murray, A.D. and Barringer, A.R. (1978). A new multi-element technique for the analysis of airborne particulates in air quality surveys. In: *Applications of Inductively Coupled Plasmas to Emission Spectroscopy* (R.M. Barns, Ed.) pp. 121–145; Franklin Institute Press, Philadelphia.

Alderton, D.M.H., Thompson, M., Rankin, A.H. and Chryssoulis, S.L. (1982) Developments in the ICP-linked decrepitation technique for the analysis of fluid inclusions in quartz, *Chem. Geol.*, **37**, 203–213.

Berman, S.S., McLaren, J.W. and Willie, S.N. (1980). Simultaneous determination of five trace metals in sea water by inductively coupled plasma atomic emission spectrometry with ultra-sonic nebulisation, *Anal. Chem.*, **52**, 488–492.

Church, S.E. (1981). Multielement analysis of fifty four geochemical reference samples using inductively coupled plasma atomic emission spectrometry, *Geostandards Newsletter*, **2**, 133–160.

Date, A.R. and Gray, A.L. (1983). Development progress in plasma source mass spectrometry, *Analyst*, **108**, 159–165.

Demers, D.R. and Allemand, C.D. (1981). Atomic fluorescence spectrometry with an inductively coupled plasma as an atomization cell and pulsed hollow cathode lamps for excitation, *Anal. Chem.*, **53**, 1915–1921.

Fuller, C.W., Hutton, R.C. and Preston,B. (1981). Comparison of flame, electrothermal and inductively coupled plasma atomisation techniques for the direct analysis of slurries, *Analyst*, **106**, 913–920.

Gray, A.L. (1975). Mass spectrometric analysis of solutions using an atmospheric pressure ion source, *Analyst*, **100**, 289–299.

Hale, M. and Thompson, M. (1983). Laser ablation-inductively coupled plasma spectrometry for geochemical exploration, *Inst. Min. Metall. Trans.*, **92,** B23–B27.

Hiraide, M., Ito, T., Baba, M., Kawaguchi, H. and Miziuke, A. (1980). Multielement preconcentration of trace heavy metals in water by coprecipitation and flotation with indium hydroxide for inductively coupled plasma atomic emission spectrometry, *Anal. Chem.*, **52**, 804–807.

McLeod, C.W., Otsuki, A., Okamoto, K., Haraguchi, H. and Fuwa, K. (1981). Simultaneous determination of trace metals in seawater using dithiocarbonate preconcentration and inductively coupled plasma emission spectrometry, *Analyst*, **106**, 419–428.

Motooka, J.M., Mosier, E.L., Stutley, S.J. and Viets, J.G. (1979). Induction coupled plasma determination of Ag, Au, Bi, Cd, Pb and Zn in geological materials using a selective extraction technique—preliminary investigation, *Appl. Spectrosc.*, **33**, 456–460.

Rankin, A.H., Alderton, D.H.M. and Thompson, M. (1982). Determination of uranium: carbon ratios in fluid inclusion decrepitates by inductively coupled plasma emission spectroscopy, *Min. Mag.*, **46**, 179–186.

Sturgeon, R.E., Berman, S.S., Desaulniers, J.A.H., Mykytiuk, A.P., McLaren, J.W. and Russell, D.S. (1980). Comparison of methods for the determination of trace elements in seawater, *Anal. Chem.*, **52**, 1585–1588.

Thelin, B. (1981). Nebuliser system for the analysis of high salt-content solutions with an inductively coupled plasma, *Analyst*, **106**, 54–59.

Thompson, M., Pahlavanpour, B., Walton, S.J. and Kirkbright, G.F. (1978). Simultaneous determination of trace concentrations of arsenic, antimony, bismuth, selenium and tellurium in aqueous solution by introduction of the gaseous hydrides into an inductively coupled plasma source for emission spectrometry, *Analyst*, **103**, 568–579 and 705–713.

Thompson, M., Rankin, A.H., Walton, S.J., Halls, C., and Foo, B.N. (1980). The analysis of fluid inclusion decrepitate by inductively coupled plasma atomic emission spectroscopy: an exploratory study, *Chem. Geol.*, **30**, 121–133.

Thompson, M., Goulter, J.E. and Sieper, F. (1981a). Laser ablation for the introduction of solid samples into an inductively coupled plasma for atomic emission spectrometry, *Analyst*, **106**, 32–39.

Thompson, M., Pahlavanpour, B. and Thorne, L.T. (1981b). The simultaneous determination of As, Sb, Bi, Se and Te in waters by an inductively coupled plasma volatile hydride method, *Water Research*, **15**, 407–411.

Thompson, M., Ramsey, M.H. and Pahlavanpour, B. (1982). Water analysis by inductively coupled plasma atomic emission spectrometry after a rapid preconcentration, *Analyst*, **107**, 1330–1334.

Thompson, M. and Walsh, J.N. (1983). *A Handbook of Inductively Coupled Plasma Spectrometry*; Blackie, Glasgow.

Walsh, J.N., Buckley, F. and Barker, J. (1981). The simultaneous determination of the rare earth elements in rocks using inductively coupled plasma source spectrometry, *Chem. Geol.*, **33**, 141–153.

Exploration Geochemistry in the Shallow Marine Environment

P. A. SMITH

Introduction

Exploration geochemistry waded offshore in the mid-1960s at a time when a growing awareness that novel sources of minerals were required to meet the potential resource demands of the future coincided with the realization that the previously unexplored ocean floor could provide attractive prospects for a wide range of mineral resources (Mero, 1965). Accordingly, groups such as the Applied Geochemistry Research Group at Imperial College, under the direction of John Webb, and Professor Robby Moore's group at the University of Wisconsin (and later at Alaska and Texas) began to test the well-established geochemical techniques and philosophy (see, e.g. Hawkes and Webb, 1962; Levinson, 1974) developed for land exploration and to thereby modify or develop new techniques appropriate for marine mineral exploration.

At the University of Wisconsin investigations were carried out on bedrock copper mineralization under Lake Superior (and although not marine, the problems are identical) (Smith and Moore, 1972; Goodden, 1974; Nebrija *et al.,* 1977) and platinum placer exploration off Alaska (Moore and Welkie, 1976; Owen, 1978, 1980).

Cassiterite in Mounts Bay in Cornwall was the AGRG's first offshore target (Tooms *et al.*, 1965; Ong, 1966; Hazelhoff Roelfzema, 1968) and at a later date placer tin exploration was extended to include north Cornwall (Yim, 1979). In the meantime, prospecting began for phosphorites in the deeper waters of the continental shelf off north-west Africa (Summerhayes, 1970; McArthur, 1974) but dived even further into deeper water for a while (manganese nodules, metal-rich sediments—see Cronan, this volume) until interest in shallow water exploration revived with an investigation into the application of geochemical techniques to explore the continental shelf of the Aegean Sea (Smith, 1979).

Applied geochemical research of this type falls into two main categories (Figure 1). Firstly, the development of geochemical surveying techniques appropriate to the marine environment to enable significant anomalies (i.e. those related to mineralization) to be recognized and dispersion pathways (or concentration gradients) delineated to pin-point the target mineralization.

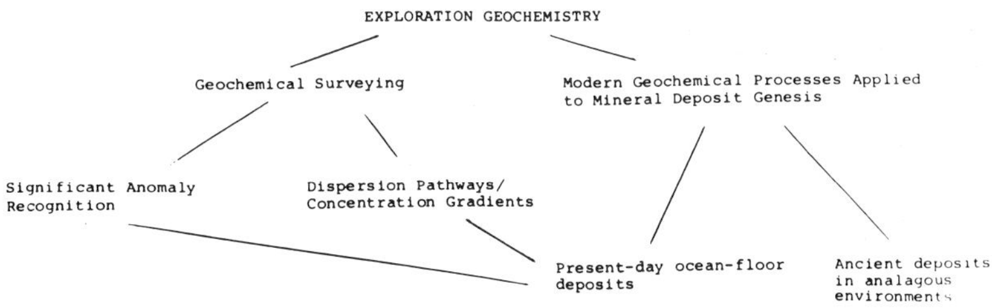

Figure 1 Exploration geochemistry in the shallow marine environment

Secondly, studies of mineral deposit genesis which are relevant not only to our understanding of the formation of present day ocean floor deposits but also to ancient deposits and associated phenomena found in analogous environments. In practice both areas of research are often part of the same investigation, particularly in the case of placer and phosphorite exploration, because our knowledge of both the most effective exploration techniques and deposit genesis is still in an embryonic stage.

In this chapter, examples of efforts made to date to apply geochemical techniques to the exploration for offshore mineral deposits are reviewed to illustrate possible techniques and strategies and, at the same time, highlight some of the problems encountered in exploring a very complex environment. Additionally, to illustrate exploration geochemical research with more of a genetic bias, the results are described of a study of shallow submarine hydrothermal activity and the formation of Fe, Mn and trace metal zoning.

Geochemical surveying

Table 1 lists the major types of mineral deposit to be found in the shallow marine environment which, for the purposes of this paper, range from the estuary to the shelf-slope boundary. Geochemical techniques are not equally appropriate to the exploration for all types of deposit and consequently an estimate is given in Table 1 of its relative importance. The reasoning behind each grade will become apparent from the discussion to follow. In sand, gravel and diamond prospecting it is difficult to envisage a role for geochemical techniques except, perhaps, as an adjunct to sediment mapping or sediment dispersal modelling.

PLACER EXPLORATION

Exploration for placers is often less a geochemical and more of a sedimentological and mineralogical problem. This is particularly true for heavy mineral deposits comprising minerals such as ilmenite, rutile, magnetite, zircon and monazite. In the case of cassiterite or chromite, however, exploration geochemical techniques are more relevant simply because it is easier to analyse chemically for Sn and Cr rather than determine

Table 1 **Mineral deposits in the shallow marine environment**

Type of deposit	Geochemistry as an exploration tool
Placers	
Heavy mineral sands	
Ilmenite, rutile, magnetite, zircon, monazite	*
Cassiterite, chromite	**
Precious metals (Au, Pt)	**
Gemstones (diamonds)	—
Bedrock mineralization (e.g. Cu, Pb, Sn)	**
Phosphorites	*
Aggregates (sand and gravel, shell sands)	—

*—geochemical techniques used mainly for mapping trends in concentration
**—all geochemical techniques are relevant

the abundance of cassiterite and chromite mineralogically. Tin (and Cr in detrital particles) is not very reactive chemically in natural environments and so modern geochemical processes are not as critical in influencing abundance as they tend to be in the case of Cu and Zn, for example. Tin concentration data are used, therefore, in the same way as mineralogical data to map concentration gradients and thereby to infer physical processes of selective dispersion and concentration.

Yim (1978) describes a geochemical exploration programme for tin placers in St. Ives Bay, Cornwall, which derive from the unusual source of mine tailings carried to the nearshore environment by streams draining the adjacent Camborne–Redruth tin mining district. On the basis of analyses carried out on different size fractions of grab samples (on a shell-free basis), the bulk of the tin was found to occur in the fine sand and silt, consistent with an origin from mine tailings in which tin losses occur in the fine fraction during the mineral processing of cassiterite mined onshore. Highest concentrations of tin were discovered off the mouths of the main rivers draining into the bay (Figure 2) but, interestingly, higher concentrations were also found just beyond the surf zone. Yim attributes this to a decrease in energy level allowing the selective concentration of cassiterite winnowed from the surf zone and suggests that this may be an attractive target zone in future exploration.

In gold and platinum placer exploration direct determination and mapping the distribution in the early stages of precious metal placer exploration is a

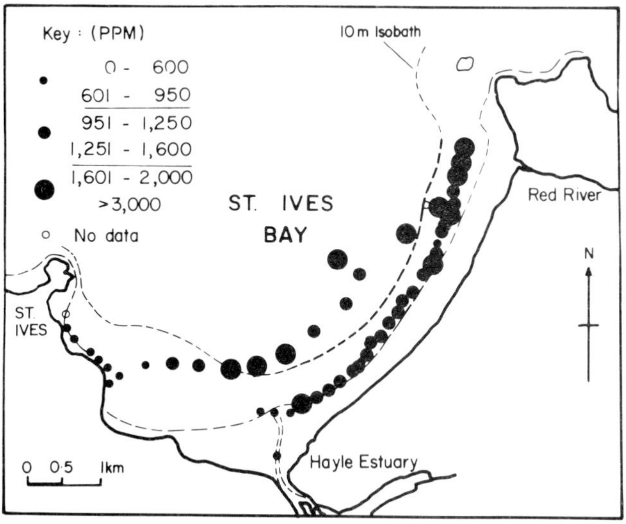

Figure 2 Distribution of tin oxide (ppm) in the -2057 μm fraction of the sediments (modified from Yim, 1979)

very costly approach because firstly, precious metals tend to be finely dispersed throughout the sediments and therefore representative samples have to be very large and difficult to analyse; and secondly, haloes of decreasing concentration gradients are often limited in extent or absent, a constraint which necessitates a very tight sampling grid to allow any chance of detecting the placer deposit. Consequently, indirect exploration methods are being developed using geochemical tracer techniques (Owen, 1978) which fully utilize geostatistical techniques to identify depositional environments suitable for placer formation (Moore and Welkie, 1976; Owen, 1978), trace sediment dispersal pathways (Owen, 1980), and identify suitable tracer elements for predicting anomalous concentrations of Pt, for example, with reasonable success (Owen, 1978). Each of these approaches is briefly reviewed below and although they were developed as part of an offshore exploration programme for platinum along the coast of the Bering Sea near Goodnews Bay, Alaska (Figure 3), where onshore alluvial platinum placers have been mined since 1934 (Moore and Welkie, 1976), they have a much broader application and are probably useful in most marine mineral exploration projects.

Exploration window

Exploration windows are defined by Moore and Welkie (1976) as a combination of geochemical and textural parameters, determined by regression analysis, which identify depositional environments where conditions are favourable for the formation of noble metal placers. The Pt placer 'window' comprises statistically significant correlations between Mn,

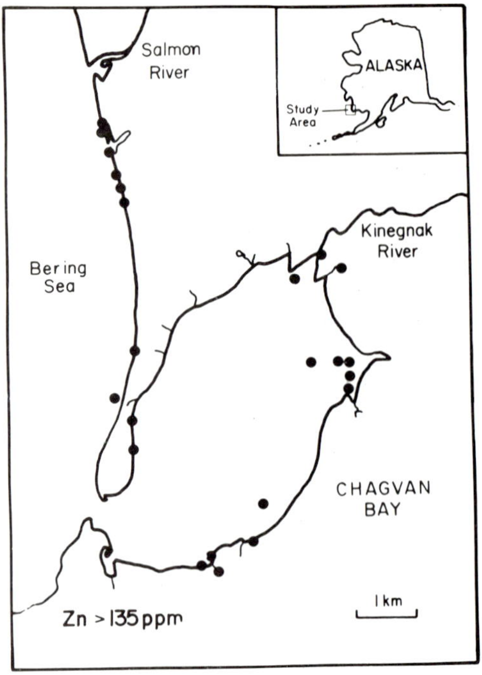

Figure 3 Distribution of samples having Zn>135 ppm, Chagvan Bay, Alaska and which have a 66.7% probability of containing anomalous Pt (modified from Owen, 1978)

Zn, Cu, Ni, V and the coarse-grained size fraction (particles 1.0 mm in diameter or greater) of the sediments. Treated in isolation such an approach has certain limitations; such geochemical associations are not unique to a given depositional environment and can be characteristic of a low energy environment, for example, where manganese scavenging is significant, or in the other extreme to an high energy environment where such an association derives from occurrence in the same minerals or rock fragments (dunite in this case). Significant in this instance is the high correlation with the coarser grain size which would appear to rule out the former option, i.e. a low-energy environment. In support of the exploration window concept, Owen (1978) discovered that in an R-mode principal components analysis of data from the same area, the first principal component in samples with background values of Pt comprised a geochemical association very similar to that proposed for the exploration window for Pt described by Moore and Welkie but importantly this component lacked a significant loading on a textural variable. Analysis of the anomalous samples, in contrast, revealed a first principal component with the same geochemical association as the background samples but, in addition, a strong negative loading on a textural parameter, the percentage of silt and clay. A combined geochemical–textural association is critical, therefore, in defining the Pt exploration window. A correlation with the coarse-grain sizes suggests furthermore, that the exploration window for Pt is more applicable to the search for high-energy placers (Owen, 1978).

Low-energy noble metal placers (Moore, 1972) may occur in environments such as Chagvan Bay (Figure 3) where fine-grained particles accumulate (see later). In the same principal components analysis of anomalous Pt samples, a second component with a significant loading on Pt, Co, Cu and Zn, plus a positive loading on silt and clay, is probably related to low-energy Pt accumulation. This suggests that the exploration window for low-energy Pt placers would comprise a Co, Cu and Zn geochemical association combined with a fine-grained texture.

The geostatistical analysis of the anomalous samples could have been carried a stage further by determining 'factor scores' which are a measure of the significance of a given factor (or principal component) on individual samples. Mapping of the factor scores for the high-energy and low-energy factors of the anomalous sample population would reveal their areal distribution and thereby help to elucidate dispersal pathways and pinpoint zones of accumulation.

A similar approach was used by Smith (1979) to map the significant underlying factors controlling the geochemistry of sediments off the island of Kithnos in the Aegean Sea. Although a different context, the principle is the same and serves to illustrate the usefulness of the technique. R-mode factor analysis of 10 variables produces a 2 factor model accounting for 71.1% of the original data variance (Table 2). Factor 1 is statistically the most dominant and is characterized by a strong association between the detrital elements Al, Fe, Ti, Cr, Zn and Ni, and an inverse loading on biogenous Mg. Factor scores

Table 2 **Rotated factor matrix for bulk chemical analyses of sediment samples from Kithnos, Aegean Sea (Smith, 1979)**

Variable	Factor		Communality
	1	2	
Fe	0.86	0.35	0.864
Al	0.90	0.35	0.936
Ti	0.84	—	0.751
Mg	-0.66	0.56	0.746
Mn	—	0.90	0.854
Cu	—	0.86	0.796
Zn	0.64	0.50	0.659
Ni	0.64	0.40	0.570
Cr	0.80	—	0.651
Pb	—	0.49	0.279

Cumulative percentage of original data variance accounted for:
Factor 1—52.7; Factor 2—71.1
Factor loadings between ±0.30 are omitted

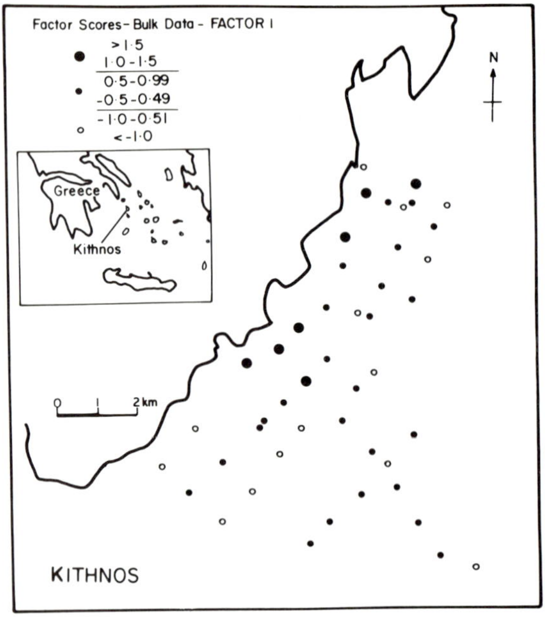

Figure 4 Distribution of factor scores of Kithnos, Aegean Sea—Factor 1 (predominantly detrital)

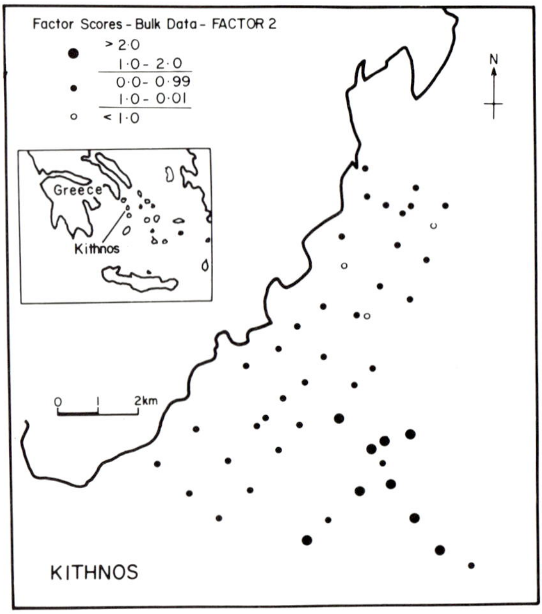

Figure 5 Distribution of factor scores off Kithnos, Aegean Sea—Factor 2 (predominantly manganese scavenging)

for this 'detrital' factor are highest in the nearshore sand lobes derived from modern terrigenous sedimentation (Figure 4). Factor 2 comprises a very strong association between Mn and Cu, with lessor loading on Mg, Zn, Ni, Fe and Al. Factor scores are highest in the offshore zone (Figure 5) where the sediment accumulation rates are lower and the scavenging action of Mn 'oxides' more significant (Mn–Cu association). The loading on Mg, Zn, Ni, Fe and Al in this factor is attributed to the presence of the chlorite-mica group of minerals which are also abundant in the offshore muds.

Zinc as geochemical tracer for platinum

To analyse for noble metals such as Pt in sample numbers adequate for exploration purposes is a time-consuming and costly process, but if it can be shown that mapping the distribution of a more readily analysed element can predict the distribution of Pt, i.e. a geochemical tracer, then the cost-effectiveness of the survey will be markedly improved. As an extension of the principal component analysis described earlier, Owen noticed that Zn was associated with Pt in both the high-energy and low-energy components of the anomalous group. On the basis of this dual Zn–Pt association, he postulated that Zn may serve as a geochemical tracer for Pt in thirds of the selected samples would contain Pt above background levels. Therefore, charting Zn data above the threshold value is an effective means of predicting zones of Pt enrichment, as illustrated in Figure 3. This map reveals two main zones of potential Pt accumulation—beach placers downdrift from the Salmon River, and fine-grained Pt derived from the Kinegnak River and other streams draining into Chagvan Bay.

Sediment dispersal modelling

Modelling sediment dispersal pathways using geochemical parameters has universal application in marine mineral exploration and one approach is to map factor scores in a similar way to that described above for Kithnos. Further studies based on the Chagvan Bay area (Owen, 1980) also illustrate how mathematical modelling techniques based on vector analysis and/or multiple regression analysis of geochemical data can help determine provenance and delineate dispersal patterns in modern sediments.

Chagvan Bay deposits derive from the three major sources: Kinegnak River and other streams, the bluffs surrounding the bay, and the littoral zone of the Bering Sea. The first task is to define sediment tracers which distinguish between each of these source areas but which also meet two additional criteria: the tracer must not undergo significant physical sorting during sediment transport, and it must be chemically conservative. In the area described by Owen five parameters were selected as sediment tracers: mean grain size, Zn, Ni, Fe and V.

In the vector analysis model each of the source areas is described by a five dimensional (corresponding to the five tracers) reference vector. The degree of compositional similarity between the bay samples and the three source areas

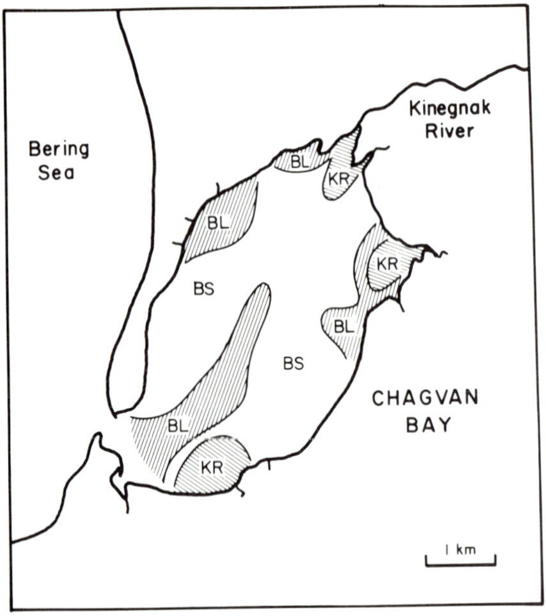

Figure 6 Dispersal pattern of source area sediments as determined by the vector analysis model (BS—Bering Sea sediment; KR—Kinegnak River sediment; BL—Bluff sediment) (from Owen, 1978)

is determined by calculating 'the coefficient of proportional similarity' (Imbrie and Purdy, 1962) which is the cosine of the angle between each bay sample vector and each of the three reference vectors. In this way it is possible to assess which of the three source areas is the dominant contributor of sedimentary material to a particular sample location. A map based on the vector analysis model (Figure 6) shows the broad pattern of sediment dispersal in Chagvan Bay.

According to Owen, the multiple regression model assumes that any sample is composed of a linear combination of sedimentary material from each of the three source areas. For example, the zinc concentration of any bay sample can be represented by:

$$S(Zn) = k_1\,(Zn)_1 + k_2\,(Zn)_2 + k_3\,(Zn)_3$$

Where S(Zn)—the zinc concentration in the sample; $(Zn)_1$, $(Zn)_2$ and $(Zn)_3$—the mean concentrations for each of the three sources areas; and k_1, k_2 and k_3—constants whose magnitude reflects the relative contribution of each source area. Similar equations can be written for each of the four remaining sediment tracers, so that a set of five equations for three unknowns (k_1, etc.) are developed and which can be solved by least squares regression analysis. Figure 7 illustrates the contour map of the present Kinegnak River sediments. Similar maps presented by Owen (1980) show the distribution of sediments derived from the remaining two source areas—the bluffs and the Bering Sea.

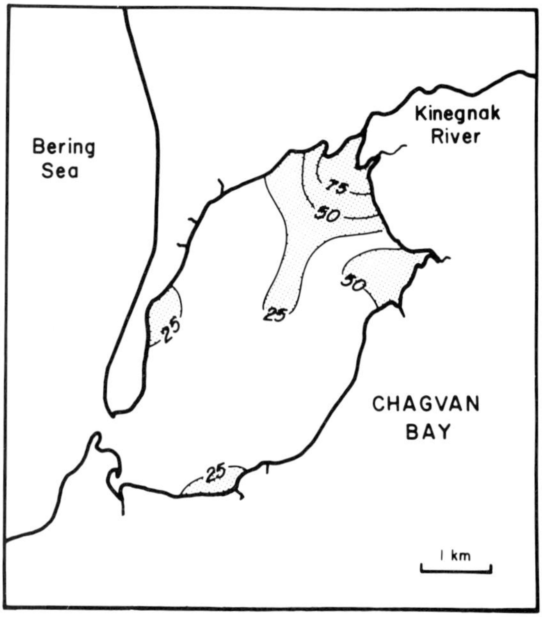

Figure 7 Contour map of percent of Kinegnak River sediments as determined by the multiple regression model (from Owen, 1978)

Both models give very similar results; the dominant contributor of sedimentary material to Chagvan Bay is the Bering Sea, and coarser sediments from the bluffs are found in the tidal channel and around the margins of the bay, whereas Kinegnak River sediments have a localized distribution at the mouth of the river.

BEDROCK MINERALIZATION

Bedrock mineralization continues to be a tantalizing prospect offshore but efforts to explore for such deposits have been tentative and will probably remain so until the first major discovery provides the necessary momentum. As far as the author is aware, no mineral deposits of this type have yet been discovered using offshore prospecting as the primary means of exploration; the known sub-seafloor deposits (e.g. Levant tin mine off north Cornwall) are simply extensions of onshore mines discovered by tracing the lodes seawards. Research into geochemical prospecting techniques for bedrock mineralization is still in an embryonic stage and as yet has not progressed much beyond an awareness of the complexity of the problem. The examples described below for tin, copper and lead/zinc exploration serve to illustrate progress to date and to highlight the main problem areas.

Tin, Mounts Bay, Cornwall

One of the first attempts to apply geochemical techniques to offshore lode exploration was in Mounts Bay, off the south coast of Cornwall, in search of tin (cassiterite) deposits (Tooms *et al.*, 1965). Tooms *et al.* discovered that the principal source of the tin in Mounts Bay sediments was streams draining into the bay—in the fine fraction of the sediments, the amount of tin decreased with increasing depth and distance from the shore. At first sight, therefore, the situation appears to be similar to that described earlier for tin in St. Ives Bay (Yim, 1979). Further work (Tooms, 1969; Hazelhoff Roelfzema, 1968), however, indicated that the distribution of tin was not quite so simple. A cubic trend surface described the decrease in tin with increasing distance from shore reasonably well, but anomalies on the trend surface were also discovered which could have been related to local sources of tin, including offshore lodes (Tooms, 1969).

As a general point, trend surface analysis is a very effective approach for anomaly recognition in marine mineral exploration. After a certain amount of trial and error, a low-order polynomial expression can usually be found which fits the data reasonably well and which also may be considered to reflect, in certain circumstances, regional trends in the data (as in the case of tin above). In a given sample, therefore, the predicted regional trend value based on the polynomial expression is subtracted from the total value and significant differences (anomalies) mapped. In this way regional trends are removed from the data which might otherwise mask significant anomalies.

Copper, Lake Superior

Copper has been mined on the Keweenaw Peninsula, on the southern shore of Lake Superior (Figure 8) since about 1843 and in the early 1900s it was one of the world's major copper producing areas. Since then onshore copper production has declined but the copper-bearing rocks are believed to extend offshore along the north shore of the eastern tip of the peninsula. Prospects for new deposits of copper in the sub-lake extensions of the Keweenawan rocks are good (Smith and Moore, 1972; Goodden, 1974) and accordingly multi-technique surveys, including geochemical sampling, have been carried out to explore for the deposits (Smith and Moore, 1972; Goodden, 1974; Nebrija *et al.*, 1976; Nebrija *et al.*, 1977). Indeed, this programme is an excellent example of the application of combined geophysical, sedimentological, mineralogical and geochemical studies in marine prospecting; each of these alone will not suffice in the marine environment.

The Nonesuch shale is one such potentially copper-bearing unit which extends offshore and Nebrija *et al.* (1977) describe a multi-technique survey used to explore its extension under Lake Superior. For geochemical and sedimentological studies, samples were collected along a series of traverses approximately at right angles to the shoreline. Results of the geochemical analysis in relation to the bedrock geology are illustrated in Figure 8, and these reveal a significant enrichment of copper in the sediments over the

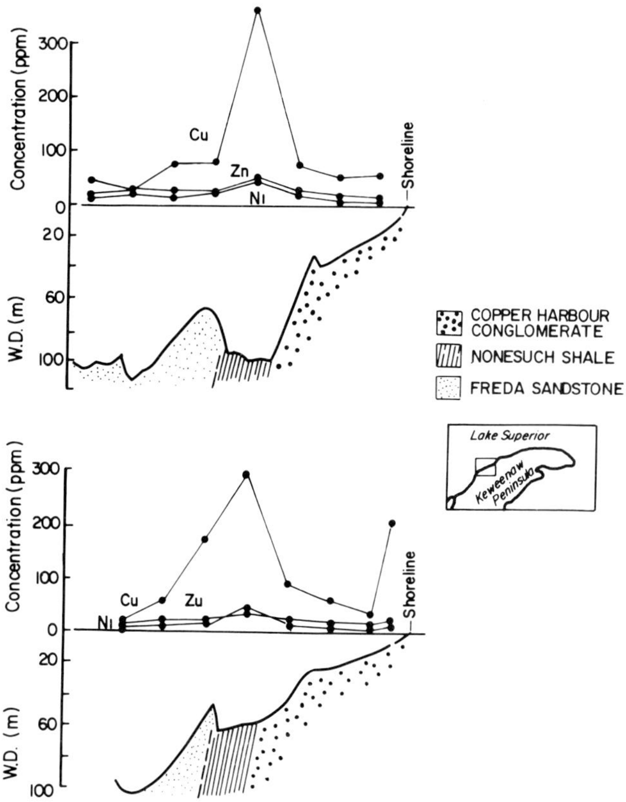

Figure 8 Profiles of concentration of Cu, Ni and Zn in relation to geology and depth along traverses perpendicular to the shoreline off the northwestern coast of the Keweenaw Peninsula, Lake Superior (modified from Nebrija *et al.*, 1977)

Nonesuch shale. After statistical analyses (correlation coefficients) of the geochemical and textural data, and mineralogical studies of the heavy mineral fraction, Nebrija *et al.* conclude that the copper is probably associated with oxide coatings (high correlation with Fe, Mn) on sand grains. The copper enrichment in sediments over the Nonesuch shale they attribute to either deposition from streams draining the mineralized areas onshore or adsorption on to Fe–Mn oxyhydroxides of dissolved Cu derived from the Nonesuch shale and transported upward by percolating groundwater in an area of increased permeability due to faulting. If the latter case could be shown to be the more likely then this would be very significant in that it would suggest that the Nonesuch shale is mineralized under Lake Superior.

Lead and Silver, Laurium, Greece

Lead, silver and zinc sulphides have been mined at Laurium, on the southern tip of the Attica Peninsula, (Figure 9) since ancient times. Exactly when

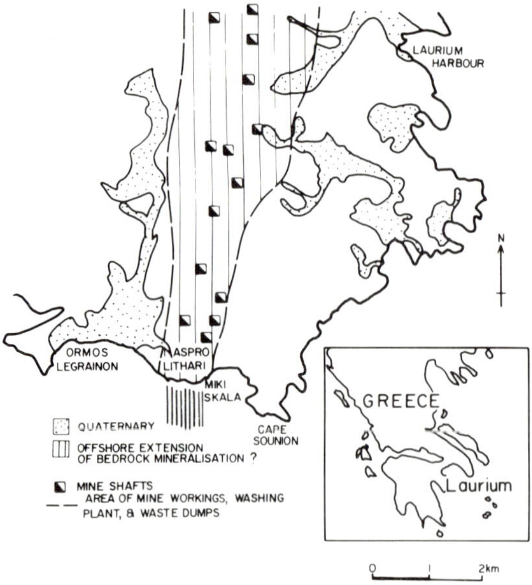

Figure 9 The Laurium mining district, Greece, showing the main area of mine workings and waste dumps (modified from Marinos and Petrascheck, 1956)

mining began is not known but it was probably before 1000 BC. The peak of activity, however, coincided with the development of the Athenian state (483 BC and onwards) and at this time the mines were of great importance—silver from Laurium is reputed to have paid for the fleet used to defeat the Persians. Eventually the rich deposits of silver-bearing galena were exhausted and so mining declined and stagnated during the middle ages. Nevertheless, large reserves of poorer ore remained and in the latter half of the nineteenth century mining was revived (Marinos and Petrascheck, 1956).

Testimony to the remarkable skills of the ancient miners are the 1000 or more mine shafts, a large number of open-cuts, and the remains of extensive metallurgical plants, all of which are still evident today throughout the Laurium district. Unfortunately, a further legacy of this intensive activity are the huge waste dumps which, in combination with additional waste from modern operations, provide an abundant source of potentially toxic material for subsequent wind or water-borne transport to the adjacent marine environment, particularly where dumps occur close to the shoreline (e.g. Mikri Skala and Laurium Harbour, Figure 9).

According to Marinos and Petrascheck (1956) the deposits trend north–south and could well extend offshore to the south of Asprolithari (Figure 9). This trend is well illustrated by the distribution of old mine workings (Figure 9). Because of the potential occurrence of ore deposits offshore, the AGRG carried out a reconnaissance survey of the Sounion area in 1972 and 1973 (Smith, 1979) in which both offshore grab and beach samples were collected to assess the general levels of Pb and Zn in the sediments and to

Table 3 **Average and range of the partial chemical composition of bioclastic sands from Kithnos, Naxos and Sounion, Aegean Sea (from Smith, 1979)**

	Kithnos (16)	Naxos (22)	Sounion (23)
Ca (%)	29.7	30.1	29.2
	27.1–33.9	27.4–33.7	27.3–32.0
Fe (%)	1.10	0.65	0.96
	0.40–1.60	0.30–1.10	0.37–1.40
Al (%)	1.00	0.98	0.81
	0.50–1.80	0.53–1.91	0.50–1.11
Mg (%)	2.04	1.81	1.59
	1.75–2.30	1.45–2.15	1.18–1.99
Ti (%)	0.08	0.09	0.08
	0.04–0.14	0.05–0.17	0.05–0.15
Mn (ppm)	290	124	235
	120–400	75–215	132–310
Cu (ppm)	12	10	11
	10–15	7–16	8–16
Zn (ppm)	32	24	88
	21–49	13–57	36–227
Ni (ppm)	17	12	18
	5–25	5–25	5–30
Cr (ppm)	38	20	32
	5–75	5–45	5–66
Pb (ppm)	26	23	112
	15–40	10–35	55–285

Number of samples given in parentheses

search for anomalies related to mineralization cropping out on the seafloor.

According to Smith (1979), most Sounion samples are bioclastic sands rich in $CaCO_3$ and undoubtedly the most significant differences between the average composition of this group and their counterparts elsewhere (Table 3) are the much greater concentration of Pb and Zn. On average Pb is five times, and Zn three times, more abundant, whereas the concentration of elements such Cu, Ni and Cr are very similar.

The distribution of Pb and Zn in the beach and offshore samples is illustrated in Figures 10 and 11, respectively and the pervasive impact of mine waste on the geochemistry of the sediments is immediately evident. One beach sample has 1% of combined Pb and Zn but in general they average about 400 ppm Pb and 400 ppm Zn (Smith, 1979). Highest levels of Pb and Zn in the offshore deposits occur off Laurium Harbour (1340 ppm Zn and 1540 ppm Pb), but in the main part of the survey area the most distinctive feature of the distribution of both Zn and Pb is the gradual decrease in concentration with

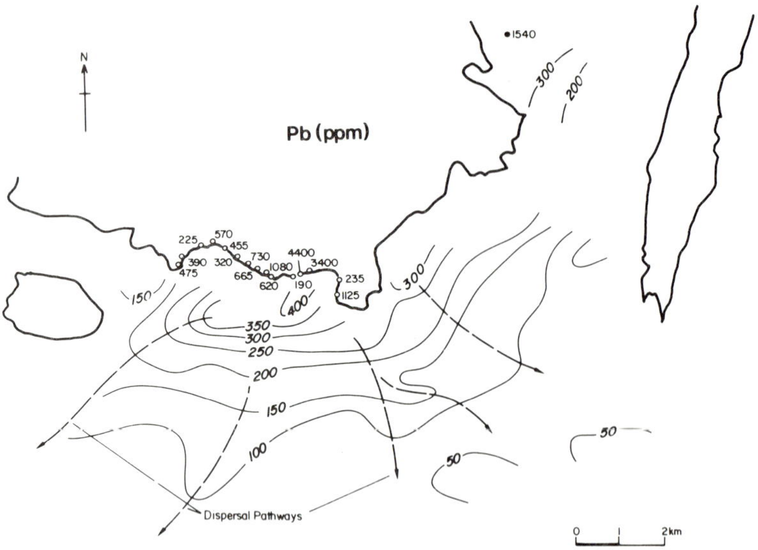

Figure 10 Distribution of total Pb (ppm) in beach and offshore samples adjacent to the Laurium mining district, Greece. The arrows indicate possible dispersion pathways

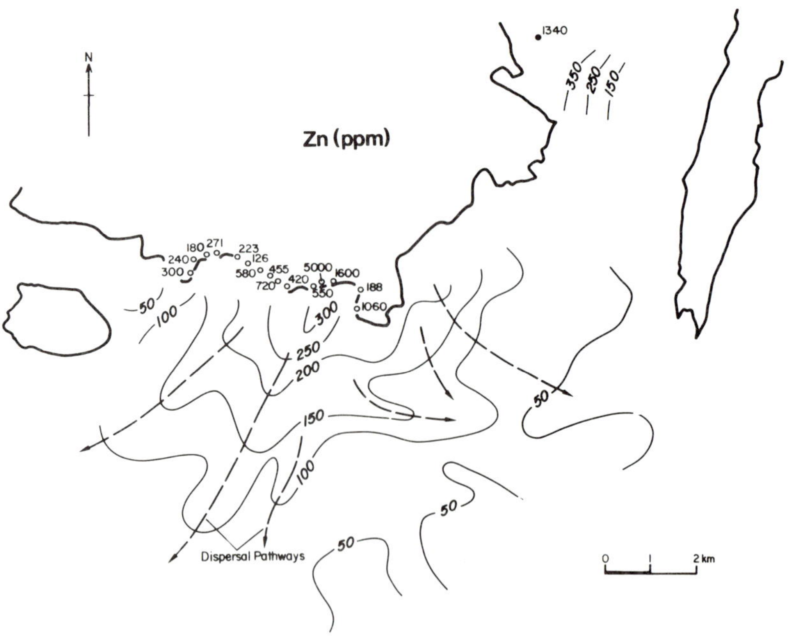

Figure 11 Distribution of total Zn (ppm) in beach and offshore samples off Laurium, Greece. The arrows indicate possible dispersion pathways

distance away from the shoreline. Off Asprolithari (Figure 9), Zn and Pb reach 322 ppm and 425 ppm but decrease to less than 40 ppm and 50 ppm respectively on the shelf edge. Such concentration gradients are characteristic of 'modern' terrigenous sedimentation. Superimposed on the overall trend is the movement of Pb and Zn, as part of the 'modern' terrigenous component of the sediment, along preferred dispersion pathways. In the western sector, in particular, lobes in the concentration isoline (e.g. 100 ppm Zn—Figure 11) coincide with valleys in the shelf edge topography (Smith, 1979). Some possible dispersion pathways are illustrated in Figures 10 and 11.

Trends in the distribution of Pb and Zn offshore and alongshore confirm that a major source of the waste material is the area around Asprolithari and Mikri Skala. This is not surprising: the area has a long history of mining activity and, more recently, construction work close to the shoreline. How the material reaches the marine environment is difficult to establish in this essentially 'static' study but the limited scale of stream input into the nearshore zone implies that windborne transport may be quantitatively significant process, especially during dry periods. In addition, it would not be surprising to discover that material also derived from direct dumping on to the beach or into the nearshore zone, either as part of the extensive construction work or by the deliberate marine disposal of mine tailings.

Although samples from closer to the shoreline have the highest concentrations, those from deeper water, particularly in the south-western sector, are also enriched in Pb and Zn (see Figures 10 and 11). (Note: the maximum values found in such deposits elsewhere are 75 ppm Zn and 45 ppm Pb—Smith, 1979). Only a few samples from the south-eastern corner have concentrations of Pb and Zn similar to those found in other parts of the Aegean. Therefore, it appears that the waste material can reach even the deepest parts of the continental margin.

Whether or not a pollution problem exists is difficult to assess simply from the composition of the sediments. In a comprehensive review of the problems involved in identifying trace metal pollution in coastal waters, Phillips (1977) points out '. . . it is conceivable that even in a situation where metal concentrations in sediments are very high, little metal is available to the biota . . . '. To take an extreme case, metals easily chelatable are likely to be more available than the same metals held within the detrital silicates. Obviously bulk analyses will shed little light on this problem.

Useful information may be gleaned from the sediments, however, by leaching them with a weak, organic chelating agent and although any chemical reagent is limited in its capacity to predict 'availability', at least this approach reveals the degree to which metals are bound in the sediment and improves on bulk analysis as a means of predicting 'availability'. Accordingly, selected samples were leached with 0.05M EDTA (Ethylenediaminetetracetic acid) for five minutes, filtered and the solutions analysed for Pb, Zn and Cu (Smith, 1979).

The results (summarized in Figure 12), reveal some interesting features:

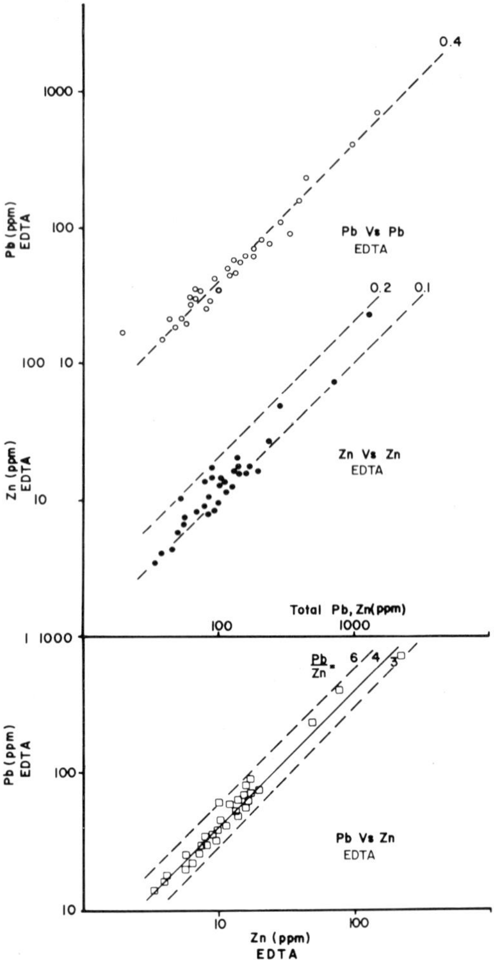

Figure 12 Relationships between the EDTA soluble and total element concentrations for Pb and Zn in sediments off Laurium, Greece

1. 40% of total Pb (Pb_{tot}) but only 10% of the total Zn (Zn_{tot}) is soluble in EDTA ($Zn_{EDTA} = 0.1\ Zn_{tot}$; $Pb_{EDTA} = 0.4\ Pb_{tot}$).
2. Pb_{EDTA} and Zn_{EDTA} covary, indicating that both occur in the same component (i.e. the mine waste), but the amount of Pb available is approximately four times the amount of Zn (i.e. $Pb_{EDTA} = 4\ Zn_{EDTA}$), even though total Pb is only 1.5 times total Zn.
3. As expected, the amount of available Cu is exceedingly small. Consequently, a potential pollution problem exists off Sounion, not because of the high total levels of Pb and Zn but because a substantial proportion of the Pb is readily extractable from the sediment and may therefore be in a form suitable for uptake by marine biota. Conversely Zn may be less of a problem.

At Sounion, analysis of the sediment only indicates that a problem may exist and the next step must be to sample and analyse the marine biota. Phillips (1977) suggests that the filter-feeding bivalve molluscs and the macroalgae are appropriate to assess the uptake of metals from solids and solution respectively (see chapter by Darracott, this volume). Such knowledge is vital if the impact of the mine waste on the marine environment of the Aegean is to be understood.

Returning to exploration problems, although in this instance the Pb to Zn ratio in the EDTA leach is fairly constant, in other areas this might be a useful approach to distinguish between different source areas, e.g. mine waste and outcropping mineralization. Smith and Moore (1972) for example, found that the EDTA Cu to total Cu ratio in sediments distinguished between source areas around the Keweenaw Peninsula (see earlier). Partition studies such as this, together with techniques described later for the hydrothermal deposits at Santorini, are potentially powerful geochemical exploration tools in their own right and an important adjunct to geostatistical studies; in isolation the latter can be misleading but partition studies can help to support, clarify or deny the geostatistical models developed.

PHOSPHORITES

At the present time, of the offshore mineral deposits phosphorites are probably the most commerically viable. The deposits can be fairly extensive and well within the reach of present dredging technology; processing of the unconsolidated deposits is relatively straightforward; and, perhaps most importantly for such a low-value commodity, in certain places (e.g. Mexico) a local demand exists (hence low transportation costs) which cannot be met from onshore resources.

The genesis of marine phosphorites is reviewed in detail elsewhere (e.g. Cronan, 1980); in continental margin areas they can occur in the form of recent unconsolidated or semi-consolidated sediments or as bedrock deposits. Recent deposits are forming at the present day close to the shelf-slope boundary off the west coast of southern Africa and South America in environments where phosphorus precipitates in the pore waters of organic-rich sediments associated with areas of upwelling, high biological productivity and low terrigenous sediment supply.

Probably more important commercially, however, are recent deposits derived from the erosion of bedded phosphorites cropping out on the continental shelf.

Summerhayes *et al.* (1972) consider erosion of bedrock outcrops to be the source for recent/Pleistocene phosphorites off the Moroccan coast of north-west Africa where Cretaceous and Tertiary phosphorites have been eroded during Pleistocene lowering of the sea-level. The distribution of P_2O_5 (Figure 13), largely in the form of sand-sized grains of phosphorite, suggests that the detrital phosphorite was redistributed by currents along the Pleistocene shoreline (Summerhayes *et al.*, 1972).

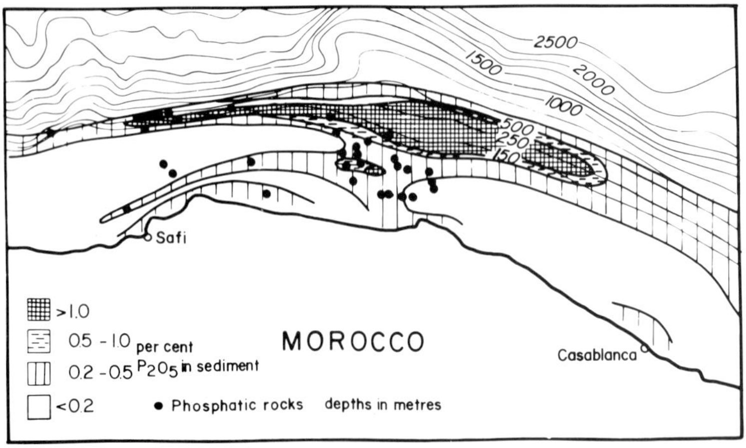

Figure 13 Distribution of phosphate (as P_2O_5) in sediments off Morocco, and location of sites where phosphatic rocks were dredged (from Summerhayes *et al.*, 1972)

Geochemical surveys are obviously a rapid and cost-effective means of exploring for unconsolidated deposits by simply analysing and mapping the distribution of phosphorus. For bedrock deposits, the erosion of phosphorite rock at outcrop and dispersion of detrital phosphorite has obvious potential in geochemical exploration but it should be borne in mind that the dispersion pattern is not straightforward and like other detrital particles, will depend on wave and current patterns, both present-day and those prevailing at lower stands of the sea-level (Cronan, 1980).

Shallow submarine hydrothermal activity and the zonation of Fe, Mn and trace metals in associated sediments

The scope of exploration geochemistry in the shallow marine environment includes research on the genesis of mineral deposits and associated phenomena such as dispersion haloes. These investigations into modern geochemical processes are useful not only to elucidate the mechanisms of formation of ancient deposits formed in analogous environments, but also modern deposits presently forming in deeper waters and which are less accessible for study than their counterparts in shallow water. One example of such research is an investigation carried out on iron-rich sediments associated with hydrothermal activity from Santorini, Greece (Smith, 1979) in which particular emphasis was placed on the extensive ferruginous horizons and the associated metal zoning which is also often found around ancient stratiform massive sulphides.

The Santorini Archipelago is situated on the Cyclades volcanic arc in the southern Aegean Sea, approximately 130 km north of Crete (Figure 14). Morphologically, it consists of a fractured Caldera open to the sea through channels between the islands of Thera, Therasia and Aspronisi. In the middle

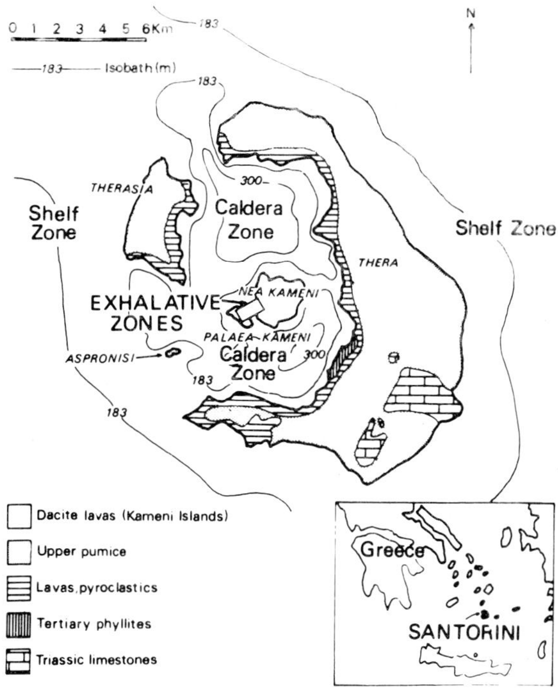

Figure 14 Simplified geological map of Santorini showing bathymetric zones (from Smith and Cronan, 1983)

of the caldera are the islands of Palaea and Nea Kameni which are the centres of the volcanic activity and around the perimeter of the Kameni islands are a series of small embayments, generally developed between diverging lava flows. The samples collected in this particular investigation were in the form of a series of traverses in two such embayments, one on Nea Kameni and one on Palaea Kameni, from the submarine fumerolic activity at the landward end of each embayment to the channel separating the two islands.

Smith and Cronan (1975) divide the embayments and channel into three separate bathymetric zones, each possessing distinct physical and chemical characteristics. The 'inner exhalative zones' are the sites of submarine hydrothermal activity and are characterized by warm, shallow water and restricted circulation. The Eh of the surficial sediments is mildly oxidizing, although reducing in patches, and the pH is slightly acidic. Inner exhalative zone sediments are the major iron deposits and include a surface layer of ferric hydroxide mud overlying ferrous carbonate (siderite) and hydroxide muds with varying amounts of pumice, vivianite and opaline silica. Iron and associated elements in the hydrothermal solutions are derived by the leaching of volcanic rocks by heated, circulating, meteoric and sea water. On entering the marine environment, a combination of the increase in pH, redox potential, and the action of iron oxidizing bacteria, cause the precipitation of hydrated ferric oxides. Local patches of sediment with negative redox potentials occur where sulphate reducing bacteria are active, and/or volcanically derived H_2S is present.

Depths range between 4 m and 40 m in the 'outer exhalative' zones which are subject to wave action and open seawater circulation and in which the pH is close to that of normal seawater. The sediments are highly oxidizing and are an admixture of ferric oxides and volcanic sands reflecting the decreasing influence of hydrothermal activity and the increasing contribution of detrital volcanic material. 'Channel' zone sediments are deposited in depths in excess of 30 m in a comparatively high energy environment and comprise bioclastic debris, organic detritus, volcanic sands and lapilli pumice.

Smith and Cronan (1983) describe in detail the geochemistry of sediments and waters from the exhalative and channel zones. Water analyses were carried out to help to answer such questions as 'what elements are contributed by the hydrothermal solution?'; 'do these elements disperse in solution or in

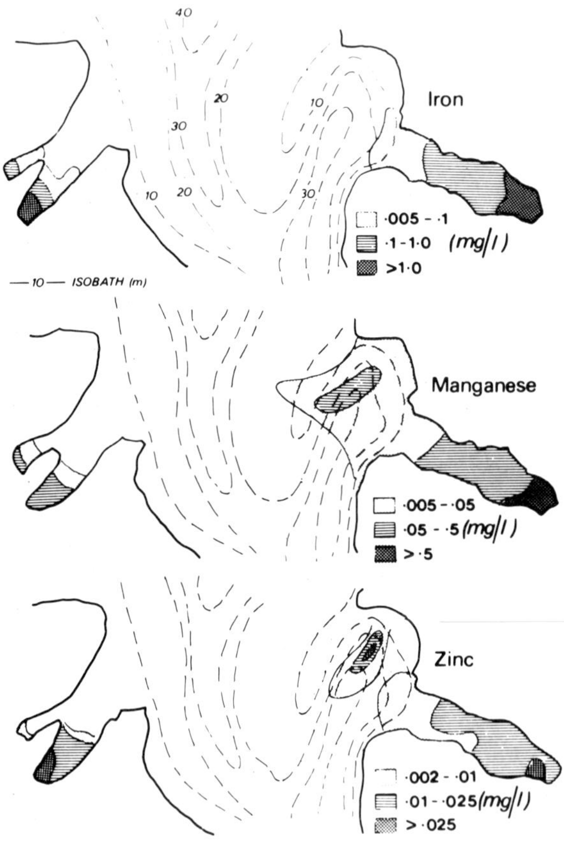

Figure 15 Trends in the concentration of Fe, Mn and Zn in seawater in the exhalative and channel zones, Santorini, Greece (from Smith and Cronan, 1983)

particulate form?'; and 'is there any evidence for the differential mobility of elements derived from the hydrothermal solution?'.

At Santorini it appears that the hydrothermal solutions supply moderate amounts of Fe, Mn and Zn and possibly Cu and Pb, but only to a limited extent. The general order of abundance is Fe>>Mn>Zn>Cu and Pb. On the basis of differences in their lateral distribution in the waters of the exhalative and channel zones (Figure 15) Mn and Zn appear to be more mobile than Fe. 'Soluble' Fe decreases rapidly to reach normal seawater levels at the boundary between the inner and outer exhalative zone of Nea Kameni. In contrast, normal seawater levels of Mn are not reached until the outer margins of the outer exhalative zones. The observed separation of Fe and Mn is related to the physiochemical gradient developed between the inner and outer exhalative zones. In the acid and mildly oxidizing conditions of the inner exhalative zone Fe oxidizes rapidly and begins to precipitate as soon as the hydrothermal fluids mix with seawater. The oxidation rate of Mn under these conditions is lower and it can remain in solution until the more alkaline conditions of the outer-exhalative zone are reached. For Zn, scavenging by iron oxides is more likely to control its mobility and for some reason, the precipitating iron oxides are unable to scavenge all of the hydrothermal Zn in the inner exhalative zone. Zn, therefore, has a greater mobility than iron. On the other hand, the limited amount of Cu and Pb supplied by the hydrothermal solution is scavenged by the iron oxides thereby restricting the mobility of these elements.

A significant geochemical characteristic of the exhalative zone deposits is the enrichment of Fe, Mn and trace metals in distinct zones. Trends in the concentration of elements in the surficial sediments (see e.g. Figures 16 and 17) illustrate how Fe and Sc and, to a lesser extent, Cu and Mo are concentrated in the inner exhalative zones, whereas Mn in particular, but also V, Cr and Zn are concentrated in the outer exhalative zone.

The interpretation of such trends in total abundance data to assess the significance of hydrothermal processes on the distribution of the elements is ambiguous and so Smith and Cronan (1983) use partition studies to discover the dominant components, i.e. hydrothermal or detrital, governing the distribution of the elements. Partition studies have been used in a large number of investigations of the geochemistry of marine sediments (see e.g. Chester and Hughes, 1967; Cronan and Garrett, 1973; Cronan, 1976a) and are based on the principle that selective chemical attacks can be used to separate the major components of the sediment. Analysis of the leach solutions or the residues reveals, therefore, the partition of the elements between the main components. In the investigation described by Smith and Cronan (1983) the samples were sequentially leached with acetic acid primarily to remove the carbonates and some colloidal iron, acetic acid/hydroxylamine hydrochloride which is a combined acid-reducing agent and which preferentially attacks Mn oxides if present, and hot HCl which will attack most components of the sediment apart from the resistant detrital minerals. The results of the partition indicate that the bulk of the hydrothermally derived iron oxides occurs in the HCl-soluble fraction, distinct Mn oxide phases exist in the outer

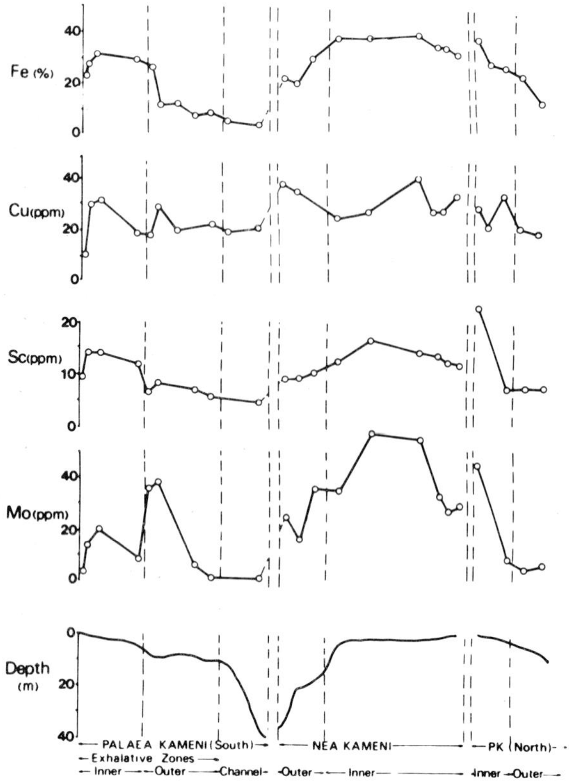

Figure 16 Trends in the concentration of Fe, Cu, Sc and Mo in sediments from the exhalative and channel zones, Santorini, Greece (from Smith and Cronan, 1983)

exhalative zones, and that the enrichment of Zn, V and Cr in the outer exhalative zones is related to scavenging by the iron oxides in that, like the iron oxides (and no other major component), a substantial proportion of their total abundance occurs in the HCl-soluble fraction.

From the point of view of ore genesis and mineral exploration, there are several features of the Santorini hydrothermal deposits that are worthy of note (Smith, 1979). Firstly they are a significant modern example of the separation of iron and manganese derived from hydrothermal solutions in the marine environment. This phenomenon was originally explained by a model formulated by Krauskopf (1957) and since invoked by many investigators of both modern and ancient environments to account for separated deposits of iron and manganese.

Secondly, an important feature of iron-rich units found in association with massive sulphide deposits is the development of distinct metal zoning (Fryer and Hutchinson, 1976). Typically, the zoning features a lateral sequence away from the sulphide deposits of Fe rich, Mn poor → Fe rich, Mn rich → Host rock Fe, anomolous Mn → Host rock Fe, background Mn (i.e. distal enrichments of

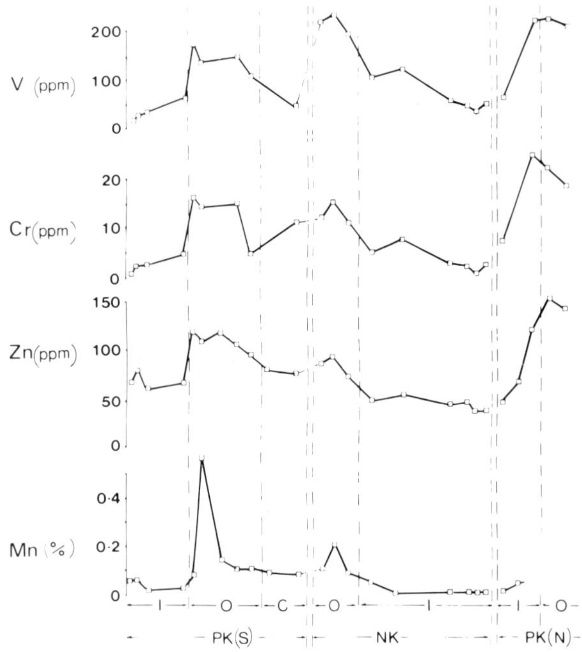

Figure 17 Trends in the concentration of V, Cr, Zn and Mn in sediments from the exhalative and channel zones, Santorini, Greece (from Smith and Cronan, 1983)

manganese in the iron-rich unit, and an extensive manganese halo).

Taylor (1974) reports such a sequence in the Hanesavo beds overlying the massive sulphides found on islands of the Florida Group, BSIP and so also do Derry *et al.* (1965) and Russell (1975) for the iron formation associated with the Tynagh massive sulphide deposit. At Meggan the iron rich unit is limited in extent but there is an extensive Mn halo (Gwosdz and Krebs, 1977). In the modern Red Sea deposits Bignell *et al.* (1976) describe a similar dispersion of manganese resulting in an halo extending up to 10 km from the deposit. At Santorini, manganese is also distally enriched in the iron-rich unit (Figure 18). This suggests that the processes described at Santorini may be used to account for the observed distribution of iron and manganese in other deposits exhibiting similar zoning and that such Mn zoning may be a common feature of the volcanogenic exhalative–sedimentary (submarine hydrothermal) ore-forming process and consequently important in exploration for such deposits (Cronan, 1976b).

Like manganese the selective dispersion and consequent zoning of heavy metals within the iron-rich unit may also be a useful exploration guide. Russell (1975), for example, describes zones of Pb and Zn enrichment in the distal manganiferous layers associated with the Tynagh massive sulphide deposit. Distal enrichment of Cu, Zn and Hg in the iron-rich unit is also a feature of the Red Sea deposits (Bignell *et al.*, 1976). Although the abundances are generally

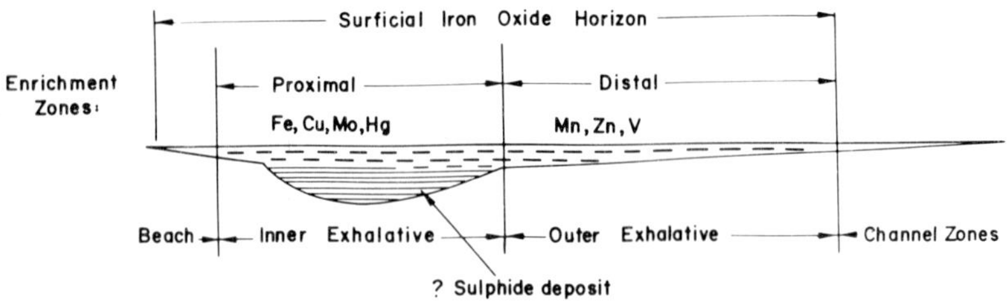

Figure 18 Schematic zoning of elements in the surficial iron-rich horizon, Santorini, Greece

much lower, a similar zoning of heavy metals related to hydrothermal activity occurs at Santorini. Figure 19 illustrates the element zoning using the HCl-soluble (i.e. hydrothermal) element against depth profiles, while Figure 18 summarizes, schematically, the element zones within the surficial iron-rich unit.

Smith (1979) proposes that heavy metal zoning related to hydrothermal activity develops because of the scavenging action of the scavenging phases

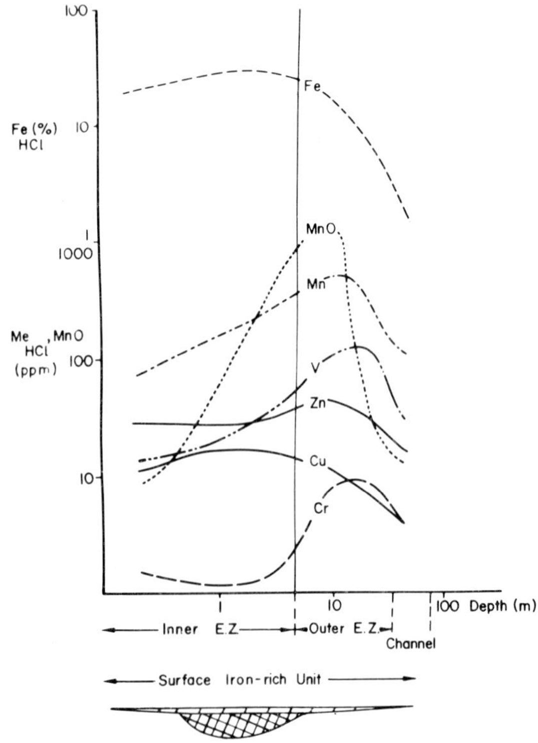

Figure 19 Summary of element zoning within the surficial iron-rich unit based on partition studies, Santorini, Greece

present. At Santorini the iron oxides are the major scavenging phase but elsewhere manganese oxides could be equally or more important. 'Proximal enrichment' occurs when the hydrothermal supply is (a) small and element mobility is limited (e.g. Cu, Mo, Pb, Sc and Hg); or (b) large, in which case solubility or oxidation kinetics rather than scavenging is the main constraint on mobility (e.g. Fe). 'Distal enrichment' occurs when there is (a) no hydrothermal influence and enrichment results simply from scavenging from normal seawater (e.g. Cr)—this feature may be particularly significant for deeper volcanogenic sedimentary—exhalative environments (e.g. the Red Sea and mid-oceanic ridge or rise systems) where the dispersion of the particulate scavenging phase may be widespread; (b) a moderate hydrothermal supply and moderate mobility (e.g. Zn and V); and (c) a substantial hydrothermal supply where, once again, solubility or oxidation kinetics are more important than scavenging but the gradual change in the physio-chemical environment and lower oxidation rates allows the element to pass to the outer zones before precipitation (e.g. Mn).

In terms of exploration for other modern deposits, these observations may have great potential value (Cronan, 1976b). In the Red Sea, for example, dispersion haloes are many times the extent of the deposits (Bignell *et al.*, 1976). Therefore regional geochemical surveys may be a useful way of exploring for volcanogenic sedimentary–exhalative deposits on the ocean floor.

General observations and conclusions

Geochemical surveys have two broad aims: to identify significant anomalies (i.e. those related to mineralization) and to delineate dispersion pathways (or concentration gradients) to point the way towards the target mineralization. It is worth noting at this point that these two aims are common to the other major area of applied geochemistry offshore, namely environmental geochemistry (dealt with elsewhere in the text) but instead of mineralization substitute pollution.

Significant anomaly recognition is particularly important in exploration programmes which use metals such as Cu and Zn as tracers (e.g. bedrock mineralization and precious metal placers—see earlier) and false anomalies can derive from a wide range of sources (see Levinson, 1974). Anomalously high concentrations of trace metals due to scavenging by manganese oxides is a well-known problem in land-based exploration and the marine environment is no exception. Off the island of Andros in the Aegean Sea, for example, high concentration of Ni in some samples relates to scavenging and not to ultrabasic provenance as was initially suspected (Smith, 1979). Other metals, particularly Cu and Zn, are similarly affected. Other factors related to the natural range in composition of marine sediments such as texture, particularly the proportion of fine-grained material, dilution by bioclastic carbonates, and the range in mineralogical composition of the detrital fraction, can all either lead to false anomalies or, alternatively, mask those which are significant (Smith, 1979).

False anomalies caused by mining contamination are the bane of exploration geochemists working in areas of past or present mining activity. Anomalies derived from previously undiscovered mineralization are very difficult to distinguish from those related to past mining activity. Ironically, continental shelves with the greatest mineral potential by virtue of the proximity onshore of extensive deposits which appear to trend offshore are also those most likely to have sediments contaminated by earlier mining operations. Sounion is a good example but the problem is common to many offshore areas with a similar potential. As discussed earlier, in St. Ives Bay the amount of tin tailings reaching the marine environment is so great that the surficial placers of cassiterite that result are mineable.

Establishing the significance of an anomaly is not particularly helpful if it is not then possible to delineate dispersion pathways and to pinpoint the potential source area. In stream sediment surveys, for example, a unidirectional flow regime simplifies the procedure considerably, but as Cronan (1980) points out, 'in an environment as complex as that found offshore it is naive to expect simple dispersion patterns'.

Reworking of the sediments by wave and current action, for example, may modify or obscure dispersion pathways. Anomalies related to relict components would be particularly difficult to decipher because they could derive from a wide variety of sources including subaerial or shoreline erosion at a lower stand of the sea-level or relict alluvial deposits. In addition, much of the relict material may have been redistributed during or since the passage of the Holocene transgressing shoreline. Often only terrigenous material derived since the last transgression appears to leave a readily detectable dispersion pathway (e.g. off Sounion). But modern terrigenous material can also derive from a variety of sources including longshore drift, local erosion and river drainage (e.g. Chagvan Bay). Indeed, one of the main applications of geochemistry in marine mineral exploration is in deciphering sediment dispersal patterns.

Obviously the problems are severe but this does not negate the use of exploration geochemistry; on the contrary, successful exploration will depend upon the effective use of the range of geochemical techniques available, integrated with geophysical, geological (and possibly oceanographic) investigations and the examples briefly reviewed above give some indication of the range of techniques available or being developed.

To conclude, to date the application of geochemical techniques to offshore mineral exploration has been on a very modest scale and mainly confined to academic research. Industrial exploration has been tentative and awaits the first major discovery to provide the requisite momentum. Exclusive Economic Zones created as part of the Law of the Sea treaty, however, may encourage exploration of the continental margins and exploration geochemistry is ready to meet the challenge.

References

Bignell, R.D., Cronan, D.S. and Tooms, J.S. (1976). Metal dispersion in the Red Sea as an aid to marine geochemical exploration, *Trans. Inst. Min. Metall.*, **B85**, 273–278.

Chester, R. and Hughes, M. (1967). A chemical technique for the separation of ferro-manganese minerals, carbonate minerals and adsorbed trace elements from pelagic sediments, *Chem. Geol.*, **2**, 249–262.

Cronan, D.S. (1976a). Basal metalliferous sediments from the eastern Pacific collected during the DSDP, *Bull. Geol. Soc. Am.*, **87**, 928–934.

Cronan, D.S. (1976b). Implications of metal dispersion from submarine hydrothermal systems for mineral exploration on mid-oceanic ridges and in island arcs, *Nature*, **262**, 567–569.

Cronan, D.S. (1980). *Underwater Minerals*; Academic Press, London.

Cronan, D.S. and Garrett, D. (1973). The distribution of elements in metalliferous Pacific sediments collected during DSDP, *Nature*, **242**, 88–89.

Derry,D.R., Clark,G.R., and Gillat, N. (1965). The Northgate base-metal deposit at Tynagh, County Galway, Ireland, *Econ. Geol.*, **60**, 1218-1237.

Fryer,B.J. and Hutchinson,R.W. (1976). Generation of metal deposits on the sea floor, *Can J. Earth Sci.*, **13**, 126-135.

Goodden,J.J. (1974). Sedimentological aspects of underwater copper exploration in Lake Superior, *MS Thesis, University of Wisconsin* (unpubl.).

Gwosdz, W. and Krebs, W. (1977). Manganese halo surrounding Meggen ore deposit, Germany, *Trans. Inst. Min. Metall.*, **86**, 73–77.

Hawkes, H.E. and Webb, J.S. (1962). *Geochemistry in Mineral Exploration*; Harper Row, New York.

Hazelhoff Roelfzema, B.H. (1968). Geochemical dispersion of tin in marine sediments, Mount's Bay, Cornwall, *PhD Thesis, London University* (unpubl.).

Imbrie, J. and Purdy, E.G. (1962). Classification of modern Bahamian carbonate sediments. In: *Classification of Carbonate Rocks* (Ed. W.E. Ham), 253-272; Amer. Assoc. Petrol. Geol., Tulsa, Oklahama.

Krauskopf, K.B. (1957). Separation of manganese from iron in sedimentary processes, *Geochim. et Cosmochim. Acta.*, **12**, 61–84.

Levinson, A.A. (1974) *Introduction to Exploration Geochemistry;* Applied Publishing Ltd., Calgary.

McArthur, J.M. (1974). The geochemistry of phosphorite concretions from the continental shelf off Morocco, *PhD Thesis, London University* (unpubl.).

Marinos, G. and Petrascheck, W.E. (1956). Laurium *Geol. Geophys. Research 4*, Inst. Geology and Subsurface Research, Athens.

Mero, J.L. (1965). *The Mineral Resources of the Sea;* Elsevier, Amsterdam.

Moore, J.R. (1972). Exploration of ocean mineral resources—perspectives and predictions, *Proc. R. Soc. Edinb.*, **72**, 193-206.

Moore, J.R. and Welkie, C.J. (1976). Metal-bearing sediments of economic interest, coastal Bering Sea, *Proc., Symp. on Sedimentation,* Alaska Geological Society, k1–k17.

Nebrija, E.L., Young, C.T., Meyer, R.P. and Moore, J.R. (1976). Electrical prospecting for copper veins in shallow water. *Proc. 8th Offshore Technology Conf.*, paper 2454, Houston, Texas.

Nebrija, E.L., Welkie, C.J. and Meyer, R.P. (1977). Exploration for copper mineralisation in an extension of the Nonesuch shale under Lake Superior, *Marine Mining*, **1**, 149–170.

Ong, P.M. (1966). Geochemical investigation in Mount's Bay, Cornwall, *PhD Thesis, London University* (unpubl.).

Owen, R.M. (1978). Geochemistry of Platinum-enriched sediments: application to mineral exploration, *Marine Mining*, **1**, 259–282.

Owen, R.M. (1980). Quantitative models of sediment disperal patterns in mineralised nearshore environments, *Marine Mining*, **2**, 231–249.

Phillips, D.J.H. (1977). The use of biological indicator organisms to monitor trace metal pollution in marine and estuarine environments—A review, *Environ. Pollut.*, **13**, 281–317.

Russell, M.J. (1975). Lithogeochemical environment of the Tynagh base-metal deposit, Ireland and its bearing on ore deposition, *Trans. Inst. Min. Metall.*, **B84**, 128–133.

Smith, P.A. (1979). Geochemical investigation of recent sediments from the Aegean Sea, *PhD Thesis, London University* (unpubl.).

Smith, P.A. and Moore, J.R. (1972). The distribution of trace metals in the surficial sediments surrounding Keweenaw Point, Upper Michigan, *Proc. 15th Conf. Great Lakes Res.*, 383–393.

Smith, P.A. and Cronan, D.S. (1975). The dispersion of metals associated with an active submarine exhalative deposit. *Proc. Oceanology International Conference, Brighton*, 111–114, BPS Exhibitions, London.

Smith, P.A. and Cronan, D.S. (1983). The geochemistry of metalliferous sediments and waters associated with shallow submarine hydrothermal activity. (Santorini, Aegean Sea), *Chem. Geol.*, **39**, 244–262.

Summerhayes, C.P. (1970). Phosphate deposits of the northwest African continental shelf and slope, *PhD Thesis, London University* (unpubl.).

Summerhayes, C.P., Nutter, A.N. and Tooms, J.S. (1972). The distribution and origin of phosphate in sediments off Northwest Africa, *Sediment. Geol.*, **8**, 3–28.

Taylor, G.R. (1974). Volcanogenic mineralization in the islands of the Florida Group, BS.I.P., *Trans. Inst. Min. Metall.*, **B83**, 120–130.

Tooms, J.S. (1969). Some aspects of exploration for marine mineral deposits. In: *Proc. 9th Commonwealth Mining and Metallurgical Congr.*; Institution of Mining and Metallurgy, London.

Tooms, J.S., Taylor Smith, D., Nichol, I., Ong, P.M. and Wheildon, J. (1965). Geochemical and geophysical mineral exploration experiments in Mount's Bay, Cornwall. In: *Submarine Geology and Geophysics* (Eds. W.F. Wittard and R. Bradshaw), *Colston Papers*, **17**, 363–39; Butterworths, London.

Yim, W.W.-S. (1978). Geochemical exploration for offshore tin deposits in Cornwall. In: *Proc. 11th Commonwealth Mining and Metallurgical Congr., Hong Kong*, Institution of Mining and Metallurgy, London.

Yim, W.W.-S. (1979). Geochemical exploration for tin placers in St. Ives Bay, Cornwall, *Marine Mining*, **2**, 59–78.

Geochemical Exploration for Deep Sea Mineral Deposits

D. S. CRONAN

Introduction

Initiated in the Applied Geochemistry Research Group by Professor Webb in the early 1960s, studies on deep sea minerals within the Group have been involved with all potentially economic varieties of the deposits. Ocean wide studies on manganese nodules in an attempt to delineate ore grade varieties (Cronan, 1967; Cronan and Tooms, 1969; Glasby, 1970; Cronan, 1975; Cronan and Moorby, 1976; Moorby, 1978; Cronan and Moorby, 1981) evolved into detailed process related investigations of these deposits in small discrete areas (Cronan and Tooms, 1967; Glasby *et al.*, 1974; Colley *et al.*, 1979) and most recently into Exclusive Economic Zone (EEZ) and Mine Site evaluation studies (Cronan, 1981; Aplin, 1983; Aplin and Cronan, 1985; Cronan, 1984). Studies on metalliferous sediments and polymetallic sulphides commenced with investigations in the Red Sea (Tooms, 1970; Holmes and Tooms, 1972; Bignell *et al.*, 1974; Bignell, 1975; Bignell *et al.*, 1976) and developed into broad ranging studies on mid-ocean ridges (Cronan and Garrett, 1973; Horowitz, 1974; Horowitz and Cronan, 1976; Cronan, 1976a, b; Cronan *et al.*, 1979; Varnavas, 1981; Shearme, 1981; Cronan and Varnavas, 1981; Skirrow, 1983; Malahoff *et al.*, 1983) and subsequently in island arcs (Smith and Cronan, 1975; Cronan, 1976a; Cronan and Thompson, 1978; Cronan *et al.*, 1982; Smith and Cronan, 1983; Cronan, 1983; Cronan *et al.*, 1984). These studies have been focussed on a number of aspects of the deposits, but always there has been the intention to develop better geochemical methods for locating potentially economic varieties of them (Cronan, 1978).

Over the twenty years in which the Applied Geochemistry Research Group has been involved in work on deep sea mineral deposits, awareness of and attitudes towards the resources have changed considerably. In the early days, the view of Mero (1965) that there was virtually unlimited wealth on the deep sea floor was widely accepted. This led to the declaration by the United Nations in 1967 that the deep ocean floor is the common heritage of mankind. Concurrent with this claim came the need to investigate deep ocean mineral resources in more detail, and this led to substantially more conservative estimates of their worth than those initially claimed (Archer, 1976; Frazer, 1977; see Cronan, 1980 for a review), while nevertheless still admitting that substantial resources do occur. At the same time, the UN Convention on the Law of the Sea (UNCLOS) deliberated on ownership and exploitation of the deposits over a ten year period (1973–1982), finally agreeing a regime which is widely regarded as unacceptable in some of the developed countries best able to recover the deposits. Partly as a result of this situation there has been a considerable lessening of activity in deep sea mineral studies by the major marine mining consortia in the UNCLOS declared International Seabed Area, but more interest in nearshore areas.

One of the decisions of the Law of the Sea Convention was to create Exclusive Economic Zones of 200 miles around Coastal States, in which the seabed resources were under the exclusive jurisdiction of the Coastal State. This has had the effect of inducing many countries to look much more closely at their immediate offshore areas than had hitherto been the case, and some international marine mining companies are now concentrating on such areas. For a long time, the potential of deep sea mineral deposits adjacent to coastal states was considered to be low, mainly because few of them had any appreciable deep sea area within 200 miles of their coasts. Placer minerals and in some cases phosphorites were known to be common in some nearshore situations (Cronan, 1980) but not manganese nodules nor metalliferous sediments. However, recent discoveries of polymetallic sulphides off the west coast of the USA and Canada (Figure 1) and of manganese nodules off some of the oceanic islands of the Pacific (Figure 2) have changed this situation. Now a number of nations such as the USA and groups of nations such as the Committee for Coordination of Joint Prospecting for Mineral Resources in South Pacific Offshore Areas (CCOP/SOPAC) are putting much effort into exploring for deep sea mineral deposits within 200 miles of their shorelines.

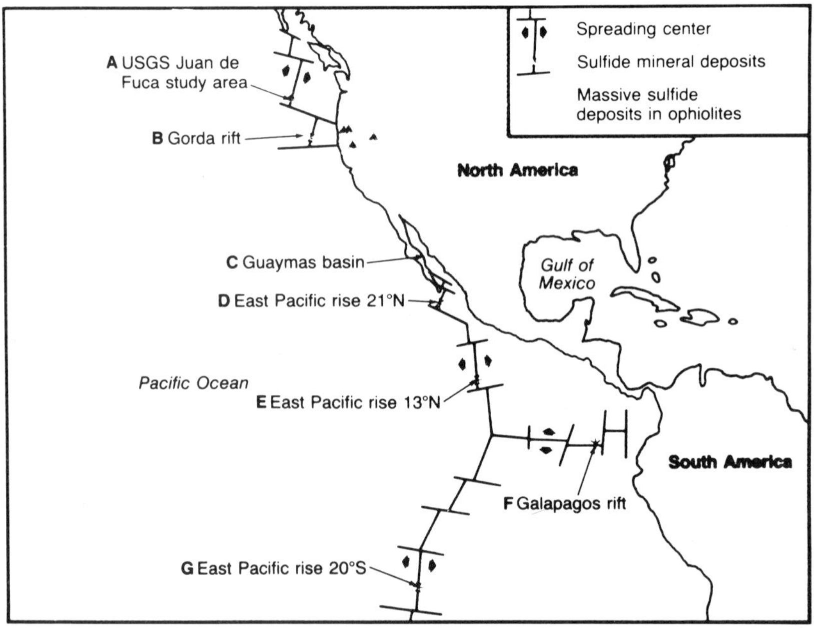

Figure 1 Polymetallic sulphide deposits in the Pacific (from Ocean Industry, 1983)

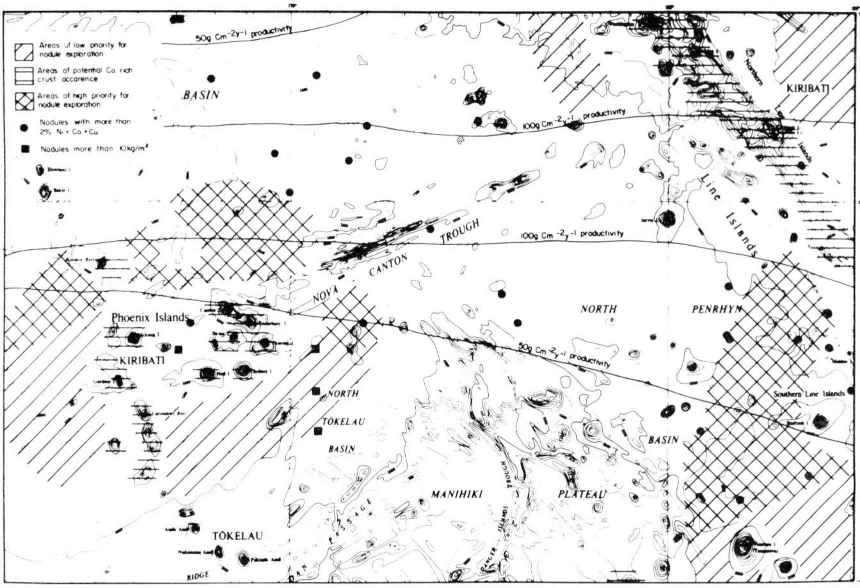

Figure 2 Areas of potentially economic manganese nodules in the East Central Pacific Basin (from Cronan, 1984)

This trend is likely to accelerate during the 1980s, until the International Seabed Authority is fully operational, and recent and continuing efforts by the Applied Geochemistry Research Group as part of this endeavour form the main bulk of the present chapter.

Metalliferous sediments

Metalliferous sediments variously include mud composed of iron and manganese oxides, iron silicates, and a variety of potentially economic base metal sulphides such as chalcopyrite and sphalerite of submarine hydrothermal origin (see Cronan, 1980 for a review of occurrences). *Polymetallic sulphides* are an important sub-class of these deposits in which the base metal sulphides are not usually in the form of unconsolidated muds, but occur as crystalline sulphide rock (Malahoff, 1982; Malahoff *et al.*, 1983).

Hydrothermal metalliferous sediments of different varieties were more or less simultaneously discovered on the East Pacific Rise spreading centre (Bostrom and Peterson, 1966) and associated with the Atlantis II Deep on the Red Sea spreading centre (Degens and Ross, 1969) but it was not until later that the genetic links between the deposits were recognized. All known occurrences of these deposits on the sea floor are now thought to comprise different members of a general submarine hydrothermal fractionation sequence (Cronan, 1976a). Sulphides on the East Pacific Rise may be regarded as the analogues of sulphides forming near to the discharge point of the hydrothermal fluids in the SW Basin of the Atlantis II Deep. Iron silicates and associated iron oxides in the FAMOUS[1] and Galapagos areas, and in the Gulf of Aden, may be regarded as the analogues of iron silicates and oxides which occur over much of the floor of the Atlantis II Deep. Manganese oxides in the TAG[2] area, although compositionally very pure, may be the equivalent of manganese oxides near the margin of the Atlantis II Deep. Finally, widely dispersed mixed Fe and Mn oxides occurring over mid-ocean ridges at large may be the equivalent of widely dispersed oxides found in a dispersion halo up to 20 km in diameter around the Atlantis II Deep (Bignell *et al.*, 1976). Also by analogy with the Atlantis II Deep, the sulphides on the East Pacific Rise at 21°N would be early formed precipitates from hydrothermal solutions, the iron rich silicate and oxide deposits from the various localities mentioned would represent intermediate formed precipitates, and the manganese oxide deposits in the TAG area would represent last formed precipitates. The widely dispersed iron and manganese-rich sediments most probably represent precipitates from iron and manganese-bearing solutions which discharged from the hydrothermal vents into the overlying sea water.

Spatially, the relationships between these different deposits will depend on the physico-chemical conditions below the sea floor and at the point of hydrothermal discharge. Where high temperatures and reducing conditions are maintained up to the point of discharge, metal rich sulphides will precipitate from black smokers on the sea floor as on the East Pacific Rise, and the remaining metals, mostly iron and manganese will escape into sea water, some to precipitate in the vicinity and the remainder elsewhere. Where downward flowing sea water mixes with and cools the ascending hydrothermal solutions, sub-surface precipitation of sulphides is likely to take place and lower temperature hydrothermal precipitates, firstly silicates and then oxides, are likely to occur on the sea floor, as in the FAMOUS and Galapagos areas and in the Gulf of Aden, their relative distribution depending, among other factors, on the Eh of the depositional environment. In such instances, manganese is likely to precipitate either above or adjacent to the iron deposits, as occurs in the Galapagos area and in the Gulf of Aden. Finally, where extensive cooling and mixing of the hydrothermal solutions takes place below the sea floor, most of the hydrothermal minerals will precipitate sub-surface and the final precipitates, manganese oxides, will precipitate on or near the sea floor, as in the TAG area.

As mentioned, the submarine hydrothermal fractionation sequence detailed above often leads to the development of dispersion haloes of Fe and Mn oxides in sediments around a potentially economic sulphide deposit, such haloes sometimes exhibiting decreasing concentrations of Fe and Mn up to several kilometres from the hydrothermal source, as occurs in the Red Sea (Bignell *et al.*, 1976). The occurrence of such dispersion haloes considerably simplifies the task of exploration for high grade metalliferous sediments, because the halo can be many times the size of the deposit of economic interest

[1]French–American Mid–Ocean Underwater Survey
[2]Trans–Atlantic Geotraverse

and thus a sampling programme can be designed accordingly. However, dispersion haloes in the open ocean are not always as clear-cut as in the Red Sea. Because of more turbulent bottom water conditions, particulate iron and manganese oxides, for example, become homogenized in the water column and get dispersed over a wide area leading to a broad zone of manganese and iron enrichment in sediments near submarine hydrothermal centres, rather than a gradient of decreasing Mn and Fe concentrations of variation in the Mn/Fe ratio away from the vents (Cronan *et al.*, 1979). In addition, prevailing bottom and mid-water currents can transport the particulate oxides for many miles from their sources to give broad zones of Mn and Fe enriched sediment, as, for example, is found on the west flank of the East Pacific Rise.

MID-OCEAN RIDGES

Most work on exploration for metalliferous sediments and polymetallic sulphides other than in the Red Sea, has been concentrated on the East Pacific Rise and adjacent spreading centres. The West German minerals industry efforts here are particularly noteworthy. Furthermore, it is on the EPR, off the western margin of the USA and Canada, that a number of polymetallic sulphide deposits have recently been found either within or close to the 200 mile EEZ (see Figure 1).

Apparently, the largest of the polymetallic sulphides associated with the eastern Pacific spreading centres found to date, is that in the Galapagos Rift at 86°W (Malahoff *et al.*, 1983). The Galapagos Rift is a medium-rate spreading centre (75 mm per year) with a well-developed rift valley 1.5 km wide on average. Submersible dives encountered massive sulphide mounds in a water depth of 2850 m which appear to be flanked east and west by hydrothermally precipitated iron and manganese oxides. The sulphide mounds seem to have developed through hydrothermal activity along the southern wall of the rift valley, and consist of chimneys 2–5 m high and 1–2 m diameter, through which black smokers would once have emerged. The chimneys would appear to have been formed as a result of hydrothermal venting along fissures formed by vertical motion along the boundary fault. On another dive, less than 2 km from where the sulphides were found, iron oxide coated hummocks and small domes containing veins of manganese oxides were found along the boundary fault. Highly ferruginous sediments were also found at this site, while in the surrounding area are calcareous and siliceous sediments containing the residual precipitates from the hydrothermal activity.

The sulphide deposits from the Galapagos Rift consist of chalcopyrite and pyrite, with minor sphalerite, and are associated with locally abundant amorphous silica (Malahoff *et al.*, 1983). Samples of the deposits contain up to 27% Cu and 31% Fe. Chalcopyrite is most abundant at the tops of the mounds, whereas pyrite is most abundant in their interiors. Alteration products of the primary sulphides include covellite and chalcanthite. Most elements determined other than Cu, Fe, Si and sometimes Zn are low in the deposits. However, an exception is silver, which is present in similar concentrations to those found in Red Sea muds.

Oxide deposits from the general vicinity of the sulphides contain up to 42.5% Mn and 33.8% Fe. Of considerable interest in these oxides is their high Zn content, up to 2.3%. This is much higher than has been found in the past in hydrothermal manganese crusts (Table 1). However, an important difference between the Galapagos Mn crusts and the others in Table 1 is that the Galapagos crusts are closely associated with sulphides whereas no close association between the other crusts and sulphides has been found. This latter group appear to be either the residual precipitates of a rather drawn out hydrothermal fractionation sequence, much of which probably has taken place below the sea floor where any associated sulphides would be most likely to occur, or have formed from hydrothermal solutions which did not first precipitate sulphides. By contrast, the hydrothermal fractionation in the Galapagos Rift has resulted in the close association of sulphides and oxides together on the sea floor, and the Zn enrichment in the latter may result from this close association. Such an association has a geochemical exploration significance in that analysis of manganese oxide crusts from submarine hydrothermal systems for Zn could indicate the likely proximity of sulphide deposits on the sea floor.

ISLAND ARCS

Notwithstanding the possible value of metalliferous sediment and polymetallic sulphide deposits in mid-ocean ridge settings, most of these deposits fall within the international seabed area. By contrast, island arcs, by their very nature, contain large areas of the sea floor within 200 miles of the islands that comprise them, and thus their sea floor mineral deposits are likely to fall largely within EEZs.

On the basis of tectonic considerations, island arcs contain two settings in which submarine hydrothermal activity and thus metalliferous sediment and polymetallic sulphide deposition of the same sort found on mid-ocean ridges

Table 1 **Composition of hydrothermal manganese crusts**

	Mn %	Fe %	Ca %	Al %	Co ppm	Ni ppm	Cu ppm	Zn ppm
Galapagos* oxide crust	42.5	33.8	1.6	0.1	122	110	1200	23 000
Average hydrothermal Mn oxide crust not immediately associated with sulphides†	38.4	4.4	2.1	0.5	38	411	130	256

*From Malahoff *et al.* (1983)

†From Cronan (1980)

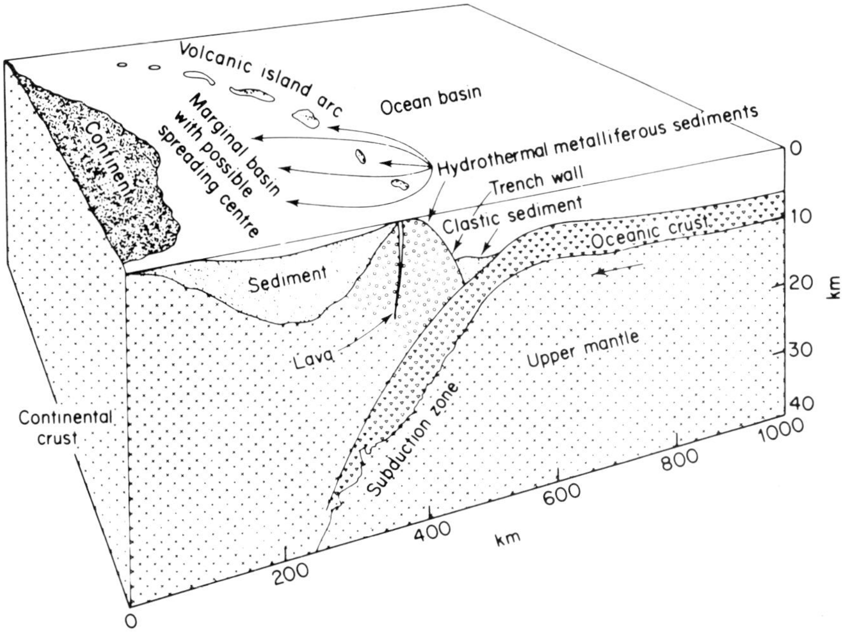

Figure 3 Settings of submarine hydrothermal activity in island arcs (from Cronan, 1980)

could occur (Figure 3). These are within the volcanic arc itself and in its marginal basins. The first of these represents the area behind the subduction zone where partial melting gives rise to surface volcanic activity, while the other is the site of secondary back arc sea floor spreading.

Kuroko deposits are well-known fossil hydrothermal ore deposits found in ancient island arc settings. Unlike fossil analogues of mid-ocean ridge deposits, these deposits contain gold and silver mineralization in addition to the other metals of interest mentioned already. Thus modern hydrothermal mineral deposits in island arcs might also be expected to contain concentrations of these metals, which would significantly enhance their economic potential (Yuasa and Yokota, 1982).

Early indications of metalliferous sedimentation associated with an island arc were noted off Santorini in the Eastern Mediterranean by Bonatti *et al.* (1972), and Smith and Cronan (1975). Using a geochemical exploration approach, Smith and Cronan (1975) successfully related the sequence of hydrothermal deposits found off Santorini to the Red Sea dispersion model of Bignell *et al.* (1976), thus demonstrating that submarine hydrothermal fractionation with all its attendant geochemical exploration implications was not just restricted to the Red Sea. However, the hydrothermal deposits off Santorini are small and of no economic value and so a larger more active island arc system that might contain potentially economic deposits was sought.

It is evident from any consideration of the worldwide distribution of island arcs that those of the western Pacific are the largest and most active of all. Accordingly, a programme of investigation into possible metalliferous sediment occurrences in the south-west Pacific was initiated in 1975 with the analysis of all previously collected sediment samples from the area that could be obtained from existing oceanographic collections (Figure 4). Preliminary results indicated several possible areas of metalliferous sediment occurrence (Cronan and Thompson, 1978) and thus the samples were augmented and subjected to a more rigorous treatment (Cronan, 1983; Cronan, in press).

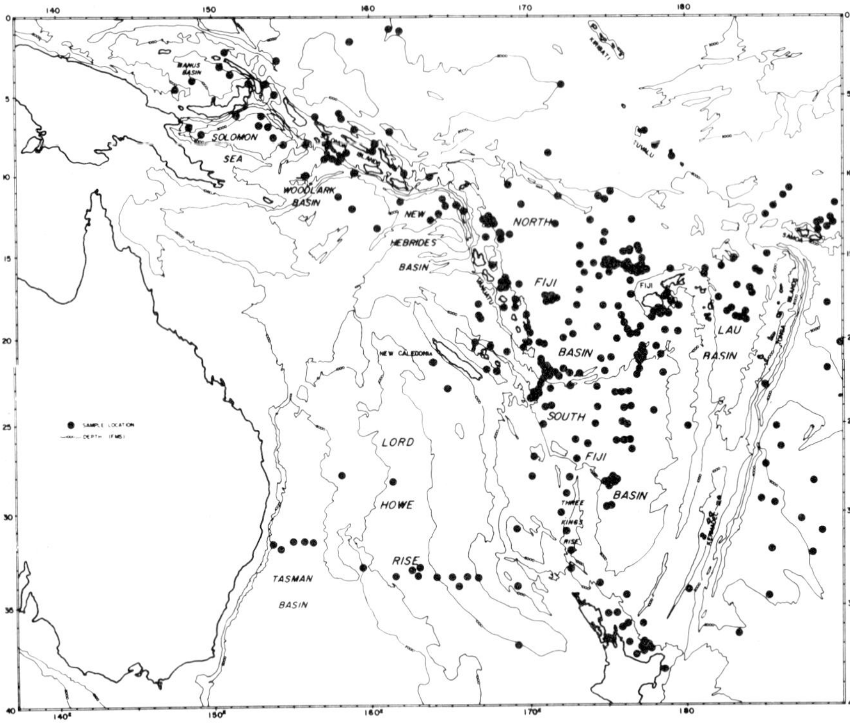

Figure 4 Distribution of samples analysed for Mn, Fe, Ni, Cu, Pb, Zn, Ca and Al in the SW Pacific Ocean (from Cronan, 1983)

Unlike most mid-ocean ridge areas where the only diluent of hydrothermal deposits is biogenic debris, in island arcs terrigenous and/or organic detritus from nearby land areas in addition to biogenic remains can bury potentially economic metalliferous sediment deposits and obscure their dispersion haloes. Furthermore, the volcanoes of the arc that rise above sea level can generate large quantities of ash which blanket the sea floor for considerable distances from the site of volcanic activity, and they can also generate turbidity slumps of volvaniclastic material (Cronan *et al.*, 1984), both of which can serve to bury or obscure hydrothermal deposits. However, these processes may be advantageous in localizing the products of submarine hydrothermal activity, many of which are dispersed over wide areas on mid-ocean ridges, possibly leading to the formation of Besshi type deposits at shallow depth under the sea floor. Because of these factors and the different tectonic settings that have a potential for metalliferous sediment formation in the SW Pacific, and the wide variety of sediment types that they contain, a multi-element approach to the development of geochemical exploration techniques for metalliferous sediments was adopted in the SW Pacific with the determination of Ca, Al, Ni, Cu and Zn, as well as Mn and Fe.

An additional reason for adopting a multi-element approach to the development of geochemical exploration techniques in the SW Pacific was that in parts of the region, unlike in the Red Sea and off Santorini, Mn could become enriched in the sediments purely by authigenic processes bearing no relation whatsoever to submarine hydrothermal activity. One way of distinguishing this enrichment from hydrothermal enrichment is by element association patterns (see below).

Analysis of Ca in sediments from an area of potential metalliferous sedimentation or polymetallic sulphide occurrence provides a measure of their $CaCO_3$ content, which acts as a diluent of any hydrothermal deposits formed above the calcium carbonate compensation depth and thus can obscure dispersion haloes. Analysis for aluminium provides a measure of the detrital content of the sediments, both terrigenous material derived from land areas, and the products of terrestrial and submarine volcanic activity such as basalt fragments, ash, etc. which can also dilute metalliferous sediments. Analysis for Cu and Zn can provide an indication of the presence of sulphide minerals such as chalcopyrite and sphalerite dispersed from 'black smoker' type hydrothermal vents. Alternatively, the sulphides may have become oxidized during particulate dispersion through sea water, but nevertheless maintain their Cu and Zn enrichments (Shearme *et al.*, 1983). In such cases, they can still provide an indication of the proximity of sulphide forming hydrothermal activity. Analysis for Ni can, in association with Mn, provide an indication of the magnitude of authigenic Mn enrichment in the sediments, the importance of which was mentioned above.

In order to outline the main controls on the geochemistry of SW Pacific sediments, the chemical data obtained on the samples analysed by Cronan and Thompson (1978) were subjected to factor analysis (Cronan, 1983). This is a statistical technique which aims to resolve the variance in a data set of 'n' variables in terms of fewer than 'n' elements, and has a considerable potential value in marine geochemical exploration. Results of this procedure (Table 2) showed that much of the variance within the data set could be explained in terms of three factors. Factor 1 was heavily loaded with $CaCO_3$, suggesting that it was a biogenic carbonate factor. Aluminium had a high negative loading on this factor, which would be expected if it were attributable to biogenic material, since Al reflects detrital influences. Factor 2 had high loadings of Fe, Mn and Cu, but a weak loading of Ni, which precluded it being an authigenic ferromanganese oxide factor because Ni is generally enriched in authigenic ferromanganese oxides. It was considered to be a hydrothermal ferromanganese oxide factor because such deposits are generally depleted in Ni but can

Table 2 **SW Pacific sediments rotated factor matrix***

	I	II	III
Ca	*.969*	.048	-.012
Mn	.148	*.659*	*.580*
Fe	-.242	*.859*	.024
Ni	.053	.216	*.889*
Al	*-.780*	.068	-.420
Cu	.313	*.751*	.305
	Bio	Hydro	Auth

*230 samples

sometimes show Cu enrichment. The remaining factor was loaded with Mn, Ni and to a lesser extent Cu, suggesting that *it* was the authigenic manganese oxide factor. The calculation of factor scores for the elements helped to clarify further their behaviour. Factor scores express the contribution made to each sample by each of the factors and thus the major source influences on each sample can be discerned.

In an additional attempt to identify metals supplied by submarine hydrothermal activity in the SW Pacific sediments, a number of samples were subjected to geochemical partition analysis, another technique that is achieving success in marine geochemical exploration. Usually in marine sediments, any one element is variably partitioned between a number of phases, although it is often more abundant in one phase than in the others. The partition method used has been described by Chester and Hughes (1967) and Cronan (1976a). Samples are first leached with acetic acid to dissolve $CaCO_3$ and associated elements, and adsorbed ions; next with a mixture of hydroxylamine-HCl and acetic acid to dissolve the above plus amorphous iron minerals and reducible ferromanganese oxides together with the elements associated with these phases; and last with hot HCl to dissolve all the above and remaining Fe oxides, and to dissolve partially detrital and authigenic silicate material. The main value of this approach in geochemical exploration for marine minerals is in determining the form in which the enriched elements are present in the sediments, and thus the most likely source of the enrichment. It can be particularly useful when a metal is present in the bulk sediment in low concentrations as a result of dilution by biogenic or detrital phases, but nevertheless in a phase indicating hydrothermal input.

Using the above techniques, together with element distribution maps for the SW Pacific (Cronan, in press) areas of possible submarine hydrothermal activity in the region have been outlined (Figure 5). It is evident that hydrothermal deposits potentially have a widespread distribution in the SW Pacific. The areas of interest in this regard include: (a) the plate boundary in the Bismarck Sea; (b) the volcanically active submarine areas off Vanuatu; (c)

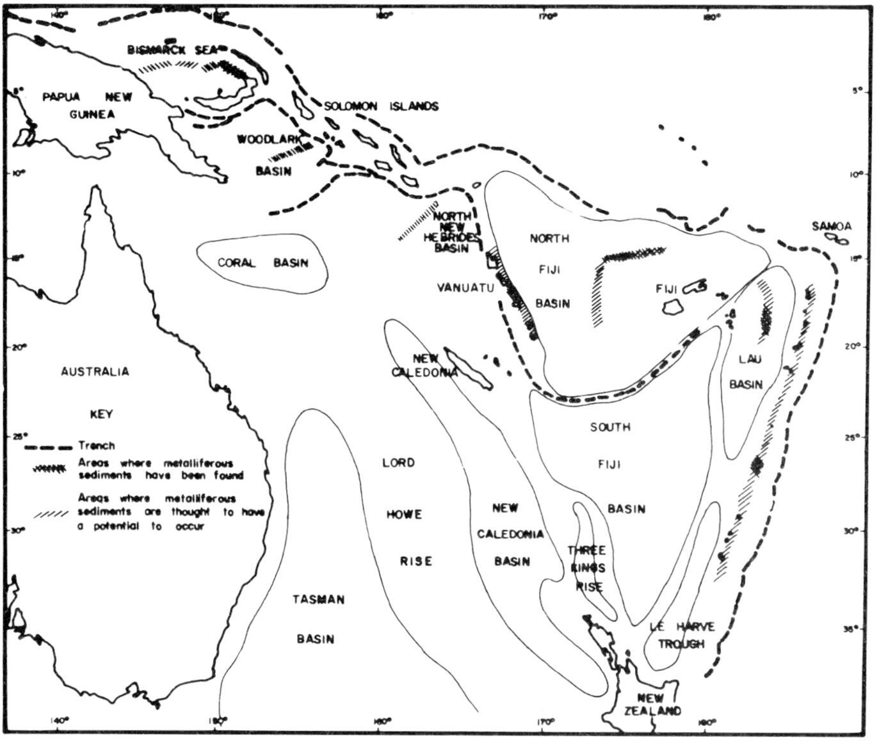

Figure 5 Areas of potential submarine hydrothermal mineralization in the South-west Pacific (from Cronan, 1983)

back arc spreading areas in the North Fiji Basin; (d) spreading areas in the Lau Basin; (e) the Tonga–Kermadec Ridge. All of these locations fall in the two tectonic settings most favourable for submarine hydrothermal activity in island arcs, either associated with spreading centres or transform faults in extensional back arc basins or associated with the volcanic arcs themselves.

As described, polymetallic sulphides are early precipitates from hydrothermal solutions, later precipitates from which are iron and manganese oxides. The hydrothermal deposits which have been located so far in the SW Pacific are all iron and manganese rich (Cronan *et al.*, 1982; Cronan, 1983; Exon and Cronan, 1983; Cronan *et al.*,) and thus represent the later stages of the hydrothermal fractionation sequence. Any sulphide minerals associated with them are likely to be present closer to the hydrothermal vents than has thus far been sampled, and it is therefore in this situation that they should be sought. Evidence in the Red Sea and elsewhere has shown that hydrothermal sulphides occur in greatest concentrations when the hydrothermal solutions discharge into depressions that can pond the precipitates. It would follow from this that the most likely setting for hydrothermal sulphide occurrences in the SW Pacific, and by analogy in other island arcs, would be within the restricted confines of submarine volcanic craters, or within downfaulted parts of axial rifts of back arc spreading centres. The closest deposits in the hydrothermal fractionation sequence to sulphides that have so far been found in the region are iron oxides occurring within the crater of a submarine volcano off Epi, Vanuatu (Exon and Cronan, 1983).

Manganese nodules and encrustations

Unlike metalliferous sediments of hydrothermal origin, nothing is 'given off' or dispersed away from manganese nodules and encrustations which can be used in exploration for them. By contrast, in exploration for potentially economic varieties of these deposits, a knowledge of the conditions under which they form will be the greatest aid in the search for them. Exploration programmes can then be directed to those environments in which such conditions are met. A deductive approach to manganese nodule and crust exploration can thus be fruitful.

Potentially economic manganese nodules and crusts comprise only a small proportion of the totality of these deposits on the sea floor. Certain cut-off grades and abundances of manganese nodules have been deemed necessary to support a first generation mining operation, but the same cannot be said for manganese crusts. The possible economic potential of these has only recently been mooted, and no reliable resource estimates are yet available for them.

It is generally agreed that the average abundance of nodules required for a first generation mine site is 10 kg m^{-2} or more, with a minimum of 5 kg m^{-2} (Archer, 1979). Compositionally, average and cut off contents of 2.27% and 1.18% combined Ni and Cu, the two elements of greatest economic importance in manganese nodules, are required to support a first generation mining operation (Kildow *et al.*, 1976). These grades and abundances occur within 200 miles of some of the islands of the South Pacific (Exon, 1983; Cronan, 1984) (see Figure 2).

The recent interest in the economic potential of manganese crusts lies principally in their Co content. The observation of Co enrichment in shallow seamount manganese crusts is not a new one. Cronan and Tooms (1969) pointed out that areas of Co enrichment in ferromanganese oxide deposits were centred on regions of elevated topography, mainly in the western and southern Pacific, where the deposits usually occur as encrustations of variable thickness on exposed rock surfaces. However, economic interest in these deposits was overshadowed by interest in the the manganese nodules, and until lately their economic potential has been considered to be low.

Several factors have led to the recent reappraisal of the economic potential of Co-rich manganese crusts. First, they tend to occur in shallow water areas, often 2000 m less than most potentially economic manganese nodule deposits, and sometimes within only a few hundred metres of the sea surface. Second, they are thought to cover extensively exposed rock surfaces on the sea floor. Third, Co is a strategic metal of high market value present in low concentrations in most Ni and Cu rich nodules, but present in maximum concentrations of up to 2% in some crusts. Fourth, by their very occurrence on seamounts and the submerged portions of volcanic islands, they will frequently fall within the EEZs of the adjacent island states and thus be subject to the jurisdiction of those states.

It is difficult to make reliable resource estimates of Co-rich crusts on the basis of the limited data available but, assuming a thickness of 2 cm and a density of 2.0 Halbach *et al.* (1982) have estimated a crust coverage of 40 kg m^{-2} on seamount slopes in parts of the Pacific. On this basis, these authors considered metal values in the crusts to be overall in the same order as in nodules from the potential mining area in the NE Pacific.

Conditions of formation of potentially economic nodules and crusts of relevance in marine geochemical exploration for the deposits

In the north-eastern equatorial Pacific, ore grade manganese nodules represent the end product of diagenetic enrichment of Ni and Cu. They contain these elements combined in excess of 3%, are rich in manganese, have a high Mn/Fe ratio, and contain the mineral todorokite as their principal mineral phase. They lie under the northern margin of the equatorial zone of high biological productivity, largely between the 50 and 100 grammes of carbon per square metre per year (g m^{-2} a^{-1}) productivity isolines, below the depth at which $CaCO_3$ dissolves, on sediment composed largely of siliceous ooze.

A number of workers have proposed models to account for the mechanism of formation of ore grade nodules in the north-eastern equatorial Pacific (see Cronan, 1980 for a review). According to these models, organic remains including faecal material containing metals extracted from the surface waters in the region of high biological productivity, sink to the ocean floor. Partial dissolution occurs within the water column, especially of calcareous organisms, but siliceous remains and faecal material, and probably a portion of the so called 'white snow' (ill-defined gelatinous organic material) reaches the sea floor. Once sedimented, the breakdown of the organic material results in diagenetic reactions liberating Mn, Ni and Cu into the interstitial waters of the uppermost sediments resulting in the formation of todorokite rich nodules and their enrichment in Mn, Ni and Cu.

On the basis of these models, it is evident that the primary requisite for ore grade nodule formation is elevated biological productivity, and thus it is in such areas elsewhere in the world oceans that ore grade nodules can most profitably be sought.

Pautot and Melguen (1979) have drawn attention to the importance of the lysocline (depth of significant increase in the rate of dissolution of $CaCo_3$) in the formation of both abundant and ore grade nodules. These workers found the preferential concentration of nodules in specific depth intervals in different basins of the South Pacific. In many cases, the maximum nodule abundance and grade was within the depth interval between the lysocline and the CCD[1] where the aggressive properties of the AABW[2] were promoting $CaCO_3$ dissolution (Aplin, 1983). Aplin and Cronan (1985) and Cronan (1984) have carried out geochemical investigations on a suite of nodules and associated sediments from the SW Pacific and found, as might be expected, that Mn, Ni and Cu rich todorokite-bearing nodules were restricted to areas underlying high productivity. They concluded that sinking faecal material was the principal route by which the enriched elements reached the sea floor. This was considered to represent a mechanism by which a mixture of oxides, metals and organic carbon is rapidly transferred from surface waters to the sediment water interface to drive the Mn oxide–organic carbon redox couple essential to the formation of todorokite. The importance of the sedimentary lysocline was believed to be that it marks a level below which sedimentation rates become sufficiently low adequately to concentrate these phases such that todorokite can form if the supply of the divalent cations Cu^{2+}, Ni^{2+}, Zn^{2+} and Mn^{2+} is sufficiently high.

It is evident from the above that in exploration for potentially economic nodules, two main environmental parameters have to be taken into account. First, deposits should be sought underlying areas of high biological productivity in order that enough organic material can be transported to the sea floor to supply the metals of interest in sufficient quantities and drive the diagenetic reactions leading to todorokite formation and Ni and Cu enrichment in the deposits. Second, deposits should be sought below the lysocline and most probably in the vicinity of and below the CCD because it is only here that sedimentation rates are sufficiently low to enable the metal rich organic phases to become sufficiently concentrated at the sediment water interface to form high-grade nodules. These reduced sedimentation rates also promote high nodule abundances on the sea floor. Thus under optimal geochemical conditions there should be a concurrence between high nodule grades and high abundances, in contrast to the often reported negative correlation between these parameters. However, even where grades are high and abundances seemingly low, because grade generally varies much less than abundance in a given area (Cronan, 1980) detailed search in that vicinity could locate areas of greater than average abundance. Grade would therefore appear to be more important than abundance in initial exploratory investigations for potentially economic nodules. Areas of potentially economic nodules near some of the islands in the equatorial Pacific based on the above considerations are shown in Figure 2.

Conditions under which cobalt enrichment in ferromanganese oxide crusts takes place are less well understood than the conditions of formation of ore grade nodules. Halbach *et al.* (1982) have noted a positive correlation between Co and Mn and the Mn/Fe ratio in the crusts that they examined, implying that Co abundance is controlled by the Mn content of the colloidal ferromanganese oxide particles that have been deposited from near bottom

[1]Carbonate compensation depth
[2]Antarctic bottom water

seawater. Cronan (1977) suggested that a reason for Co enrichment in such deposits was their slow growth rates and concommitant increased scavenging ability. Halbach *et al.* (1982) support this conclusion and go further in suggesting that only Co of the economically important minor metals in oceanic ferromanganese deposits increases in concentration as the growth rates of the deposits decrease.

Klinkhammer and Bender (1980) have noted an oxygen minimum in water depths between 900 m and 1300 m in the Central Pacific. These workers also pointed out that this is closely associated with an enrichment of dissolved Mn in seawater. Halbach *et al.* (1982) suggest that manganese diffuses from this zone into deeper water, resulting in the precipitation of Mn on substrates located close beneath the oxygen minimum zone. These precipitates have large surface areas enabling them to be efficient scavengers of transition metals such as Co.

The observations made by Halbach *et al.* (1982) have been largely confirmed in a suite of Line Islands crusts by Aplin (1983) and Aplin and Cronan (1985) who have provided more detail on the possible mechanisms leading to Co enrichment in the deposits, and consider that the dominant controls on the minor element composition of the crusts are the surface properties and crystal chemistry of the Mn and Fe oxide floculates.

It is evident from the above that in exploration for Co-rich manganese crusts, the position of the oxygen minimum zone in the water column is critical. Highest Mn/Fe ratios and thus Co contents, are associated with depths where Mn is enriched in waters containing between 0.09 and 2.2 ml l^{-1} dissolved oxygen (Aplin, 1983). Delineation of such areas should thus enable prediction of where the Mn/Fe ratio is likely to be high and thus where Co will be highest. It is evident from Figure 6 that the North Pacific is depleted in dissolved oxygen relative to the South Pacific. Values of less than 2.2 ml l^{-1} are rare in the Pacific south of 20°S and thus Co-rich Mn crusts are also likely to

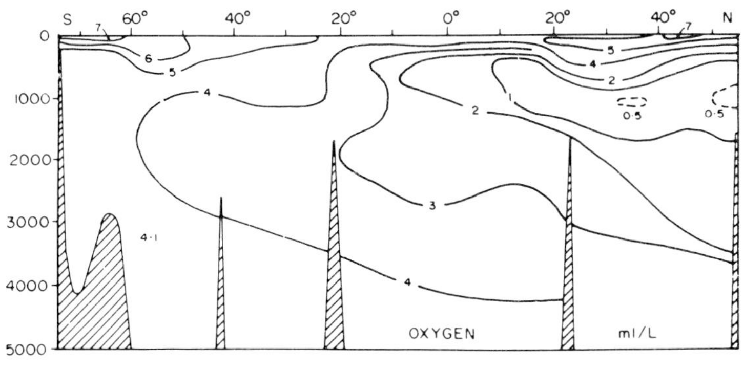

Figure 6 Distribution of dissolved oxygen along 175°W in the Pacific Ocean (from Pickard, 1975)

be rare south of this latitude. Horizontal advective-diffusion processes could lead to modification of this conclusion. Over much of the remainder of the Pacific, an oxygen minimum of less than 2.2 ml l^{-1} occurs between about 200 and 1500 m, deepening northwards (Figure 6). One would therefore predict the greatest probability of finding high Mn/Fe and high Co crusts in the Pacific is north of about 15°S and at depths between 300 and 2500 m. Some areas of the equatorial Pacific where these conditions may be met are shown in Figure 2.

Summary and conclusions

It is evident from the material presented in this chapter that conditions conducive to the formation of potentially economic deep sea mineral deposits such as metalliferous sediments, polymetallic sulphides and manganese nodules can occur within 200 miles of land, as well as in the deep ocean basins far from land.

Geochemical exploration for metalliferous sediments and polymetallic sulphides is best conducted in mid-ocean ridge and island arc areas using a multi-element approach. Manganese and iron dispersion haloes can define areas of submarine hydrothermal activity, and can be detected in sediments using factor and partition analysis even when the metal anomalies are present in very low concentrations due to dilution by calcareous and aluminosilicate debris. Zinc anomalies can possibly more closely pinpoint areas of sulphide occurrence. Nickel analyses enable non-hydrothermal authigenic manganese enrichments to be identified and disregarded. Favoured locations for high grade potentially economic submarine hydrothermal mineral deposits are in volcanic craters and downfaulted parts of back arc and oceanic spreading centres.

Geochemical exploration for potentially economic manganese nodules and encrustations is best centred on a deductive approach based on a knowledge of the environmental conditions under which such deposits form. Exploration can then be directed to such environments. Pre-requisites for high grade and abundant nodules occurrences are elevated biological productivity and a sea floor depth near or below the calcium carbonate compensation depth. Turbidite sedimentation is deleterious to the occurrence of both high grade and abundant nodules. Pre-requisites for the occurrence of Co rich manganese encrustations would appear to be a depth less than 2000 m and a dissolved oxygen minimum in the water column of less than 2.2 ml l^{-1}. Such conditions are widespread in the island and seamount groups of the central Pacific north of 15°S.

References

Aplin, A. (1983). The geochemistry and environment of deposition of some ferromanganese oxide deposits from the South Equatorial Pacific, *PhD Thesis, London University*.

Aplin, A. and Cronan, D.S. (1985). Ferromanganese oxide deposits from the Central Pacific Ocean, Geochim. et Cosmochim. Acta, **49**, in press.

Archer, A. (1976). Prospects for the exploitation of manganese nodules: the main technical economic and legal problems, *UNESCAP CCOP/SOPAC Tech. Bull 2* (Eds. G.P. Glasby and H. Katz), 21–39.

Archer, A. (1979). Resources and potential reserves of Ni and Cu. In: *Manganese Nodules: Dimensions and Perspectives*, pp. 71–82; D. Reidel Publishing Co., Dordrecht, Holland.

Bignell, R. (1975). The geochemistry of metalliferous brine precipitates and other sediments from the Red Sea, *PhD Thesis, London University*.

Bignell, R., Tooms, J.S., Cronan, D.S. and Horowitz, A. (1974). An additional location of metalliferous sediments in the Red Sea, *Nature*, **248**, 127–128.

Bignell, R., Cronan, D.S. and Tooms, J.S. (1976). Metal dispersion in the Red Sea as an aid to marine geochemical exploration, *Inst. Min. Metall. Trans.*, **85**, B273–B278.

Bonatti, E., Hônnorez, J., Joensuu, O. and Rydell, H. (1972). Submarine iron deposits from the Mediterranean Sea. In: *The Mediterranean Sea* (Ed. D.J. Stanley), pp. 701–710; Dowden, Hutchinson & Ross.

Bostrom, K. and Peterson, M.N.A. (1966). Precipitates from hydrothermal exhalates on the East Pacific Rise, *Econ. Geol.*, **61**, 1258–1265.

Chester, R. and Hughes, M.J. (1967). A chemical technique for the separation of ferro-manganese minerals, carbonate minerals, and adsorbed trace elements from pelagic sediments, *Chem. Geol.*, **2**, 249–262.

Colley, N., Cronan, D.S. and Moorby, S.A. (1979). Some geochemical and mineralogical studies on newly collected ferromanganese oxide deposits from the north-west Indian Ocean. In: *Sur la Genese des Nodules de Manganese* (Ed. C. Lalou), pp. 13–21; Colloque Internat. du CNRS, No. 289.

Cronan, D.S. (1967). The geochemistry of some manganese nodules and associated pelagic deposits, *PhD Thesis, London University*.

Cronan, D.S. (1975). Manganese nodules and other ferromanganese oxide deposits from the Atlantic Ocean, *J. Geophys. Res.*, **80**, 3831–3837.

Cronan, D.S. (1976a). Implications of metal dispersion from submarine hydrothermal systems for mineral exploration on mid-ocean ridges and in island arcs, *Nature*, **262**, 567–569.

Cronan, D.S. (1976b). Geochemistry of metalliferous sediments from the Eastern Pacific, *Bull. Geol. Soc. Am.*, **87**, 928–934.

Cronan, D.S. (1977). Deep Sea Manganese Nodules: Distribution and Geochemistry. In: *Marine Manganese Deposits* (Ed. G.P. Glasby), pp. 11–44; Elsevier, Amsterdam.

Cronan, D.S. (1978). Recent advances in geochemical exploration for undersea mineral deposits, *Inst. Min. Metall. Trans.*, **A87**, 143–146.

Cronan, D.S. (1980). *Underwater Minerals*; Academic Press, London.

Cronan, D.S. (1981). Local variability in the grade and abundance of manganese nodules: implications for manganese nodule mine site evaluation, *Proc. ECOR Symposium*, London.

Cronan, D.S. (1983). Metalliferous sediments in the CCOP/SOPAC region of the Southwest Pacific with particular reference to geochemical exploration for the deposits, *CCOP/SOPAC, Tech. Bull*, **4**, Suva.

Cronan, D.S. Regional geochemistry of sediments from the Southwestern Pacific. In: *Sedimentation and Mineral Deposits in the Southwest Pacific Ocean* (Ed. D.S. Cronan); Academic Press, London, in press.

Cronan, D.S. (1984). Criteria for the recognition of areas of potentially economic manganese nodules and encrustations in the CCOP/SOPAC area of the central and southwestern Pacific, *South Pacific Marine Geology Notes*, 1984, **3**, 1–17.

Cronan, D.S. and Tooms, J.S. (1967). Geochemistry of Mn nodules from the NW Indian Ocean, *Deep Sea Res.*, **14** ,239–249.

Cronan, D.S. and Tooms, J.S. (1969). The geochemistry of Mn nodules and associated pelagic deposits from the Pacific and Indian Oceans, *Deep Sea Res.*, **16**, 335–359.

Cronan, D.S. and Garrett, D.E. (1973). Distribution of elements in metalliferous Pacific sediments collected during the Deep Sea Drilling Project, *Nature*, **242**, 88–89.

Cronan, D.S. and Moorby, S.A. (1976). Preliminary results of renewed investigations on manganese nodules and encrustations in the Indian Ocean, *UNESCAP CCOP/SOPAC Tech. Bull.*, **2**, (Eds. G.P. Glasby and H. Katz), 118–123.

Cronan, D.S. and Moorby, S.A. (1981). Manganese nodules and other ferro-manganese oxide deposits from the Indian Ocean, *J. Geol. Soc. Lond.*, **138**, 527–539.

Cronan, D.S. and Thompson, B. (1978). A preliminary regional geochemical reconnaissance survey for metalliferous sediments in the southwestern Pacific, *Inst. Min. Metall. Trans.*, **87**, B87–B90.

Cronan, D.S. and Varnavas, S.P. (1981). Hydrothermal and dissolution related geochemical variations in sediments from an East Pacific Rise Fracture Zone at 9°S, *Oceanologica Acta. S. P.*, 47–58.

Cronan, D.S., Rona, P.A. and Shearme, S. (1979). Metal enrichments in sediments from the TAG Hydrothermal Field, *Marine Mining*, **2**, 79–89.

Cronan, D.S., Glasby, G.P., Moorby, S.A., Thompson, J., Knedler, K. and McDougall, J.C. (1982). A submarine hydrothermal manganese deposit from the south-west Pacific island arc, *Nature*, **298**, 456–458.

Cronan, D.S., Moorby, S.A., Glasby, G.P., Knedler, K., Thompson, J. and Hodkinson, R. (1984). Hydrothermal and volcaniclastic sedimentation on the Tonga-Kermadec Ridge and in its adjacent marginal basins, *J. Geol. Soc. Lond. Spec. Publ. 16* (Eds. B.P. Kokelaar and M.F. Howells), 137–149.

Degens, E. and Ross, D. (Eds.) (1969). *Hot Brines and Recent Heavy Metal Deposits in the Red Sea*; Springer Verlag, New York.

Exon, N.F. (1983). Manganese nodule deposits in the central Pacific Ocean and their variation with latitude, *Marine Mining*, **4**, 79–107.

Exon, N.F. and Cronan, D.S. (1983). Hydrothermal iron deposits and associated sediments from submarine volcanoes off Vanuatu, Southwest Pacific, *Marine Geol.*, **52**, M43–M52.

Frazer, J. (1977). Manganese nodule reserves, an updated estimated, *Marine Mining*, **1**, 103–124.

Glasby, G.P. (1970). The geochemistry of manganese nodules and associated pelagic deposits from the North-west Indian Ocean, *PhD Thesis, London University.*

Glasby, G.P., Tooms, J.S. and Howarth, R.J. (1974). Geochemistry of manganese concretions from the NW Indian Ocean, *NZ J. Sci.*, **17**, 387–407.

Halbach, P., Manheim, F.T. and Otten, P. (1982). Co-rich ferromanganese deposits in the marginal seamount regions of the Central Pacific Basin—results of Midpac '81, *Erzmetall*, **35**, 447–453.

Holmes, R. and Tooms, J.S. (1972). Dispersion from a submarine exhalative ore body. In: *Geochemical Exploration 1972*, pp. 193–202; Inst. Min. Metall., London.

Horowitz, A. (1974). Geochemical investigations of sediments associated with the Mid-Atlantic Ridge, *PhD Thesis, London University.*

Horowitz, A. and Cronan, D.S. (1976). Geochemistry of basal Deep Sea Drilling Project sediments from the North Atlantic, *Marine Geol.*, **20**, 205–228.

Kildow, J.T., Bever, M.B., Dar, V.K. and Capstaff, A.E. (1976). *Assessment of Economic and Regulatory Conditions Affecting Ocean Minerals Resource Development*; Prepared by MIT for US Dept. Interior, Ocean Mining Admin.

Klinkhammer, G.P. and Bender, M.L. (1980). The distribution of manganese in the Pacific Ocean, *Earth Planet. Sci. Lett.*, **46**, 361–384.

Malahoff, A. (1981). Comparison between Galapagos and Gorda spreading centres, *Proc. Offshore Tech. Conf. OTC 4129*, 115–120.

Malahoff, A., Embley, R.W., Cronan, D.S. and Skirrow, R. (1983). The geological setting and chemistry of hydrothermal sulphides and associated deposits from the Galapagos Rift at 86°W, *Marine Mining*, **4**, 123–137.

Mero, J.L. (1965). *The Mineral Resources of the Sea;* Elsevier, Amsterdam.

Moorby, S.A. (1978). The geochemistry and mineralogy of some ferromanganese oxides and associated deposits from the Indian and Atlantic Oceans, *PhD Thesis, London University.*

Pautot, G. and Melguen, M. (1979). Influence of deep water circulation and sea-floor morphology on the abundance and grade of central South Pacific manganese nodules. In: *Marine Geology and Oceanography of the Pacific Manganese Nodule Province* (Eds. J.L. Bischoff and D.Z. Piper), pp. 621–652; Plenum Press, New York.

Pickard, G.L. (1975). *Descriptive Physical Oceanography;* Pergamon Press, Oxford.

Shearme, S. (1981). The geochemistry of sediments from the Atlantic and Pacific Oceans, *PhD Thesis, London University.*

Shearme, S., Cronan, D.S. and Rona, P.A. (1983). Geochemistry of sediments from the TAG Hydrothermal Field, MAR at latitude 26°N, *Marine Geol.*, **51**, 269–291.

Smith, P.A. and Cronan, D.S. (1975). Dispersion of metals associated with a submarine exhalative deposit, *Proc. Oceanology Int. Brighton, 1975*, 111–114.

Smith, P.A. and Cronan, D.S. (1983). The geochemistry of metalliferous sediments and waters associated with shallow submarine hydrothermal activity (Santorini, Aegean Sea), *Chem. Geol.*, **39**, 241–262.

Tooms, J.S. (1970). Metal deposits in the Red Sea, their nature, origin and economic worth, *Underwater Sci. and Technology Journal*, **March**, 28–33.

Varnavas, S.P. (1981). Geochemical investigations on sediments from the eastern Pacific, *PhD Thesis, London University.*

Yuasa, M. and Yokota, S. (1982). Hydrothermal manganese and ferromanganese concretions from sea floor of the Ogasawara arc-trench region, northwestern Pacific, *UNESCAP/CCOP Tech. Bull.*, **15**, 51–64.

Geochemistry and Animal Health

G. LEWIS

Copper deficiency in cattle and sheep

The first suggestion that deficiency of copper occurs naturally in grazing animals was made in the early 1930s in reports from America and Holland (Neal *et al.*, 1931; Sjollema, 1933). In 1937 Bennetts and Chapman in Western Australia showed that enzootic ataxia, a disease of lambs, could be prevented by copper supplementation of the ewe during pregnancy. This nervous disorder in lambs, characterized by incoordination of movement, has long been recognized in many parts of the world and given various local names. In Britain for example the condition is commonly known as 'swayback' or 'warfa'.

All these conditions appear to be pathologically similar and the disease as described by Innes and Shearer (1940) consists of lesions of demyelination of the spinal cord involving mostly the ventro and dorsal aspects of the white matter. Diffuse symmetrical demyelination occurs in the cerebral white matter of acute cases, ranging from slight involvement to gross degeneration and liquifaction of almost all the white matter. This gross lesion, with cavities in the brain, is not uncommon in acute neo-natal cases of swayback in Britain but is rarely seen in cases of neo-natal enzootic ataxia in Australia.

In delayed swayback, which is essentially similar to enzootic ataxia, lambs appear clinically normal at birth but develop signs of the disease from a few days to 6 months of age. The essential lesion consists of degeneration and necrosis of the motor neurones of the brain stem and spinal cord and degeneration of the nerve fibres of the spinal cord. It has been shown that the lesion of delayed swayback can develop after birth in copper deficient lambs (Lewis *et al.*, 1974).

The disease has been reported in most parts of Britain and is endemic in certain areas. Incidence varies from year to year but is likely to be lower in a hard winter than a mild one. The disease only occurs when copper concentrations are low in the tissues of a ewe and its lamb. However not every copper deficient ewe will give birth to a swayback lamb and not every copper deficient lamb will develop swayback. Lambs may be affected at birth or the clinical signs may appear a few days to several months after birth. Incoordinated movements of the hind legs, a stiff, staggering gait and swaying of the hind quarters are shown progressively as the disease develops until locomotion finally becomes impossible. Mild cases may also occur in which the ataxia is only apparent when lambs are hurried or driven some distance. Appetite usually remains unimpaired and the lambs grow well until the final stages when they are unable to feed.

A copper deficiency disease called 'peat scours' in cattle on peat land in New Zealand was reported by Cunningham (1946). The most characteristic symptoms were loss of condition and, in dairy cattle, persistent severe diarrhoea which occurred whenever there was flush growth of pasture. Blood and liver copper levels of cattle grazing were low compared with values in cattle grazing healthy areas. The first suspicion of copper deficiency in cattle in Britain was reported, also in 1946, by Jamieson and Russell on land newly reclaimed from heather in Aberdeenshire. Diarrhoea, emaciation and anaemia were features but no blood or liver copper values were given to confirm the condition. However a single analysis of herbage showed a low copper content of 2.5 mg kg^{-1}. In the same year, Allcroft (1946) demonstrated low blood copper levels in a small dairy herd suffering a severe scouring disorder on peat land on the border of Shropshire and Cheshire. Subsequently Allcroft and Parker (1949) showed, on a larger neighbouring dairy farm in the same area, that daily administration of 2 g copper sulphate incorporated in a high protein cake over a period of 14 months increased blood copper levels to normal and resulted in a marked improvement in clinical condition with control of the previous persistent diarrhoea. The copper content of the herbage appeared adequate (mean 14.2 mg kg DM; range 8.7–22.7) but molybdenum levels were higher than normal (mean 4.4 mg kg^{-1} DM; range. These preliminary observations indicated that although the copper content of the pastures was normal, a low copper status of the animals was a factor in the disease syndrome and administration of copper produced a marked improvement in the condition.

The calves on the Shropshire farms were stunted and did not do well and in Caithness a survey of 123 herds showed that 65% had hypocupraemic disorders. The beef breed calves, mainly Aberdeen-Angus crossbred and crossbred Highland calves, born in April/May and then pastured with their mothers till autumn, showed symptoms about a month after turning out to grass. The calves showed a stilted gait, progressive signs of malnutrition and a rough dull and discoloured coat. In some cases there was depigmentation, so the black hair round the eyes changed to grey giving the calves a 'spectacled' appearance. Diarrhoea was not a constant feature and anaemia was not found. Investigating this 'copper pine' in calves Jamieson and Allcroft (1950) showed that small doses of copper given either orally or intravenously produced a dramatic improvement in bodily condition and weight gain. The quality of the pastures was good and the pining was not associated with lack of food, the copper content of the herbage was normal and molybdenum values varied from 1.1–19.5 mg kg^{-1} DM.

Ferguson *et al.* (1943a) showed that the disease of cattle, known locally as 'teart', which occurred on the Lower Lias clay outcrops in Somerset and small areas of Gloucester and Warwick was due to a high molybdenum content of soil and pasture. 'Teart' which had been recognized for many years affects mainly cows in milk and young stock, severe diarrhoea may commence within two days of putting cattle on to affected pastures in Spring. The animals develop rough, staring, discoloured coats and rapidly lose condition; milk yield drops. Ferguson *et al.* (1943b) showed that the diarrhoea could be controlled and the animals maintained in good health on 'teart' pasture by the oral administration of copper sulphate. They did not, at that time, examine the copper status of cattle in the 'teart' area. It was fortuitous and with hindsight perhaps very fortunate that, in those days, copper sulphate because of its astringent properties was the common remedy for any scouring disorder whatever the cause.

The scouring effect of 'teart' pastures has been regarded as a straightforward molybdenosis (Allcroft, 1952) since it could be produced in a time which seemed too short to allow depletion of the animals copper reserves to deficiency levels. However if the intake of molybdenum is prolonged, even at lower dietary concentrations where diarrhoea is not a marked effect, the copper status of the animal will fall.

Work by Dick in Australia (1953a, b) has shown that the limiting effect of molybdenum on copper storage in sheep could be influenced by the inorganic sulphate content of the diet. When this was low, sheep stored more copper than those fed a diet of similar copper and molybdenum content but of higher sulphate concentration. However, analysis of herbage from areas of England where copper deficiency was either prevalent or uncommon did not reveal any marked differences in sulphate content and all values were on the high side in relation to Dick's work (Allcroft and Lewis, 1957). It was concluded at that time, that variation in the sulphate content of herbage was not a major factor in predicting whether copper deficiency might or might not occur in animals grazing a particular sward.

Two early studies of blood copper levels carried out in conjunction with Veterinary Investigation Officers in England, Wales and Scotland showed that the distribution of hypocupraemia in cattle was widespread (Allcroft, 1952; Allcroft and Uvarov, 1959). Although the condition was found in varying degrees of severity in 'teart' and peaty areas it was by no means confined to them and there was generally little correlation with the geological map of England and Wales.

The analyses of stream sediment samples carried out by Professor Webb's group at Imperial College and, initially, production of maps of selected areas using the geochemical reconnaissance data particularly in relation to molybdenum, provided a valuable tool for the investigation of the distribution of copper deficiency in grazing ruminants.

Clinical copper deficiency had been recognized for many years on a number of farms within an area of 10 square miles in North Staffordshire and appeared to be related to a dietary excess of molybdenum, confirmed by analysis of soil and herbage. However, geochemical reconnaissance in 1965 delineated areas totalling 60 square miles in Derbyshire and North Staffordshire (including the above area) in which molybdenum occurred in above normal concentrations in stream sediments, rocks, soils and pasture. The source of the molybdenum was found to be black marine shales of Carboniferous age. Within the areas the presence of molybdeniferous soils and pasture is related to local variation in geology and parent material and is particularly affected by drift cover.

Analysis of blood copper on 350 cattle from 26 herds (excluding those in which copper deficiency had been demonstrated previously) indicated hypocupraemia in 64% within the geochemically defined molybdeniferous area compared to 35% in nearby control areas. On the whole, deficient animals showed no marked clinical signs. However, copper supplementation of young dairy animals grazing within the molybdenum rich areas produced mean live weight gains of 10–70% representing mean herd increases of 30–70 lb per animal over a 6 month grazing period. Control untreated animals on these farms, with few exceptions, showed no clinical signs of hypocuprosis in spite of their low blood copper levels (Thornton *et al.*, 1972).

In another study (Thompson *et al.*, 1972), stream sediments were sampled systematically in nine areas totalling 1100 square miles underlain in part by marine black shales. Regional patterns of molybdenum in the sediment were related to the underlying bedrock. Soils developed from the black shale parent material in eight of the areas contained above normal concentrations of molybdenum (>3 ppm). The molybdenum content of herbage was higher in these areas (up to 7.2 ppm/DM) of black shale parent material than in surrounding areas underlain by other rocks.

Blood copper results, on diagnostic samples submitted to the Central Veterinary Laboratory through the Veterinary Investigation Service over a number of years, were mapped for the relevant areas. There was a good correlation between sites where hypocupraemia had been confirmed by blood analysis and the outcrops of marine black shales, particularly in the area stretching from Dorset, Somerset and the Severn Valley across Buckinghamshire and Oxfordshire to the Wash, Lincolnshire and the northeast coast of England. For the first time we were able to correlate the presence or likely presence of hypocupraemia in grazing cattle with a geological map, which also delineated areas where agricultural practices such as liming might increase the incidence of livestock disorders.

Swayback, a condition affecting the nervous system of lambs, occurs widely in Britain. The incidence varies from year to year but is usually most severe in lambs born in the Spring after a mild winter. It is related to a copper deficiency in the ewe and its lamb and can be prevented by administration of copper to ewes in susceptible flocks during pregnancy. The relationship between copper and molybdenum in the diet of the grazing sheep and the aetiology of the disease has not been as clear cut as the role of molybdenum in inducing copper deficiency in the bovine.

Early work in Derbyshire, where the disease is endemic, showed that herbage copper concentrations were adequate. It was not considered that sufficient molybdenum was present in the diet, on farms with a history of swayback, to contribute to the disease under natural conditions (Allcroft and Lewis, 1957; Suttle, 1970). However, the disease has been produced under experimental conditions by maintaining ewes, of low copper status, on diets of high molybdenum and sulphate content (Fell *et al.*, 1961; Suttle and Field, 1968a, b).

Barlow *et al.* (1960) concluded that in southeast Scotland a dietary deficiency of copper was the underlying factor and the disease has been produced experimentally by feeding a semi-purified diet of low copper with no added molybdenum (Lewis *et al.*, 1974).

Alloway (1973) made a study of trace elements in soils and herbages from farms in three major swayback areas, north Derbyshire, Northamptonshire and Denbighshire, as part of the programme on agricultural applications of geochemical reconnaissance data. Control samples from farms with no history of the disease were also collected in Northamptonshire.

Copper: molybdenum ratios in the samples of winter herbage were calculated and were lowest for the most severely affected areas, Derbyshire and Northamptonshire, where both swayback and bovine hypocupraemia are endemic. Swayback incidence is less severe and more spasmodic in Denbighshire and appears to be worse on recently limed and reseeded pastures. This was reflected in the Cu: Mo ratios which had a higher mean value and wider range than for the other two areas. The Northamptonshire swayback farms had a much lower mean ratio (4.7) than the controls (8.8) and included some very low values. The mean molybdenum content of herbage on the swayback farms was double that of the controls.

The molybdenum content of the herbage is becoming increasingly important with the widespread use of single species swards, particularly of ryegrass. The copper content of leys cut at a moderately mature stage is usually below 10 mg kg^{-1} DM and frequently well below the 8 mg kg^{-1} DM which the Agricultural Research Council (1980) calculated to be the dietary concentration the ruminant may often need to maintain an adequate copper status. Modern fertilizer practice and grassland husbandry has tended to increase the molybdenum content and lower the copper content of herbage, the sulphur content may be increased by such practices particularly on improved grazings in hill areas.

Over the last decade there has been an increasing reliance on fresh herbage and silage in the diet and it has been shown that copper in fresh grass is less readily absorbed than from hay, with the availability from silage intermediate (Suttle, 1982). Other changes may have given rise to an increase in the incidence of copper deficiency throughout the UK. Trends have been towards higher stocking rates, herd sizes have risen and more cows are calving off grass in the autumn, a time when liver reserves in the dam (particularly first calving heifers) may be low. The calves will then start life with a low copper status, and the effects exacerbated by the increase in the production of larger, more rapidly growing calves with a greater demand for copper.

These effects were reflected in the dramatic increase in the number of blood and tissue samples received by the Veterinary Investigation Service for determination of copper. Results have shown that hypocupraemia especially in young stock is common. In many instances it is not until marked clinical signs appear in calves, that farmers first appreciate that they have a copper deficiency problem which may have resulted from changes in management and husbandry made over the last 5–10 years (Lewis, 1980). The distribution of bovine hypocupraemia in England and the Welsh borders, based on blood samples analysed by the MAFF from 1977–1980, has been mapped (Leech *et al.*, 1982) and the incidence discussed in relation to the regional distribution of elements published in the *Wolfson Geochemical Atlas*.

It is apparent that several hypocupraemic regions fall outside the molybdeniferous areas and so must be attributed to other causes including a low intake of copper, further multi-element interactions involving copper and management effects. The geochemical atlas will provide valuable data for further studies particularly in those areas where multi-element interactions may be involved.

Cobalt

Within the British Isles it had been reported since the late 18th Century that a 'pining' condition occurred in ruminants on certain grazings and the worst effects could be avoided by moving the animals periodically to 'healthy' grazings in other areas. The first evidence that cobalt is a dietary essential was obtained almost 50 years ago when Underwood and Filmer (1935) and Filmer and Underwood (1937) showed that two naturally occurring debilitating diseases of sheep and cattle in Australia could be cured and prevented by small oral doses of cobalt salts. Subsequently 'pining' in sheep and cattle in Britain could be associated with soils of low cobalt content derived from granites and Old Red Sandstone.

Later cobalt was shown to be an essential constituent of vitamin B_{12} and in 1951 Smith and others effected complete remission of all signs of cobalt deficiency in lambs by injection of vitamin B_{12}. A unique nutritional situation was disclosed, in which animals utilize a trace element solely as an integral part of a vitamin and are dependent upon rumen bacteria for synthesis of their supply of the vitamin. For this reason cobalt given by injection is ineffective for treatment since it does not reach the rumen organisms; it must be given orally and regularly. The cobalt status of ruminant diets is therefore crucial to growth and health of the animals.

In both cattle and sheep severe cobalt deficiency is characterized by loss of appetite, a progressively developing anaemia, ill-thrift and loss of weight leading to emaciation and death if treatment is not given. Sheep may show a varying degree of lachrymation, cattle may develop a rough coat and depraved appetite.

Sub-clinical deficiency, especially in lambs, may be the most important economically. No marked signs are seen but the animals do not thrive. The poor weight gain in lambs may not be appreciated until the animals fail to reach market weight at the expected time in the autumn.

Cobalt deficiency, resulting from a low cobalt intake by grazing ruminants, may be caused in three ways, inherent, developed or induced. Inherent deficiencies arise where the parent materials are so impoverished with respect to cobalt that the soils derived from them are of low cobalt content and the herbage grown on them will not contain sufficient cobalt to meet the requirement of the grazing animal. Developed deficiencies result from such soil-forming processes as podzolization where cobalt is leached from the surface horizons and is not available for assimilation by plants. Geological and geochemical maps based on stream sediment analysis indicate areas where deficiencies from these causes may be expected.

Farming practices, such as the use of lime to raise the pH of the soil, reduce the availability of trace elements, particularly cobalt and manganese, to the herbage and hence may induce a cobalt deficiency. The trace element content of plants varies with species, growth rate and stage of maturity. When hill land is improved lime and fertilizers are applied and the botanical composition of the sward is altered. Mitchell (1971) has shown that renovating and reseeding may markedly reduce the cobalt content of pastures under these conditions. Increased stocking rates may exacerbate the problem in livestock. The geochemical map will indicate areas where changes in farming practice, particularly liming, are most likely to potentiate the risk of inducing an inadequate cobalt intake by grazing ruminants, and where cobalt supplements or cobalt fertilizers may need to be used to maintain a cobalt status in the stock.

Selenium and vitamin E

In Britain the primary selenium deficiency disease is white muscle disease (WMD), an acute myodegeneration of calves and lambs, also seen less frequently in yearlings and infrequently in adults. Ill-thrift and infertility in sheep, although seen in Australia and New Zealand, have not been confirmed during recent investigations in Britain.

Selenium deficiency is accompanied by low activity of the selenium-containing enzyme glutathione peroxidase (GSH–Px). This enzyme is present in many tissues of the body and its low activity in muscle is regarded to be at least part of the aetiology of white muscle disease. With the development of methods for the rapid and simple measurement of selenium (as GSH–Px) in blood, widespread hyposelenaemia has been demonstrated in both cattle and sheep in Britain (Anderson *et al.*, 1979). This condition is not necessarily synonymous with disease but indicates areas of potential risk. As with other trace elements, in general these are present as herd or flock problems where individuals succumb with clinical signs when subjected to greater stress. This may be nutritional, climatic or physiological, as in late pregnancy, at parturition or during periods of rapid growth.

Selenium and vitamin E are interrelated. Many of the selenium-responsive diseases can be treated with selenium or vitamin E, the vitamin will spare the requirement of selenium and vice versa. They have been shown to act at different points in the metabolic pathway. Unsaturated fatty acids from dietary lipid may be oxidized to lipid peroxides, accumulation of peroxides within cells is potentially dangerous and can cause disruption of cell membranes and eventual cell death. In the presence of adequate vitamin E, an anti-oxidant, formation of lipid peroxides is restricted or prevented. The selenium-containing enzyme, glutathione peroxidase, acts in the conversion of the lipid peroxides to harmless alcohols. Vitamin E and selenium (as GSH–Px) therefore act together to prevent the accumulation of lipid peroxidases–vitamin E by restricting their formation and GSH–Px by accelerating their elimination.

A good source of vitamin E is not necessarily a good source of selenium and vice versa. Fresh grass, dried grass, silage and kale are all good sources of the vitamin. Cereal grains, maize silage and good quality hay have a moderate content, while straw, root crops and inferior quality hay are poor sources, However, the vitamin E content of certain feedstuffs can be seriously depleted under some storage conditions, for example when cereal grains are stored 'wet', either in sealed silos or after treatment with proprionic acid. The vitamin E content of grass falls with maturity, so the vitamin E content of silage may vary greatly depending on the stage of growth from which it is made.

In contrast, the concentration of selenium in feedstuffs is principally related to the geographical source, much less to the type of feed and even less to storage. Although there are regional differences in Britain, none have very high soil selenium concentrations. At any given location the variation in selenium concentration between home grown feedstuffs such as hay, silage and cereal grain is usually small. The selenium content of oil seed cakes and proprietary protein concentrates is usually high and when fed to milking cows will maintain an adequate selenium status, however grazing animals, particularly sheep, dairy heifers and beef cattle are more dependent on home grown feeds.

A survey by Anderson *et al.* (1979) of blood selenium levels (as GSH–Px) in sheep on 329 farms in Britain showed that 47% of these farms were unable to provide sufficient selenium to maintain blood selenium levels greater than 0.075 $\mu g\,ml^{-1}$ which is regarded as 'marginal'. A large proportion of farms with the lowest levels were to be found in Wales and bordering countries, southern and southeast England. In Scotland most of the samples were taken in the Grampian region where 60% of the farms had low levels. This supports the studies of Blaxter (1962) on the occurrence of vitamin E or selenium-responsive nutritional myodegeneration of cattle in northeast Scotland. Soils in these areas were deficient in selenium.

The present evidence on soil selenium content does not suggest that there are areas of marked soil deficiency in England and Wales or that selenium content of herbage can be directly correlated with soil type. However it is evident that in many areas home grown feedstuffs supply insufficient selenium to maintain an adequate selenium status in certain classes of stock. Anderson *et al.* (1976) suggested that the increase in white muscle disease in Britain during the previous five years could be related to modern farming practices which tend to produce combined vitamin E and selenium deficiencies at the end of the winter feeding period.

Conclusions

Most of the sheep and a large proportion of the cattle population in Britain are heavily dependent on home grown feeds, often only grass and grass products for their dietary intake of the trace elements necessary for health, reproduction and development. It has been estimated that grassland provides almost three quarters of the total crude protein for ruminants in Britain and cereals provide another eighth. We must assume that home grown feeds are a major contribution to animal diets and that without supplementation the intake of trace elements is largely dependent on the concentration in herbage. Time has permitted consideration of only three of these elements, but it is evident that diets based largely on grass and home grown cereals are unlikely to meet the requirements for many trace elements particularly copper, cobalt, selenium and iodine at all stages in the animals' development and physiological cycle. The interrelations and interactions of elements in affecting availability from soil to plant, and plant to animal are becoming of increasing interest and importance, particularly with changes in farming practice and increased productivity both arable and animal.

Great progress has been made but increasing knowledge of geochemical, soil, plant and animal trace element relationships will enable: (a) a better definition of areas where problems may be encountered in crops and animals; (b) better understanding of husbandry practices likely to reduce or increase the risk of such problems; and (c) better advice to be given so that, where necessary, trace element supplements may be used safely, efficiently and economically to provide adequate intakes for the health of farm animals at all physiological stages.

References

Agricultural Research Council (1980). *Nutrient Requirements of Ruminants*, (2nd edition); CAB Farnham.

Allcroft, R. (1946). Hypocupraemia in cattle, *Nature*, **158**, 796.

Allcroft, R. (1952). Conditioned copper deficiency in sheep and cattle in Britain, *Veterinary Record*, **64**, 17–24.

Allcroft, R. and Lewis, G. (1957). Copper nutrition in ruminants. Disorders associated with copper–molybdenum–sulphate content of feeding stuffs, *J. Sci. Food Agriculture*, **8**, S96–S104.

Allcroft, R. and Parker, W.H. (1949). Hypocupraemia in dairy cows, *Brit. J. Nutrition*, **3**, 205–217.

Allcroft, R. and Uvarov, O. (1959). Parenteral administration of copper compounds to cattle with special reference to copper glycine (copper aminoacetate), *Veterinary Record*, **70**, 797–810.

Alloway, B.J. (1973). Copper and molybdenum in swayback pastures, *J. Agricultural Sci., Cambridge*, **80**, 521–524.

Anderson, P.H. and Berrett, S. and Patterson, D.S.P. (1976). Some observations on "paralytic myoglobinuria" of cattle in Britain, *Veterinary Record*, **99**, 316–318.

Anderson, P.H., Berrett, S. and Patterson D.S.P. (1976). Some observations on "paralytic stock in Britain as indicated by sheep erythrocyte glutathione peroxidase activity, *Veterinary Record*, **104**, 235–238.

Barlow, R.M., Purves, D., Butler, E.J. and MacIntyre, I.J. (1960). Swayback in southeast Scotland. 1. Field aspects, *J. Comparative Pathology*, **70**, 396–410.

Bennetts, H.W. and Chapman, F.E. (1937). Copper deficiency in Western Australia: A preliminary account of enzootic ataxia of lambs and an anaemia of ewes. *Australia Veterinary J.*, **13**, 138–149.

Blaxter, K.L. (1962). Muscular dystrophy in farm animals: its cause and prevention, *Proc. Nutrition Soc.*, **21**, 211–216.

Cunningham, I.J. (1946). Copper deficiency in cattle and sheep on peat lands, *New Zealand J. Sci. Technol.*, **27**, Sect. A. 372–381.

Dick, A.T. (1953a, b). The effect of inorganic sulphate on the excretion of molybdenum in sheep. The control of copper storage in the liver of the sheep by inorganic sulphate and molybdenum. *Australian Veterinary J.*, **29**, 18–26 and 233–239.

Fell, B.F., Williams, R.B. and Mills, C.F. (1961). Further studies of nervous tissue degeneration resulting from "conditional" copper deficiency in lambs, *Proc. Nutrition Soc.*, **20**, xxvii.

Ferguson, W.S., Lewis, A.H. and Watson, S.J. (1943a, b). The teart pastures of Somerset. I. The cause and cure of teartness. IV. The effect of continuous administration of copper sulphate to dairy cows, *J. Agricultural Sci., Cambridge*, **33**, 44–51 and 116–118.

Filmer, J.F. and Underwood, E.J. (1937). Enzootic marasmus. Further data concerning the potency of cobalt as a curative and prophylactic agent, *Australian Veterinary Journal*, **13**, 57–63.

Innes, J.R.M. and Shearer, G.D. (1940). "Swayback": A demyelinating disease of lambs with affinities to Schilder's encephalitis in man, *J. Comparative Pathology and Therapeutics*, **53 (1)**, 1–39.

Jamieson, S. and Allcroft, R. (1950). Copper pine of calves, *Brit. J. Nutrition*, **4**, 16–31.

Jamieson, S. and Russell, F.C. (1946). Suspected copper deficiency in cattle in Aberdeenshire, *Nature*, **157**, 22.

Leech, A., Howarth, R.J., Thornton, I. and Lewis, G. (1982). Incidence of bovine copper deficiency in England and the Welsh Borders, *Veterinary Record*, **111**, 203–204.

Lewis, G. (1980). Copper deficiency in cattle, *MAFF Animal Disease Report*, **3**, 6–9.

Lewis, G., Terlecki, S. and Parker, B.N.J. (1974). Observations on the pathogenesis of delayed swayback, *Veterinary Record*, **95**, 313–316.

Mitchell, R.L. (1971). Trace elements in soils. In: *Trace Elements in Soils and Crops*, pp. 8–20; Technical Bulletin 21, MAFF London, HMSO.

Neal, W.M., Becker, R.B. and Shealy, A.L. (1931). A natural copper deficiency in cattle rations, *Science*, **74**, 418–419.

Sjollema, B. (1933). Kupfermangel als ursache von Krankheiten bei pflanzen und tieren, *Biochemische Zeitschrift*, **267**, 151–156.

Smith, S.E. (1951). The response of cobalt deficient lambs to liver extract and vitamin B_{12}, *J. Nutrition*, **44**, 455–464.

Suttle, N.F. (1970). Discussion in *Trace Element Metabolism in Animals* (Ed. C.F. Mills), p. 113; WAAP/IBP International Symposium, Aberdeen, 1969, Edinburgh and London, Livingstone.

Suttle, N.F. (1982). Predicting the effects of molybdenum and sulphur concentrations on the absorbability of copper in grass and forage crops to ruminants. In: *Proceedings of the Fourth International Symposium on Trace Element Metabolism in Man and Animals* (TEMA 4), Perth, Western Australia, 1981, pp. 545–547; Springer-Verlag, Berlin Heidelberg, New York.

Suttle, N.F. and Field, A.C. (1968a, b). Effect of intake of copper, molybdenum and sulphate on copper metabolism in sheep. I. Clinical condition and distribution of copper in blood of the pregnant ewe. II. Copper status of the new born lamb, *J. Comparative Pathology and Therapeutics*, **78**, 351–362 and 363–370.

Thompson, I., Thornton, I. and Webb, J.S. (1972). Molybdenum in Black Shales and the incidence of bovine hypocuprosis, *J. Sci. Food Agriculture*, **23**, 879–891.

Thornton, I., Kershaw, G.F. and Davies, M.K. (1972). An investigation into copper deficiency in cattle in the Southern Pennines. I. Identification of suspect areas using geochemical reconnaissance followed by blood copper surveys. II. Response to copper supplementation, *J. Agricultural Science, Cambridge*, **78**, 157–163 and 165–171.

Underwood, E.J. and Filmer, J.F. (1935). Enzootic marasmus. The determination of the biologically potent element (cobalt) in Limonite, *Australian Veterinary Journal*, **11**, 84–92.

Webb, J.S., Thornton, I., Thompson, M., Howarth, R.J. and Lowenstein, P.L. (1978). *The Wolfson Geochemical Atlas of England and Wales*; Oxford University Press.

The Interaction Between Geochemical and Pollutant Metal Sources in the Environment: Implications for the Community

I. THORNTON, P. W. ABRAHAMS, E. CULBARD, J. A. P. ROTHER AND B. H. OLSON

Introduction

Cannon and Anderson (1971) in a paper entitled 'The Geochemist's Involvement with the Pollution Problem' draw attention to the kinds of information that the geochemist can contribute to studies on environmental health where pollution has a role. These are:

1. Data showing the background amounts of each chemical element that are to be expected in a natural uncontaminated geologic setting.
2. Information on anomalously high metal contents in soils and vegetation that are due to metal-rich rocks and transported materials derived from them.
3. The effects of air pollution on background levels of elements in soils and plants.
4. The geochemistry of soils and soil solutions and the migration of elements and their compounds through surface and ground waters.
5. The chemical composition of mining and milling wastes and the solubilities of components which result in either their retention at the site or their dispersion through the local drainage system.

Contamination of soils, vegetation, surface waters and drainage sediments arising from mining and smelting operations has long been of concern to those undertaking geochemical prospecting for metals (Webb, 1958; Hawkes and Webb, 1962). There is also a long historical record of metal pollution of the environment and of the implications of mining, processing and utilizing metals to the health of workers in industry and those affected by environmental exposure (Thornton and Abrahams, 1984). Borlase (1758) referring to early metal mining in Cornwall states 'the sterility (of the soil) may be great where metals and minerals abound, of which no place perhaps afford more frequent instances than Cornwall; for here the coarsest grounds abound most in metals, and on the other hand there is the greatest plenty of corn, grass, plants and trees, where no metals or minerals have ever appeared'. Beach sands were commonly added as soil ameliorants in Cornwall, but Borlase mentions that these could not be used if they contained mine waste as 'their acrimonial mineral juices are mortal enemies to vegetation'. Farey (1811), writing about mining and agriculture in Derbyshire, describes early smelting techniques and mentions noxious fumes emitted from Cupolas or low arched reverbatory furnaces which were used from about 1747. These fumes poisoned herbage for ¼ mile or more around the Cupolas, and owners were obliged to pay a rent to the farmers for the damage caused to their land. Farey (1817) described poisoning of cattle in Derbyshire, locally known as Belland, caused by grazing too long in the vicinity of the lead smelter furnaces. Lead poisoning could also affect sheep, poultry, and small animals such as dogs. Human lead poisoning was quite common in people working in the smelters and was also recorded in people living near the Cupolas. However, like many of the early records, there are no statistics attached to any of this information, and no measurements were made of levels of lead or other metals in the environment.

Losses from smelters worldwide from the 1750s to the 1960s have been estimated by Murozumi *et al.* (1969), who quote as much as 2% Pb converted in the course of smelting to aerosol in the period 1750–1880, falling to 0.06% in the 1960s due to more efficient processing. Studies on Greenland snow profiles clearly show increasing global lead pollution due to industry since 800 BC and point to the fact that lead concentrations in rainfall have risen in recent times. Arctic snows contained 10 times the amounts of lead than those in the Antarctic, reflecting greater uses of lead in the northern compared with the southern hemisphere.

Industrial exposure in those operating mercury mines has been recognized since Roman times (D'Itri and D'Itri, 1977). It has been suggested that the Roman authorities closed mines at Monte Amiata in northern Italy and imported mercury from Spain (Almaden), perhaps to protect Italian workers from mercury poisoning or to prevent the devastation of the Italian landscape. In 1921, at Almaden, 2100 miners and 400 furnace men were affected by mercury poisoning due to vapour escaping from the furnaces; environmental consequences are not recorded. However, mining at Idria in what is now Yugoslavia, led to mercurialism in all inhabitants of the town in 1893 due to a fire in the mine tunnel; poisoning was also recorded in farm livestock.

Present day effects of historical pollution have been reported by numerous authors (including Davies, 1980; Thornton, 1980). Effects of pollution from lead mining on river fauna and flora in the Aberystwyth area of Wales were studied as long ago as 1919 by Carpenter (1924, 1925, 1926), who provides detailed descriptions of fauna in several rivers in relation to concentrations of lead both in water and in bottom sediments. Cases of lead poisoning in cattle at times leading to death are still occasionally reported around historical mining sites in Derbyshire, when animals stray on to mine spoil; nowadays the old lead rakes are sometimes intentionally grazed.

Research at the interface of geochemistry and pollution has had a significant role at Imperial College since John Webb founded the Applied Geochemistry Research Group (AGRG) in the early 1950s. Specific research on metal contamination of the environment and its possible implications to plant, animal and human health commenced in the 1960s and in particular developed as a result of the first trial geochemical surveys by AGRG in 1965 over 2800 square miles in England and Wales (Nichol *et al.*, 1970a, b, 1971). These and subsequent surveys based on stream sediment sampling in part met the requirement for systematic data on the distribution of trace elements and heavy metals on a regional and national scale. The foresight of Webb in recognizing the national requirement for geochemical atlases led in due course to the compilation and publication of the *Wolfson Geochemical Atlas of England and Wales* (Webb *et al.*, 1978). This valuable collection of computer-generated colour maps for 24 elements, including several such as Pb (Plate 3), Cd (Figure 1) and As (Plate 4) of prime concern to the environmental and pollution scientist, has led to a considerable expansion in research effort at AGRG over the past decade directed at their interpretation and evaluation in relation to agriculture, water supplies, estuarine fisheries, urban pollution and public health. Studies have also been initiated into the effects of natural and anthropogenic sources of metals on soil bacteria and on the quality of the indoor environment.

Figure 1 The distribution of Cd in stream sediments over England and Wales (ppm)

This chapter describes the major areas of work developed over this period and in particular focusses on current interests and future requirements.

Natural and man-made inputs of metal in the environment

Before attempting to assess the degree and extent of environmental metal pollution as a result of man's activities, it is essential to have reliable baseline information on the amounts present from natural geological sources. All of the potentially toxic elements of concern to man are present in small amounts in soil, water, food and air. Sometimes the natural geochemical source may be large, for example Cr and Ni in basaltic igneous rocks, Cd and Mo in some marine black shales (Table 1); soils developed from these parent materials tend to reflect their extreme geochemical composition (Table 2). Thus, the role of the geochemist is clearly recognized in planning an environmental impact study, whether concerned with past or present day industrial activity or with problems associated with hazardous waste management (Thornton, 1982). In this respect it is important to remember that the natural composition of both soils and waters may vary appreciably as a result of rock weathering and soil forming processes, together with factors influencing the secondary dispersion of chemical elements in the surface environment.

Both natural and anthropogenic inputs of elements may compliment one another and act in an additive fashion as a source of material to the biosphere. It is frequently difficult to distinguish or apportion metals from these different sources, though a knowledge of their precise forms and chemical species is useful. Unfortunately, although the initial mineral form or nature of the polluting compound is often known, changes due to weathering in the soil and drainage channel, mobilization and reconstitution as secondary minerals, adsorbed species, organo-mineral complexes, etc. often lead to marked chemical changes over time; old, transient and new species are difficult to quantify and apportion as to source. Recent approaches on chemical speciation have been reviewed by Sposito (1983) and will be discussed later.

In the context of this chapter, geochemical sources of metals are considered to be those derived from natural bedrock and earth materials, including mineral ores, and also metalliferous contaminants arising from the mining and smelting of these ores. Anthropogenic or man-made metal sources to the environment include those from other industrial emissions and effluents, the disposal of domestic and industrial wastes, sewage sludge and other agricultural wastes, fertilizers, ameliorants and pesticides, those arising from urban development and from vehicle emissions.

Research conducted by AGRG at the interface of geochemistry and pollution is described under the major application areas and their implications to the community. These comprise (a) agriculture, (b) soil microbiology, (c) water resources, (d) urban pollution and (e) human health.

Table 1 **Range and mean concentrations of some elements in igneous and sedimentary rocks (ppm)**

Elements	Ultramafic igneous	Basaltic igneous	Granitic igneous	Shales and clays	Black shales	Limestones	Sandstones
Arsenic	0.3–16	0.2–10	0.2–13.8	—	—	0.1–8.1	0.6–9.7
	3.0	2.0	2.0	10		1.7	2
Barium	0.2–40	20–400	300–1800	460–1700	70–1000	—	—
	1	300	700	700	300	10	20
Cadmium	0–0.2	0.006–0.6	0.003–0.18	0–11	<0.3–8.4	—	—
	0.05	0.2	0.15	1.4	1.0	0.05	0.05
Chromium	1000–3400	40–600	2–90	30–590	26–1000	—	—
	1800	220	20	120	100	10	35
Cobalt	90–270	24–90	1–15	5–25	7–100	—	—
	150	50	5	20	10	0.1	0.3
Copper	2–100	30–160	4–30	18–120	20–200	—	—
	15	90	15	50	70	4	2
Fluorine	—	20–1060	20–2700	10–7600	—	0–1200	10–880
		360	870	800		220	180
Iron	94 000	86 500	14 000–30 000	47 200	20 000	3800	9800
Lead	—	2–18	6–30	16–50	7–150	—	<1–31
	1	6	18	20	30	9	12
Mercury	0.004–0.5	0.002–0.5	0.005–0.4	0.005–0.51	0.03–2.8	0.01–0.22	0.001–0.3
	0.1	0.05	0.06	0.09	0.5	0.04	0.05
Molybdenum	—	0.9–7	1–6	—	1–300	—	—
	0.3	1.5	1.4	2.5	10	0.4	0.2
Nickel	270–3600	45–410	2–20	20–250	10–500	—	—
	2000	140	8	68	50	20	2
Selenium	—	—	—	—	—	—	—
	0.05	0.05	0.05	0.6		0.08	0.05
Vanadium	17–300	50–360	9–90	30–200	50–1000	—	—
	40	250	60	130	150	20	20
Zinc	—	48–240	5–140	18–180	34–1500	—	2–41
	40	110	40	90	100	20	16

Adapted from table compiled by M. Fleischer and H.L. Cannon (Cannon *et al.*, 1978)

Table 2 **Trace elements and metals in soils derived from normal and geochemically anomalous parent materials in Britain**

	Typical Normal range in soil (ppm)	Metal-rich soils (ppm)	Sources	Possible effects
As	<5–40	up to 2500 up to 250	Mineralization Metamorphosed rocks around Dartmoor	Toxicity in plants and livestock; excess in food crops
Cd	<1–2	up to 30 up to 20	Mineralization Carboniferous black shale	Excess in food crops
Cr	15–300	up to 3500	Ultrabasic rocks in Scotland	No known effect
Cu	2–60	up to 2000	Mineralization	Toxicity in cereal crops
F	20–500	1% or more	Mineralization	Fluorisis in livestock
Hg	0.008–0.19	1–7	Mineralization	No known effect
Mo	<1–5	10–100	Marine black shales of varying age	Molybdenosis or molybdenum induced hypocuprosis in cattle
Ni	2–100	up to 8000	Ultrabasic rocks in Scotland	Toxicity in cereal and other crops
Pb	10–150	1% or more	Mineralization	Toxicity in livestock; excess in foodstuffs
Se	<1–2	up to 7 up to 500	Marine black shales in England and Wales Namurian shales in Ireland	No effect Chronic selenosis in horses and cattle
Zn	25–200	1% or more	Mineralization	Toxicity in cereal crops

Agriculture

Investigations aimed at the interpretation of geochemical maps for crop and animal health and production have been partly focussed in those areas of Britain known historically for the mining and smelting of metals.

COPPER, TIN AND ARSENIC IN SOUTHWEST ENGLAND

Historically, southwest England was the premier metalliferous mining centre of the world. In Cornwall, more than 600 mines have operated in the past. mainly within a belt some 75 miles long and 10 miles wide. Although tin, copper, and arsenic were the main metals extracted, W, Pb, Zn, U, Fe, Sb, Ag and Mn were also economically recovered, with small amounts of Bi, Ni and Co. Total output has been estimated at 2 million tons of metallic tin and 920 000 tons of metallic copper (Dines, 1956). From about 1860 to 1900, Cornwall was the world's major producer of arsenic and at the height of the industry also imported arsenical ores for smelting. The geochemical reconnaissance map for arsenic (Plate 4) clearly shows the anomalous arsenic province in southwest England, reflecting not only the mineralized area but widespread contamination resulting from mining and processing.

This history of mining and smelting in Cornwall has left behind a legacy of dereliction and pollution, with mine tailings and other wastes abundant.

Table 3 **Ranges and mean total concentrations (ppm) of arsenic and selected heavy metals in agricultural soils (0–15 cm) from the Tamar Valley, Cornwall**

	No. of samples	As	Cu	Pb	Zn
Alluvium contaminated downstream of Devon Great Consuls Mine	12	309 90–900	620 35–2000	435 60–1008	368 114–1020
Alluvium upstream of mine	5	25 20–30	44 36–40	85 72–96	121 67–180

Data from Colbourn *et al.* (1975)

Arsenic occurs as the ore arsenopyrite, which was crushed and calcined and the vapour passed through chambers and flues where arsenious oxide accumulated as a grey powder known as 'arsenic soot' or 'crude arsenic'. Actual amounts lost and the extent and degree of environmental contamination was not recorded at this time, though it has been estimated that the average loss of tin in the course of mining and processing was in the order of 30% or more. Devon Great Consuls mine in the Tamar Valley, separating Devon and Cornwall, was at one time Europe's richest copper mine and from 1845 to 1901 produced some 70 000 tons of copper and 72 000 tons of refined arsenic. Alluvial soils downstream of this mine still contain much larger

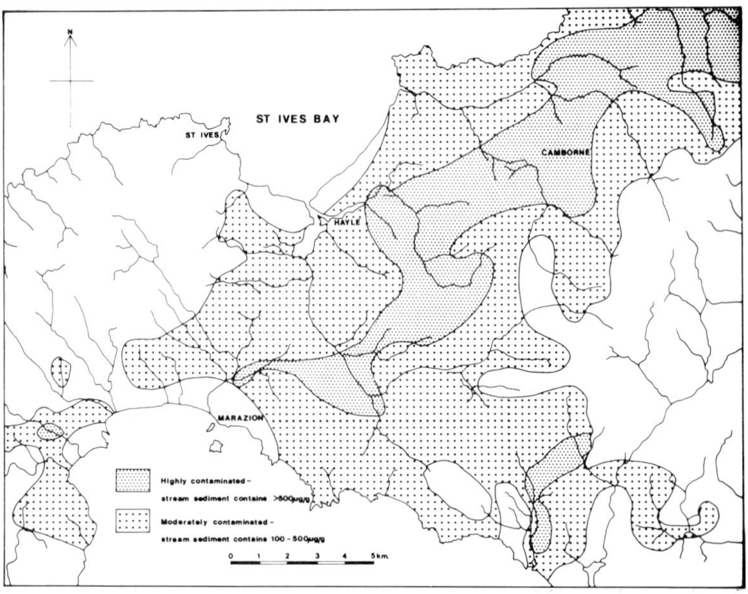

Figure 2 The distribution of As in stream sediments in the Cornish peninsula, England

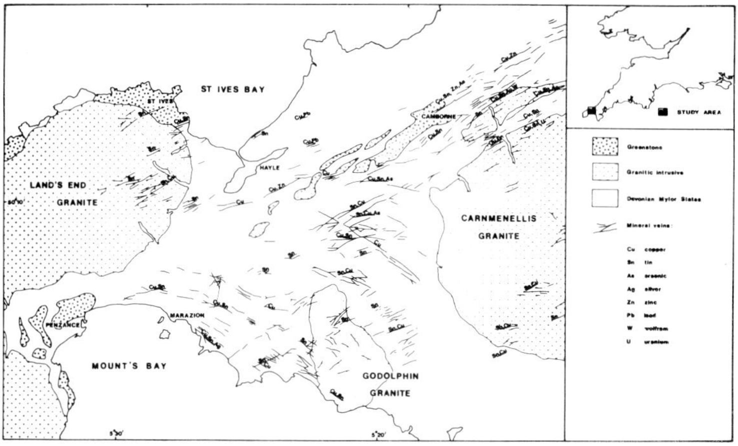

Figure 3 Mineral occurrences in western Cornwall, England

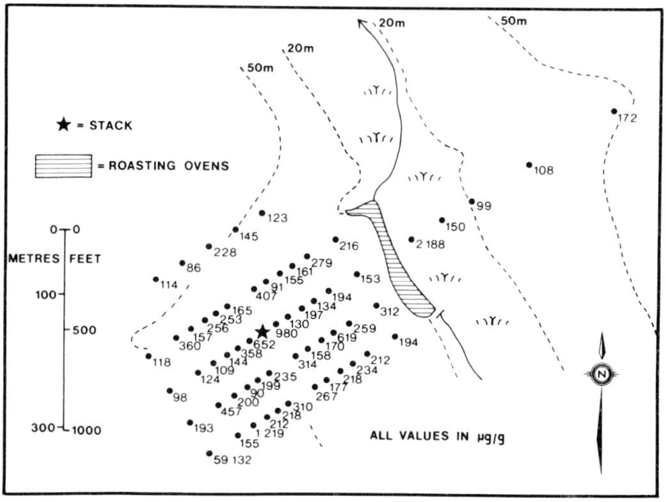

Figure 4 Concentrations of As in surface soils (0–15 cm) around an old calcining works at Newmill, Cornwall

concentrations of arsenic, copper, lead and zinc than those upstream (Table 3), due to downwash of metal-rich mine waste and suspended solids in adit waters and runoff from contaminated upland soils (Thornton, 1980).

In the western Cornish peninsula, present-day anomalies for arsenic (Figure 2), copper, lead, zinc and cadmium, in surface stream sediments extend for some 250 km^2 with peak values of 14 000 ppm As and 8000 ppm Cu. These anomalies reflect the intensity of mineral occurrences as shown in Figure 3. Extensive contamination of agricultural soils has been documented by Abrahams (1983) and is illustrated by present-day levels of arsenic in surface soils (mostly farmed) around the stack of the arsenic calcining works

Table 4 **Mean contribution of soil to total daily intake of As and Cu in eleven herdsof cattle in Cornwall, England**

Farm type (based on soil As)	Mean % soil ingested	Soil mg kg^{-1}		Total daily intake mg d^{-1}		% element ingested as soil	
		As	Cu	As	Cu	As	Cu
20–50 mg kg^{-1} As n = 3	1.4	24	26	7	140	67	4
55–140 mg kg^{-1} As n = 3	1.4	85	93	21	175	76	10
160–259 mg kg^{-1} As n = 5	1.1	202	199	52	190	58	16

at Newmill, Cornwall which ceased production in 1890 (Figure 4).

Within the Hayle/Camborne area (Figures 2 and 3) the levels of As and Cu found in surface (0–15 cm) soils across the Peninsula ranged up to 730 ppm As and 560 ppm Cu. Amounts in subsoils (30–45 cm) were appreciably lower at a number of locations confirming the widespread surface contamination of the area during the course of mining and smelting operations. The agricultural significance of these findings was in part assessed by determining the uptake of arsenic and copper into pasture herbage and the intage of these elements by grazing cattle on 11 farms in the region, selected on the basis of the arsenic content of their soils. Neither arsenic nor copper are readily taken up and translocated to the aerial parts of grasses, and it has been clearly shown that accidentally ingested soil comprises a major exposure route to the grazing animal (Table 4; Thornton and Abrahams, 1983). The amount of soil ingested (calculated on the basis of the Ti content of faeces) varied appreciably between seasons of the year and in this study ranged from 0.2 to nearly 18% of the animal's dry matter intake. In the spring when there is little grass cover, this could account for around 90% of the arsenic and 50–60% of the copper intake of grazing cattle. No adverse effects on the health or production of cattle have as yet been reported in this area.

The impact of arsenic and copper contamination of soils on the composition of pasture and food crops has been studied in an area of about 140 km^2 centred on the Tamar Valley (Thoresby and Thornton, 1980), where soils ranged up to 980 ppm Cu and 1200 ppm As. Although amounts of Cu and As were higher in spring herbage in the Tamar than in control areas selected for comparison in Kent (mean 16 ppm As and 12 ppm Cu in soil), there was little relationship between amounts of either element in herbage and soil within the Tamar area (Figure 5c). It is possible that herbage in the area is contaminated with both elements as a result of soil splash, as there is no wholly satisfactory way of removing this by simple washing procedures. Levels of arsenic in barley grain tended to increase with soil content (Figure 5d) and also increased in both lettuce and strawberries, though concentrations were very small, peaking at 1.4 ppm and 0.7 ppm DM respectively (Figures 5a and b).

It is seen from these studies in southwest England that the aerial parts of the

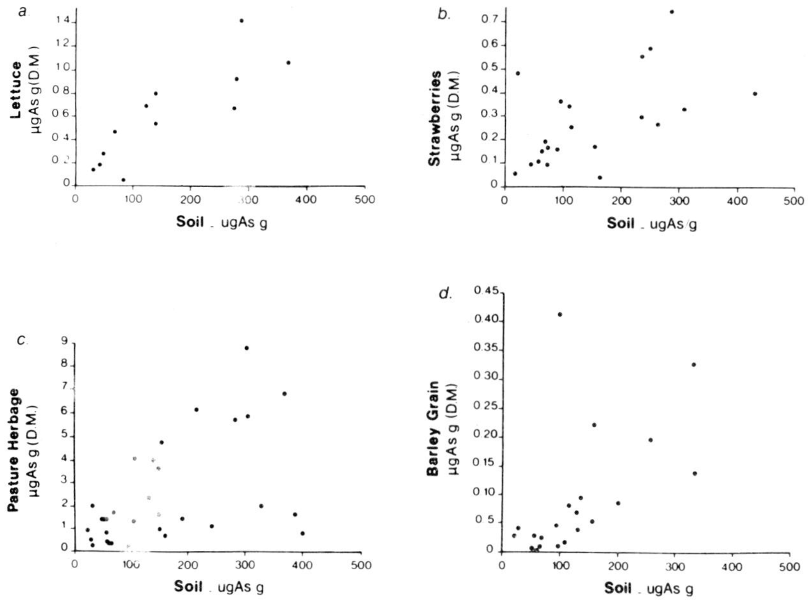

Figure 5 Relationship between the As content of: (a) lettuce; (b) strawberries; (c) pasture herbage; and (d) barley grain (ppm, dry matter) and the As content of soil in the Tamar area of south-west England

plants sampled reflect to only a small extent the large differences found in the 'total' metal contents of the soils. Of course not all the metal present in the soil is in a soluble or 'available' form for uptake into the plant root and frequently only a small amount of that taken into the root is translocated to the shoot. Investigations undertaken under controlled solution-culture conditions have shown that a large proportion of copper (up to 96%) added to perennial ryegrass is retained in the root, mostly associated with the cell wall fraction, and it has been concluded that this root process restricts the movement of copper along the soil–plant–animal pathway (Jarvis, 1978). In southwest England, it is apparent that little soil arsenic passes into the shoots of normal pasture plants or into the edible parts of food crops, though its forms in the soil and mechanisms of translocation in the plant are not yet fully understood.

LEAD, ZINC AND CADMIUM

The geochemical map showing lead distribution in England and Wales (Plate 3; Webb *et al.*, 1978) shows anomalously high patterns reflecting metal contamination on a regional scale from past and present mining and smelting and from industrial and urban sources. The majority of the anomalies occur in the major areas of lead–zinc mineralization, including those in southwest England, the Mendips, the southern and northern Pennines, central and north

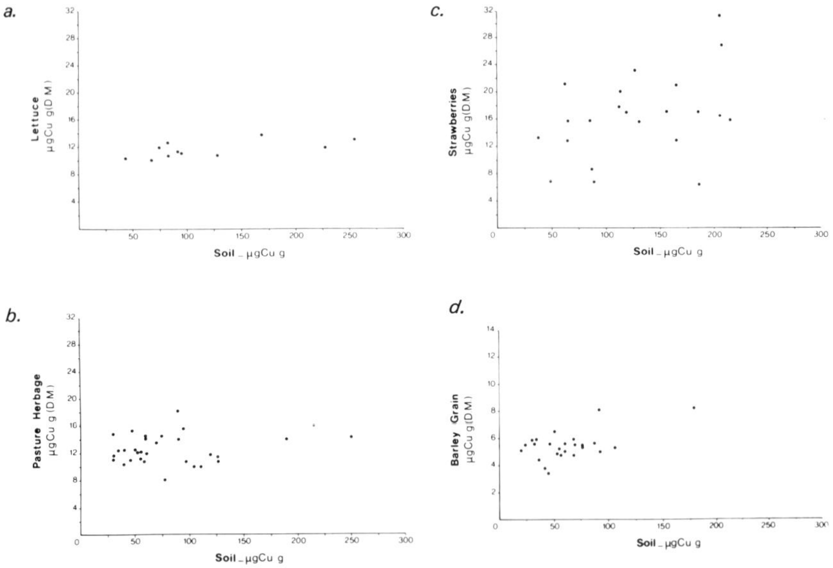

Figure 6 Relationship between the Cu content of: (a) lettuce; (b) strawberries; (c) pasture herbage; and (d) barley grain (ppm, dry matter) and the Cu content of soil in the Tamar area of south-west England

Wales. The main questions to be answered are (a) do these anomalies indicate the presence of large areas of contamination to the surface environment, particularly agricultural soil and (b) if so, what effect if any do they have on crop and animal production.

A major investigation was mounted in the southern Pennines, centred in Derbyshire, in which soils were systematically sampled across the geochemically anomalous areas and specific studies conducted into the soil–plant–animal system on 11 livestock farms. Concentrations of lead were found to range from several hundred to several thousand parts per million in the dry soil, with up to 3% Pb near old workings and smelters; the total area affected has been estimated to be *ca.* 250 km^2 and comprises the mineralized part of the Carboniferous Limestone and includes neighbouring areas where smelters were located to the east (Colbourn, 1976; Colbourn and Thornton, 1978). The lead content of uncontaminated UK soils falls within the range 10–150 ppm Pb and normally ranges from 10–70 ppm. Some agricultural soils within this Derbyshire mineralized area are also enriched with zinc and cadmium (which occurs as an impurity in zinc ores), with concentrations ranging up to 8000 ppm Zn and 34 ppm Cd (Marples, 1979).

Lead mining in Derbyshire commenced in Roman times and reached a peak in the latter part of the 19th century. Today there are no active lead mines though old waste tips are currently worked for fluorspar. Deaths of grazing livestock are rare and are due to the ingestion of metal-rich waste materials while grazing, or to soil-contaminated fodder or silage. Chronic effects are not usually recognized. On the 11 farms tested, which were chosen on the basis of

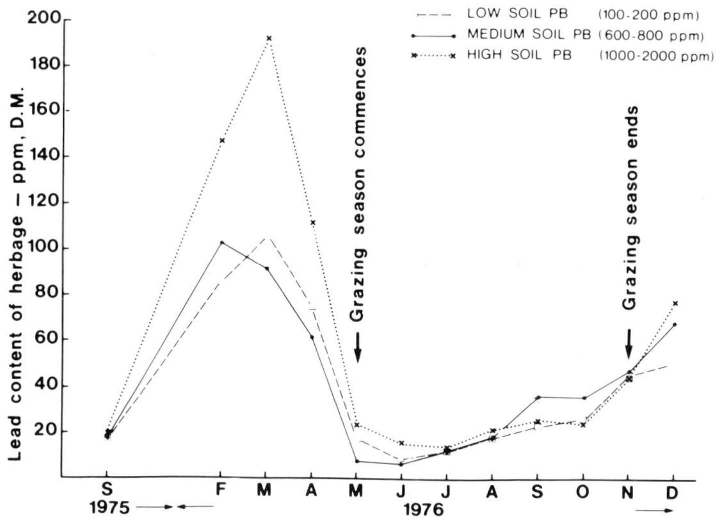

Figure 7 Seasonal variations in the Pb contents of pasture herbage growing in contaminated and uncontaminated soils in Derbyshire

their soil lead content, amounts of lead in pasture herbage mostly ranged from 6 to 25 ppm DM in the spring and 15 to 30 ppm in the autumn and only reflected the composition of the soil to a small degree (Figure 6). Low uptake of lead by pasture plants has been well documented and is restricted by barriers at the soil : root and root : shoot interfaces (i.e. Jones and Clement, 1972). The marked seasonal influence on plant lead content (Figure 7) is of significance and is thought to be partly due to soil contamination combined with aerial deposition of lead-rich particulates in the winter months, dilution associated with rapid plant growth in the spring and early summer and possibly physiological and biochemical changes within the plant at times of senescence.

Blood and faecal samples were taken from 15 cattle on each of the 11 farms in early summer and autumn. In general, blood lead values were higher in

Table 5 **Mean contribution of soil to total daily intake of Pb and Zn in eleven herds of cattle in Derbyshire, England**

Farm type (based on soil Pb)	Mean % soil ingested	Soil content mg kg^{-1}		Total daily intake mg d^{-1}		% element ingested as soil	
		Pb	Zn	Pb	Zn	Pb	Zn
100–500 mg kg^{-1} Pb n = 4	2.9	276	87	289	590	40	6
600–1000 mg kg^{-1} Pb n = 4	3.7	856	311	621	860	71	19
1000–4000 mg kg^{-1} Pb n = 3	2.9	2942	287	1700	714	66	15

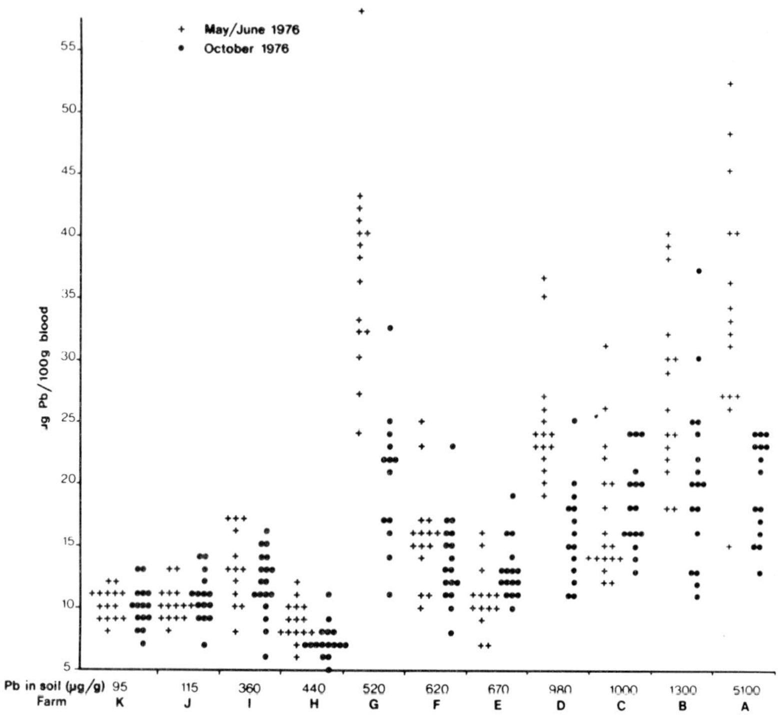

Figure 8 Within-herd variations of bovine blood Pb concentrations on 11 farms in Derbyshire. The farms are arranged in ascending order of soil lead concentrations

cattle grazing contaminated land (farms A–G, Figure 8) than uncontaminated land (farms H–K), with herd means ranging in the early summer from around 10 µg Pb/100 ml blood on the background farms to 30 µg Pb/100 ml on the most contaminated land. Faecal lead content reflected soil lead and as with the case of As in southwest England, accidental ingestion of metal-rich soils was again shown to be a primary exposure route for lead (Table 5; Thornton and Kinniburgh, 1978; Thornton *et al.*, in press).

The map showing the distribution of cadmium in stream sediments in England and Wales (Figure 1) in general mirrors that for lead, showing anomalies due to three major sources of the metal, mineralization and mining, recent and present-day smelting and other industrial activities and the presence of underlying metal-rich rocks. Where cadmium is associated with zinc ores, such as sphalerite (ZnS) and smithsonite ($ZnCO_3$), agricultural soils often contain up to 30 ppm Cd, compared with most rocks and soils which contain 1 ppm or less. At Shipham in the Mendip Hills in Somerset, where smithsonite was mined from ca. 1700 to 1850, cadmium is present in reclaimed soils at concentrations ranging from 40 to 600 ppm Cd in association with extremely large amounts of zinc (usually at Zn : Cd ratio of *ca.* 100 : 1) (Matthews and Thornton, 1980, 1982). Studies on the uptake into pasture herbage in this area have shown that little of the soil cadmium is

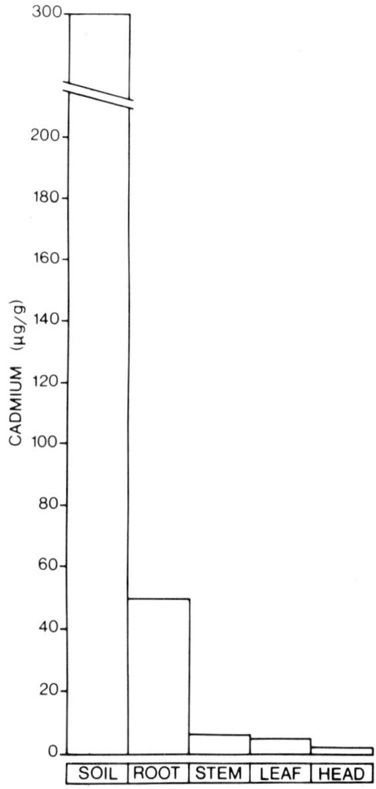

Figure 9 An example of the Cd content of soil and the root, stem, leaf and seed head of *Holcus lanatus* at Shipham, Somerset, an area heavily contaminated by past Zn mining (from Matthews and Thornton, 1982)

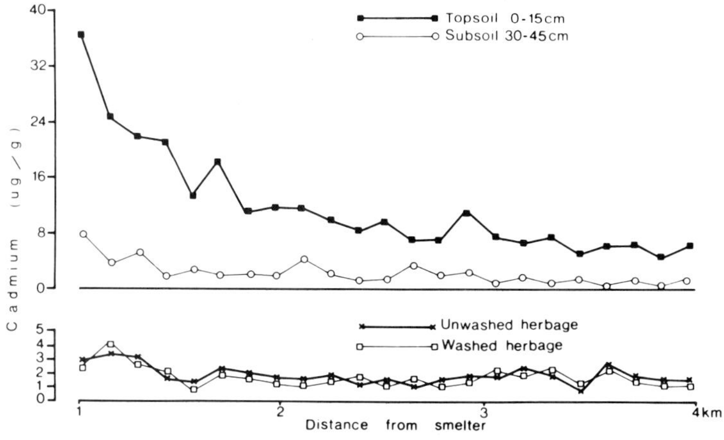

Figure 10 Distribution of Cd in soils and pasture herbage downwind from the Avonmouth smelting complex, near Bristol (from Matthews, 1982)

transferred to the aerial parts of the pasture herbage (Figure 9), and that uptake, as well as showing marked seasonal differences, varies appreciably between plant species with some of the weed species such as daisy and dandelion containing some 20–40 times more Cd than grass species growing alongside them (Matthews and Thornton, 1982).

Anomalies associated with recent and present-day smelting of zinc and lead, include those of the Swansea Valley in South Wales and around the Avonmouth smelting complex near Bristol. At the latter site, the dispersion of cadmium and zinc in surface soil (0–15 cm) falls off rapidly with distance downwind from the industrial area (Figure 10; Matthews, 1982). Little of the metal is leached into the subsoil due to the impervious nature of the estuarine clay parent material; Zn : Cd ratios of 76 : 1 were found in the topsoil compared with 83 : 1 in the subsoil (30–45 cm). Larger concentrations of metals in pasture plants in winter months compared with summer corresponded with larger proportions that could be removed by washing and it was concluded that near the smelter the main source of metal was due to aerial deposition and soil splash (Matthews, 1982).

The third type of cadmium anomaly was associated with metal-rich bedrock. For example, dolomitic conglomerate, the host-rock of the smithsonite mineralization at Shipham, weathers to give rise to soils ranging from 10 to 30 ppm Cd over an area of *ca.* 8 km^2. Carboniferous black shales in Derbyshire and North Staffordshire contain up to 200 ppm Cd (see Table 2). Studies on a catenary sequence of soil in this latter area demonstrated the mobility of cadmium under poorly drained acid soil conditions (pH 4–5) and an accumulation together with iron at the base of the soil profile in organic waterlogged sites in the valley bottom (Marples, 1979; Matthews, 1982).

THE SOIL–PLANT–ANIMAL RELATIONSHIP

From the above sections, it is seen that large areas of Britain have been contaminated as a result of the country's long history of metalliferous mining and smelting. Estimates of agricultural land affected range from in excess of 4000 km^2 in England and Wales (Thornton, 1980, 1983) to over 8000 km^2 (Davies, 1980). It is difficult to determine the extent of metal contamination from other sources, many of which may cause pollution on a local scale too small to be picked up by widescale reconnaissance mapping techniques. Close-spaced soil and plant sampling is necessary to assess the extent and degree of contamination in these cases.

It has been estimated that the annual maximum lead emission from vehicles in the UK is around 7500 tonnes Pb per year (calculated from data published by Associated Octel, 1983) of which around 10% is deposited on and within 100 m of roads (Little and Wiffen, 1978). On a wider basis, Chamberlain (1983) has calculated that a cumulative fallout of 3.4 g Pb m^{-2} (3 g from industrial sources since 1700 and 0.4 g from vehicle emission since 1946) would increase the lead concentration in the surface 15 cm of soil by 20 ppm of which 3 ppm could be attributed to lead from vehicle exhaust. The amounts involved are extremely small compared with the large concentrations found in association with mineralization and mining and in the immediate vicinity of industrial sources, particularly smelters.

It must be emphasized that the chemical and physical forms of an element may vary both in relation to their source and as a result of the physical and biological components of the environmental system in which they occur. The actual forms and chemical species present in a soil will depend on the nature of the source, soil properties such as reaction (pH) and redox potential (Eh), as well as the nature of other soil components, including reactive constituents such as hydrous oxides of iron and aluminium and complexing organic ligands. The subject of the chemical forms of trace metals in soils has been comprehensively reviewed by Sposito (1983) and Mattigod and Page (1983), and that of soil and plant factors influencing their mobilization in soils, availability and uptake by plants and subsequent translocation within the plant by Mitchell and Burridge, 1979, West, 1981 and Tinker, 1981.

The intake of metals by grazing animals depends on the nature of their diet and is affected by species composition of the sward and season of the year, amount of supplementary feed provided, selectivity of grazing and the availability of the element and its absorption within the alimentary tract. There are still large gaps in the knowledge of factors influencing trace metal availability to the animal, particularly in relation to different sources and forms of the metal and interactions with other dietary components.

In summary, it is evident that both crop plants and livestock can in general be farmed successfully in areas where both natural geochemical and man-made sources of metals are large, due to the barriers to uptake, translocation and absorption described above. In extreme cases, death of plants and animals may occur but these cases are rare. Less is known about subclinical effects that may influence production and health. The ingestion of soil by grazing animals is an important biogeochemical pathway which may by-pass barriers at the soil : plant and plant : animal interfaces; this may amount to as much as 10–20% of the dry matter intake of cattle and as much as 30% or more in sheep. However, the availability and absorption of potentially toxic metals from this source is still far from clearly understood.

Soil microbiology

Vernadsky (1944) in integrating the fields of biology, geology and chemistry in the concept of biogeochemistry, concluded that biogeochemistry should include the study of the role of micro-organisms in the dissemination of chemical elements in the biosphere. Through the study of such microbially mediated cycles as these of N and S, the concept of biogeocycling has become well established.

Three reasons have been proposed as to why micro-organisms are important in the cycling of minerals: (a) micro-organisms comprise most of the earth's biomass and have rapid generation times; (b) micro-organisms occupy the widest diversity of habitats; and (c) micro-organisms colonized the earth far earlier than higher forms of life (Trudinger *et al.*, 1979).

Bacteria and fungi can alter the valence state of metals in the course of methylation, chelation, complexation, absorption, oxidation and reduction. Thus micro-organisms can affect the bio-availability and dispersion of metals in both soil and aquatic ecosystems and influence the movement of metals into and up the food chain. Metals may be solubilized by microbially mediated oxidation, as in the case of U or by reduction as in the case of Fe and Mn. All micro-organisms form some metal-organic complexes in the course of their biochemistry; many metals essential to biochemical processes are actively sought in the metabolism of micro-organisms.

The subject of microbially mediated metal cycling has recently been reviewed by Olson (1983), focussing attention in particular on As, Cd, Hg and Pb in the biosphere and their significance to health. Mechanisms of detoxification and metal resistance have been determined (Chopra, 1971; Summers and Silver, 1972), and have obvious application in our understanding of the behaviour of both naturally occurring metals and of metal contaminants in the soil environment and on the effects of these metals on the functional role of soil micro-organisms.

Studies on the 'geochemical ecology' of soil micro-organisms have constituted a major research area for many years at the Vernadsky Institute of Geochemistry and Analytical Chemistry, USSR Academy of Sciences (Kovalsky and Letunova, 1964). It has been shown that soil micro-organisms adapt to different contents and ratios of trace elements in the soil; examples have been documented for the elements Mo, B and Se (Letunova, 1970).

EFFECTS OF METAL CONTAMINANTS ON NITROGEN CYCLING BACTERIA

Following an exchange visit to the Vernadsky Institute in 1974 (by Iain Thornton) studies were initiated at Imperial College to investigate possible effects of the large amounts of metal contaminants in the UK metalliferous mining areas on nitrogen cycling in agricultural soils. A preliminary investigation found that free-living nitrogen-fixing bacteria present in Derbyshire soils contaminated by mining and smelting were able to adapt to high concentrations of heavy metals (Brierly and Thornton, 1979).

More detailed studies were conducted on agricultural soils from Derbyshire (with up to 5 ppm Cd, 8000 ppm Pb and 500 ppm Zn (total concentration of metal)) and Somerset (with up to 400 ppm Cd, 6000 ppm Pb and 34 000 ppm Zn). Control sites were also examined with soils containing less than 2.5 ppm Cd, 150 ppm Pb and 260 ppm Zn. Nitrogen fixation was estimated using the acetylene reduction assay, mineralization by the breakdown of added peptone, and nitrification by the oxidation of ammonium–nitrogen and nitrite–nitrogen. The activity of the free-living nitrogen-fixing bacteria was determined as described by Rother *et al.* (1982a). The nitrogen fixing activity of plants of white clover (*Trifolium repens*) in a mixed clover-grass sward was measured using an *in situ* incubation technique (Rother *et al.*, 1983). The three stages of the mineralization of nitrogen were studied using a soil perfusion apparatus under laboratory conditions (Rother *et al.*, 1982b). Samples of soil were collected, and *in situ* incubations performed, at different times of the year, to determine if there was any seasonal effect on the cycling of nitrogen.

In all the contaminated sites, a proportion of the total content of heavy metals were extractable in distilled water. For Cd, the most 'mobile' of the elements studied, the water soluble fraction ranged from 1.0 to 1.5% of the total; Pb (the least soluble) from 0.1 to 0.6%. Zn was intermediate at 0.2 to 0.6%. The activities of the free-living, heterographic, nitrogen-fixing micro-organisms showed little difference between the various sites at any one time. However, there was a seasonal pattern, with higher activities during the winter and spring (20–40 μm $C_2H_4m^{-2}d^{-1}$) than during the summer (3–15 μm $C_2H_4.m^{-2}d^{-1}$). Of the various factors investigated, including heavy metal content, inorganic nitrogen and phosphorus, and moisture content, only the latter was found to be correlated with the seasonal change of activity. When the soils were dry, the bacteria were less active (Rother *et al.*, 1982a).

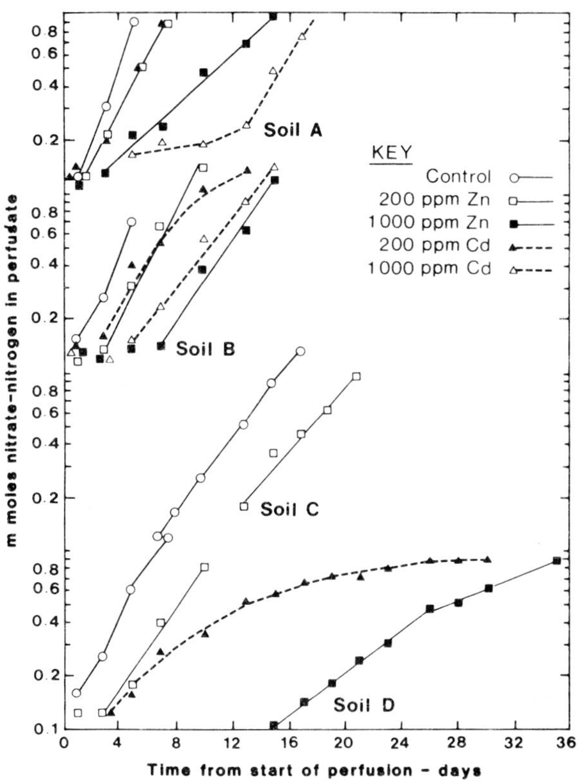

Figure 11 The effects of the addition of Cd and Zn on the nitrification of ammonium-N in soils from Shipham, Somerset; soils A and B are from a heavily contaminated area, soils C and D from an uncontaminated area

The amount of nitrogen fixed by the free-living bacteria, in common with other published findings, was small: 0.3 to 0.5 g N ha^{-1} d^{-1}. In contrast, the amount of nitrogen fixed by *Trifolium repens* (white clover) was much greater: 80 g N $ha^{-1}h^{-1}$ in the spring. Assuming that there was no marked diurnal variation and an active season (spring and early summer) of 100 days, this would represent an input of 200 kg N $ha^{-1}a^{-1}$: again, there were no consistent differences between the sites, even where the sward was established on a thin layer of organic soil (roots and dead plant material) covering the freely-draining spoil heaps (Rother *et al.*, 1983). The plants were healthy, with abundant and pink root nodules. However, in one of the contaminated Somerset sites, the clover plants were chlorotic during the spring, an effect which may have been due to Zn toxicity. This did not significantly affect microbial activity. The seasonal pattern of activity (maximum during the spring, falling to half of the maximum rate by the summer and to less than a tenth during the autumn) was consistent at all the sites.

A comparatively limited range of soil micro-organisms are capable of the chemo-autotrophic and stoichiometric oxidation of NH_4^+–N to NO_3^-–N (e.g. Nitrobacter). This, together with the fact that the product of the reaction (NO_3^-–N) can be continuously monitored in the leachate from the soil, makes nitrification an ideal system for experimental manipulation. When a sufficiently high concentration of Cd or Zn (Figure 11) or Pb was added to soils at the beginning of an experiment, a clear difference emerged between the soils studied (Rother *et al.*, 1982b). Soils from the contaminated areas (A and B) showed little or no lag in the development of activity, and even when a short lag (1–2 days) occurred the rate of activity was generally unchanged. The soils from the non-contaminated sites (C and D), in contrast, had longer lag phases, lower rates of increase (i.e. slower growth of bacteria) or a complete inhibition of activity or growth, (e.g. soil C, with 1000 ppm Zn, 200 ppm Cd and 1000 ppm Cd added). The addition of lower concentrations of metals than those used here had no effect on any of the soils, as all the metal was absorbed by the soil within minutes of addition.

When peptone (as a source of organic nitrogen) was added to the soils there was a rapid production of NO_3^-–N in all cases (Rother *et al.*, 1982b). The addition of Cd and Zn of up to 5000 ppm and of Pb up to 10 000 ppm was required before there was any inhibition. Differences between the soils were slight and were not always consistent with their origins. This was attributed to the much wider range of micro-organisms (which includes fungi) which can break down peptone.

It is probable that, in soils which are contaminated with heavy metals, tolerant strains of nitrogen-cycling bacteria are selected and develop into populations which are as active and as capable of rapid growth as those in non-contaminated soils. In the studies illustrated above, there was little difference in nitrogen fixation between the sites tested, although only the Clover–Rhizobium symbiosis would contribute significant amounts of nitrogen to the soil. Metal-tolerant populations were demonstrated for the bacteria involved in nitrification under laboratory conditions. Although tolerant populations developed eventually in the non-contaminated soils in response to the addition of heavy metals, their rate of growth and specific activity was lower than those from high-metal soils during the period of the experimental incubations. This indicates that although tolerance may arise rapidly, a longer time is required before the newly-tolerant populations become as metabolically efficient as the parent strain or as populations which have been exposed to heavy metals for decades or centuries.

RESISTANCE PATTERNS TO METALS OF POPULATIONS OF BACTERIA IN CONTAMINATED AND NATURALLY METAL-RICH LAND

The resistance of bacteria to heavy metals has been documented as a widespread phenomenon under a variety of laboratory and environmental conditions. An example is shown by the tolerance of nitrogen cycling organisms described in the previous section. Resistance of bacteria to heavy metals may result in biological transformation of certain metals, uptake and accumulation of metals by bacteria, and the production of compounds which may in some cases complex with heavy metals. Various physico-chemical factors in the environment may mitigate the effects of heavy metals on populations of bacteria, including pH, synergistic or antagonistic ion interactions, and the binding of metallic ions to other environmental constituents (Gadd and Griffiths, 1978). The effects of environmental pollution on bacterial populations have also been studied showing the emergence of both heavy metal and antibiotic resistant bacteria in contaminated environments (Timoney *et al.*, 1978; Nelson and Colwell, 1975), and variation in resistance patterns with mode of exposure (Bewley, 1979, 1980). Effects of metal pollution on exposed populations have shown that some members of the bacterial community are more sensitive than others to metals, especially to Cd (Houba and Remacle, 1980). Most previous work has overlooked the effects of physico-chemical variables in the soil environment, such as pH and Eh, on metal–microbial interactions.

Specific mechanisms of bacterial resistance to metals have been identified through detailed genetic and biochemical studies (Summers and Silver, 1978). A collaborative programme of research between environmental microbiologists at the University of California, Irvine and environmental geochemists at Imperial College commenced in 1979 to test the feasibility of using the resistance to metals of naturally occurring populations of soil bacteria as a quantitative measure of the amounts of pollutants in the soil environment and their potential bio-availability.

Initial studies were undertaken on soils from the historical mining areas of Shipham, Somerset and of Derbyshire. The soils tested covered a wide range of concentrations of Cd, Zn and Pb; metal resistance was determined using a dilution technique (in 0.3 M phosphate buffer) with agar media supplemented with $CdCl_2$, $ZnCl_2$ and $PbNO_3$ to give a range of metal concentrations (see Troyer *et al.*, 1980). Later the work was extended to cover soils contaminated with metals from past and present-day smelter emissions (Swansea and

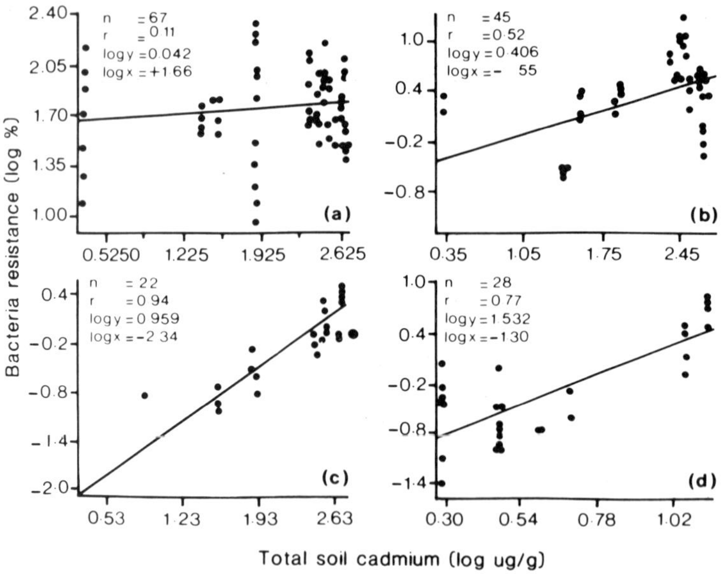

Figure 12 (a–c) Percentage of bacteria from Shipham soils resistant to Cd in media amended with 10 μg/g, and 200 μg/g respectively; (d) bacteria from Derbyshire soils resistant to Cd in media amended with 100 μg/g

Avonmouth respectively), by sewage sludge (Thames Valley) and enriched in metals due to the natural weathering of metal-rich source rocks (Carboniferous black shales in Derbyshire). Further old mining sites were included in Cornwall and the assay method extended to cover Cu and As.

These studies have shown:

1. Metal-resistant populations of bacteria occur in the soils of each of the locations tested; in the historical mining areas in Somerset and Derbyshire, patterns of resistance were quantitatively related to total concentrations of Cd, Zn and Pb in the surface soil (Olson and Thornton, 1982). For example, tolerance to Cd was greatest in soils from Shipham with high concentrations of Cd; bacteria from soils with relatively low concentrations of Cd from Derbyshire and Shipham demonstrated little or no resistance. The quantitative nature of this relationship is shown in Figure 12. The bacterial resistance to Cd at Shipham where soils contain up to 450 $\mu g\ g^{-1}$ Cd was most clearly shown ($r = 0.94$) with 200 $\mu g\ g^{-1}$ Cd in the medium (Figure 12c); resistance in Derbyshire with soils ranging up to 13 $\mu g\ g^{-1}$ Cd was best seen ($r = 0.77$) when medium containing 100 $\mu g\ g^{-1}$ Cd was used (Figure 12d). Although an arbitrary range of metal concentrations in the media is used in this assay the relationships between bacterial tolerance and soil Cd concentrations are clearly demonstrated over a range of both polluted and unpolluted soils. Results further suggest that the bacterial populations could withstand a small further input of Cd (several $\mu g\ g^{-1}$) into the environment without showing a significant

change in numbers. Also it can be concluded that bacteria from the highly polluted Shipham soils would be less sensitive to Cd additions than isolates from Derbyshire soils with a relatively low level of contaminating Cd.

Strong correlations were also obtained between the proportion of the populations resistant to Zn and the Zn concentrations in the soil. Lead resistance in the bacterial populations, though present, was not as easily demonstrated by this method as was resistance to cadmium and zinc (Olson and Thornton, 1982).

2. Metal resistance was much greater in bacteria from soils receiving present-day smelter fallout at Avonmouth than in those in which smelter contamination had ceased 10 years previously at Swansea (Olson *et al.*, 1983). Significant relationships between resistance to Cd, Cu, Pb and Zn and both 'total' and extractable (NH_4OAc, pH 7) metal concentrations were found at Avonmouth but not at Swansea (Table 6). It has been suggested that resistance reflects the bio-availability of the metal. Decrease in resistance with time could be due to changes in metal bio-availability brought about by weathering and pedogenetic process in the soil eventually leading to changes in the chemical forms and species of the contaminating metals and perhaps their incorporation into secondary soil minerals.
3. There are preliminary indications that on sludge amended soils, bacterial populations rapidly adapt to increasing metal concentrations in the soil.

Table 6 **Relationships between log percent bacterial resistance on amended media and log total metal soil levels for current and recent smelting locations in Britain**

Locations	Metals tested	Concentration range (ppm)	r value* Media concentration of metal μg/g† A	B	C	D
Smelting, current						
Avonmouth	Cd	5–50	0.34(55)	0.70(47)	0.85(39)	NA
	Zn	380–3540	ND	0.75(55)	0.68(60)	0.57(33)
	Cu	28–259	−0.47(52)	0.07(41)	0.53(28)	NA
	Pb	106–1622	0.09(58)	0.55(52)	0.50(41)	NA
Smelting, recent						
Swansea	Cd	3–81	0.28(36)	0.28(36)	0.18(30)	NA
	Zn	540–2300	ND	0.24(44)	0.14(39)	0.63(26)
	Cu	84–452	0.10(57)	−0.39(35)	0.08(30)	NA
	Pb	120–1552	0.08(58)	0.21(38)	−0.03(22)	NA

Data from Olson *et al.* (1983)
*r—correlation between log percent bacterial resistance and log total soil metal concentration ($P<0.05$)

†A:	Cd = 10 μg/g;	Cu = 100 μg/g;	Pb = 500 μg/g;	Zn = 200 μg/g
B:	Cd = 100 μg/g;	Cu = 300 μg/g;	Pb = 1000 μg/g;	Zn = 250 μg/g
C:	Cd = 200 μg/g;	Cu = 400 μg/g;	Pb = 2000 μg/g;	Zn = 500 μg/g
D:	Cd = NA;	Cu = NA;	Pb = NA;	Zn = 650 μg/g

Number of samples in brackets
NA—not applicable; ND—not determined

A comparison between sludge amended sites in the Thames Valley sampled in January 1981 and August 1982 showed metal resistance in soil bacteria increased for Cd and Zn at the one site where sludge had been further applied between the two sampling occasions; in this case, the proportion of Cd resistant bacteria on a 100 $\mu g\,g^{-1}$ Cd assay plate increased from <0.001% to 3.9%. At other sites, where no further sludge had been added, resistance patterns decreased, suggesting that either seasonal factors were affecting resistance or that 'availability' of the metal was decreasing with time.

4. Comparisons between contaminated and matched control sites demonstrated a marked reduction in species diversity in the resistant populations, even though total numbers of bacteria were similar. For example, bacteria from Shipham soils isolated on medium with no metals added were identified as belonging to the genera *Alcaligenes*, *Bacillus*, *Arthrobacter Pseudomonas*, *Flavobacterium*, *Chromobacterium* and *Klebsiella*, while isolates from media amended with 200 $\mu g\,g^{-1}$ Cd were often limited to species of two or four genera: *Flavobacterium*, *Pseudomonas*, *Acinetobacter* and *Alcaligenes*. Because of the historical nature of the polluted sites reported on here, it cannot be determined if the species composition of bacterial populations in these soils rapidly altered at the onset of mining over a century ago, or evolved gradually as weathering changed the bioavailability of the metals.

Further research is clearly necessary to understand fully the relationship between metal resistance in bacteria and the nature of contaminants in the soil. Seasonal influences, including the degree of wetness and temperature of the soil, may be important, affecting the bio-availability of metals—both to the microbiological and to the plant community. The effect of time on the ongoing availability of new metal additions to soils is still far from clearly understood, and will almost certainly be related both to the nature and chemical form of the metal input and to the physico-chemical soil environment.

In conclusion, the relationship between metal resistance and antibiotic resistance (Timoney *et al.*, 1978) has intriguing possibilities in those areas of Britain (and elsewhere) where soils are heavily contaminated with metals and in which bacterial metal resistance has been demonstrated. Reservoirs of soil-borne bacteria in which antibiotic resistance has been triggered by the presence of metal contaminants are a possible legacy from our forebears. However, the development of linked antibiotic and metal resistance in response to environmental pressure in the soil still needs to be assessed.

Water resources

The heavy metal content of river waters distributed for potable water supplies may sometimes exceed medically recommended limits, such as those required by the Council of European Communities of 50 $\mu g\, l^{-1}$ Pb and 5 $\mu g\, l^{-1}$ Cd (CEC, 1980). The monitoring of river waters is both costly and time consuming due to the diurnal and seasonal variations in soluble and insoluble chemical constituents. In the early 1970s, research commenced into the possible use of data based on the composition of river/stream sediments as an ancillary aid to the prediction of water quality and to the mapping of potential water resources. A collaborative programme involving AGRG and the Cornwall River Authority (now part of the Southwest Water Authority) focussed on some rivers in Cornwall where natural weathering processes, together with past and present mining activities, have given rise to high levels of metals such as Cu, Pb, Zn, Cd, Sn and As in the waters and sediments of local streams draining mineralized areas. A comparison of mineralized and unmineralized tributory drainage for three rivers examined (Table 7) showed large trace element concentrations in waters, reflecting their associated sediments; this demonstrated the potential usefulness of geochemical reconnaissance data based on stream sediment sampling as a preliminary indicator of metal concentration ranges in the waters (Aston *et al.*, 1974; Thornton and Webb, 1977).

As a result of this work, it was proposed that the trace element composition

Table 7 **Comparison of the average trace element concentrations in the waters and sediments of mineralized and unmineralized tributaries**

River		Sediment (ppm)		Water ($\mu g l^{-1}$)	
		Contaminated	Uncontaminated	Contaminated	Uncontaminated
Red	Mn	1020	657	720	270
	Cu	706	563	9	25
	Zn	1590	358	180	70
	Pb	775	133	18	7
	Cd	6	4	5	2
	Fe	57 000	39 000	990	490
Carnon	Mn	1054	964	830	350
	Cu	3000	416	1080	1080
	Zn	1839	391	8150	3060
	Pb	540	402	44	37
	Cd	7	3	ND	ND
	Fe	67 000	48 000	5600	4100
Gannel	Mn	3165	1000	650	87
	Cu	313	46	16	11
	Zn	3995	255	800	59
	Pb	2670	89	250	21
	Cd	21	4	ND	ND
	Fe	46 000	42 000	6670	1620

ND—not determined

of stream sediments could provide a more stable and convenient indication of the probable trace element status of associated waters than one-off sampling of the waters themselves, and at the same time could be useful in the selection of critical sites for routine water sampling. Further studies by AGRG in southwest England led to suggested tentative threshold concentrations of Cu, Pb, Zn, Cd, Fe, Mn and As in sediments above which their concentrations in unfiltered waters could exceed the recommended limits for drinking waters proposed by the World Health Organization (1971) (Aston and Thornton, 1977).

Such an empirical approach does not take into account the importance of the physical and mineralogical composition of the sediment, water pH, hardness, alkalinity and organic content on the solubility of respective elements, and is only applicable within an area of relatively uniform physico-chemical conditions. In many situations, water may be directly contaminated with elements from sewage effluents and from industrial and urban wastes; in these cases the chemical composition of the sediment is often of little value in predicting water quality.

In the light of the unique data bank on sediment composition for around 50 000 locations obtained in the course of compiling the *Wolfson Geochemical Atlas of England and Wales* (Webb *et al.*, 1978), a detailed programme of research into the applications of geochemical reconnaissance to the water industry was commenced in 1977, sponsored by the UK Department of the Environment. Studies were conducted into relationships between the concentrations of heavy metals in sediments and waters in selected rivers in England and Wales in which metal-contaminated sediments had been found; in addition this research sought to identify some of the processes governing the dispersion of metals in the dissolved and particulate phases of the river waters.

The main emphasis of this work was again in areas of historical mining and smelting where acid mine drainage together with clastic erosion of waste materials and leachates from wastes continue to provide on-going sources of metal contaminants to both river sediments and waters. Investigations in the River Hayle, Cornwall (Cu : Sn mineralization, soft water), River Gannel, Cornwall (Pb : Zn mineralization, soft water) and Minsterley Brook, Shropshire (Pb : Zn mineralization, hard water) have been described by Metcalfe (1983); these provide excellent examples of physico-chemical processes governing metal dispersion in the dissolved and particulate phases of the water that are of prime importance in the interpretation of the stream sediment–water relationship.

In the course of this work, four major sources of metals to each river were identified: (a) discharge from mine drainage adits; (b) action of subsurface drainage and groundwater on metalliferous mine spoil; (c) action of surface drainage on mine spoil; and (d) outwash of metalliferous material from river banks and spoil tips. For example, adit drainage into the River Hayle accounted for 10% of the river flow and as much as 20–60% of the metal loading in the water column at some points of discharge. Downstream of adit

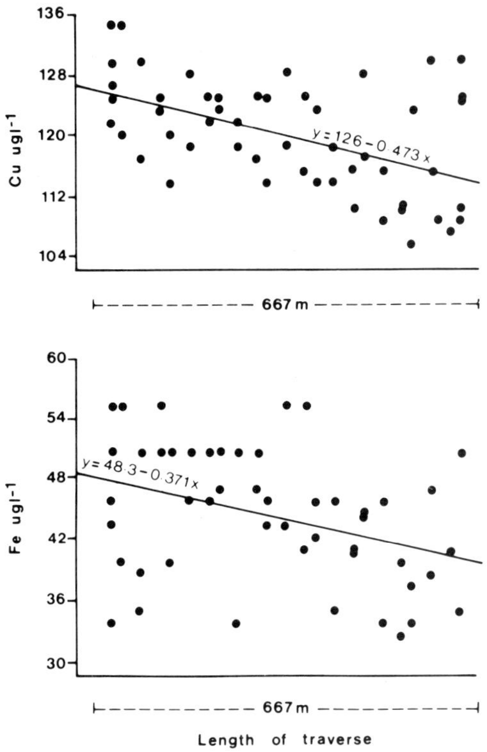

Figure 13 Dissolved Cu and Fe in waters of the River Hayle, Cornwall downstream of an input from an old mine (from Metcalfe, 1983)

input, Fe and Cu were removed from solution due to precipitation and adsorption on to particulates and sediments. This was clearly illustrated in a 700 m stretch of this river below Godolphin within which discharge was constant and some 10% of the dissolved Cu and 20% of the dissolved Fe were lost from solution either to the sediments or to the particulate phase (Figure 13). It is thought that on mixing with the more aerated waters of the mainstream, Fe(11) in the adit water is oxidized and precipitated as hydrous Fe-oxides, which are a suitable substrate for the adsorption of dissolved Cu at the mean water pH of 6.2.

The main source of metals to the sediments of each of the rivers was erosion of mine spoil. In the majority of situations studied, there was little or no evidence of solution of metals from sediments; this was confirmed in laboratory 'flume' experiments (Metcalfe, 1983). In the three rivers studies, Cu, Pb and Fe were the major metal constituents measured in the particulate phase of the waters. Below pH 6.5, there was little Zn, Cd and Mn in the particulate material; amounts increased as pH rose, with as much as 30% of the Mn associated with particulates at pH 7.5 (Metcalfe, 1983). This is illustrated by the observations made in the River Gannel (Figure 14), where

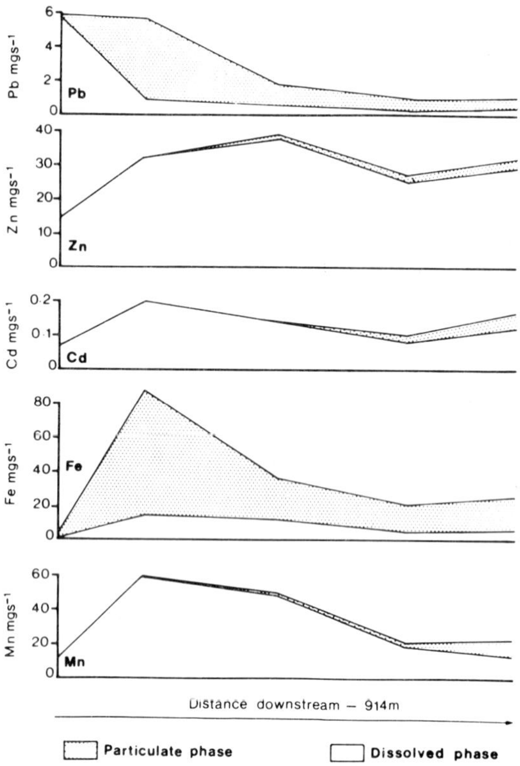

Figure 14 Distribution of Pb, Zn, Cd, Fe and Mn in dissolved and particulate phases of waters in the River Gannel, Cornwall (from Metcalfe, 1983)

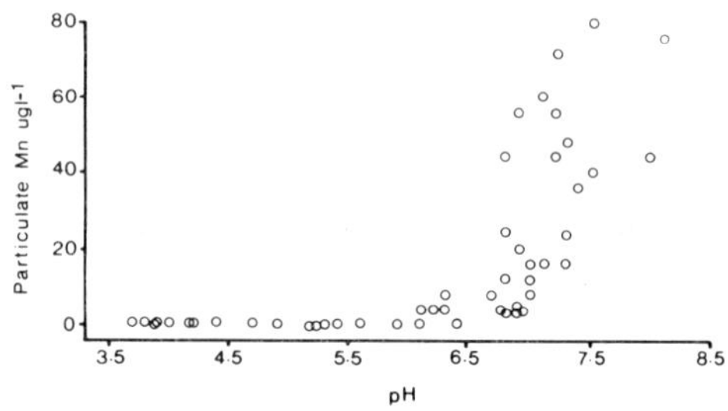

Figure 15 Relationship between concentrations of particulate Mn and water pH in the lower stretches of the River Gannel, Cornwall (from Metcalfe, 1983)

both Pb and Fe in water were mainly present as particulates, and particulate Mn increased markedly above pH 6.5 in the lower stretches of the river (Figure 15). Zinc and Cd were lost from solution in these stretches most probably due to adsorption onto the large amounts of Mn and Fe oxides in the sediment material.

In the much harder waters of Minsterley Brook at pH 8.0, dissolved Pb, Zn,

Cd, Fe and Mn were lost from solution downstream from mine wastes and were probably precipitated from solution as carbonates (Metcalfe, 1983). Where metals were being lost from solution, relatively large proportions of these metals in the sediments occurred in 'non-residual' forms such as water-soluble, exchangeable, organically bound or in association with carbonates or Fe– or Mn–oxides (these were determined with the use of weak chemical extractants, see Metcalfe (1983)). It was concluded that 'a high proportion of a metal in the non-residual phase of the sediment may indicate elevated concentrations in the water column' and that 'the concentrations of non-residual metals in river sediments may be used to infer concentrations in the waters during reconnaissance geochemical surveys providing the water pH is not anomalously low' (Metcalfe, 1983).

Clearly, where geochemical surveys show anomalously high concentrations of metals in stream sediments in drainage used for potable supplies or for irrigation purposes, a programme of water sampling should be carried out, taking into account seasonal and diurnal influences relating to rainfall, etc. The actual relationship between metals in the sediment and in the water column will depend on the nature of the metal sources and inputs into the system, the nature and amount of suspended material, water pH and hardness, and the kinetics of precipitation and adsorption reactions.

Urban pollution

Natural and anthropogenic sources of heavy metals at elevated concentrations in air, soil, water and food constitute major pathways of transfer to man, and as such present a potential risk to human health. The possible health hazard represented by elevated concentrations of metals in dust has hitherto received only limited recognition (Barltrop *et al.*, 1975; Royal Commission on Environmental Pollution, 1983). However, the discovery of increased concentrations of lead and other toxic metals within the home (Thornton *et al.*, 1980; Moorcroft *et al.*, 1982; Culbard *et al.*, 1983), suggests that housedusts could represent an important metal source, especially to young children who have a habit of sucking playthings and carrying out excessive hand to mouth activity. Data from a selection of published studies have led to the proposal of a quantitative relationship relating the concentration of lead in dust or soil to which the child is exposed to incremental changes in blood lead concentration (Duggan, 1980). The question as to whether or not a relationship exists between elevated blood Pb concentrations and detrimental effects on childhood IQ and behaviour still remains to be satisfactorily resolved.

Prior to 1981, information on the levels of lead and other toxic metals to be found within United Kingdom households was largely inadequate. In order to assess the possible importance of dust as a route of metal exposure, it was necessary to investigate the range of metal concentrations in dusts and soils to which the population are generally exposed. A nationwide survey of metals in dusts and soils was commenced by AGRG for the UK Department of the Environment in 1981. Extensive reconnaissance studies have now been completed in 53 city boroughs, towns and villages in England, Scotland and Wales, selected to represent industrial, residential and rural communities and proportionately weighted to match overall population density. Six geochemical 'hotspots' clearly defined by the *Wolfson Geochemical Atlas of England and Wales* as areas where the local community are exposed to elevated metal concentrations were included for comparison. The locations studied were:

Group A (Population >1 million)	—London, Birmingham
Group B (Population >500 000)	—Glasgow, Leeds, Liverpool, Sheffield
Group C (Population >250 000)	—Edinburgh, Bristol, Cardiff, Nottingham, Hull, Stoke-on-Trent
Group D (Population >100 000)	—Plymouth, Swansea, Portsmouth, Blackpool, Luton, Northampton, Southend, Brighton, Ipswich, Reading
Group E (Population<100 000)	—York, Preston, Gloucester, Darlington, Rotherham, Carlisle, Scunthorpe, Merthyr Tydfil, Shrewsbury, Dunfermline, Falkirk, Camborne/Hayle, Hawick, Newmarket, Leominster
Group F (Population <5000)	—Horncastle, Shaftesbury, Llanrwst, Coombe Martin, Alston, Grassington, Derbyshire Villages

The sampling strategy adopted was suitable for a large-scale multi-element reconnaissance survey requiring a rapid, low cost and publicly acceptable method of sample collection. At each location, housedust (from the vacuum cleaner) and garden soil (0–5 cm) was collected from 100 households. Soils from vegetable gardens and dusts from roads, school playgrounds and amenity areas were also taken. Methods of sample collection, preparation and analysis have been described by Culbard *et al.* (1983).

The concentration ranges of lead, cadmium, zinc and copper in internal floordusts and in surface garden soils are large (Table 8) and exceed those for rocks and for uncontaminated agricultural soils (Archer, 1971; Bowen, 1979; Davies, 1983). Average lead concentrations in garden soils are appreciably higher than in rural agricultural soils (usually within the range 10–150 ppm) suggesting anthropogenic inputs and confirming previous reported levels (Purves, 1967; Davies, 1977). The relationship between lead concentrations in housedusts and garden soils for the total data set is highly significant (Figure 16; $r = 0.531$, $n = 4512$). With the exception of the Derbyshire mining area, the concentration of lead in housedusts is significantly greater than that of the

Table 8 **Concentrations of lead, cadmium, zinc and copper in housedusts (vacuum cleaner) and garden soils (0–5 cm) collected in the UK nationwide survey**

	Housedust				Soil			
	Pb	Cd	Zn	Cu	Pb	Cd	Zn	Cu
7 London boroughs	210–12400	2–30	404–5950	46–1408	109–3318	1–5	112–1528	24–294
	1007	8	1324	208	647	1	424	73
	(683)	(683)	(683)	(683)	(579)	(579)	(579)	(579)
40 other city boroughs, towns and villages	106–4177	2–34	348–3850	47–1880	36–1656	1–4	62–1205	14–237
	507	7	1055	204	230	1	260	53
	(3953)	(3955)	(3954)	(3955)	(3550)	(3549)	(3550)	(3551)
Derbyshire Pb mining area 'geochemical hotspot'	728–6500	5–45	669–2684	84–568	1704–17900	6–53	419–4710	37–188
	1870	12	1317	203	5605	17	1483	102
	(100)	(100)	(100)	(100)	(100)	(100)	(100)	(100)

Concentration in ppm, 95% range, geometric mean; number of samples in brackets

surface garden soil (Table 8), and indicates the possibility of internal sources of Pb from within the home supplementing those from outside and/or to the selective movement of fine (metal-rich) particles which could be transported from a variety of potential sources including fine soil, atmospheric aerosol, etc. into the home. Concentrations of Pb in housedust and soil were larger in the 7 London Boroughs tested (geometric means 1007 and 647 ppm, respectively) than elsewhere (means 507 and 230 ppm) (Table 8), with the exception of the 'geochemical hotspot' of Derbyshire mining villages (1870 and 5605 ppm Pb) where levels were far greater.

Housedusts containing in excess of 2000 ppm Pb were found in 9.5% of homes. Lead concentration in both soils and dusts is related to the age of the property with the majority of high lead concentrations associated with older homes (Figure 17). Houses under 30 years of age comprise the majority of

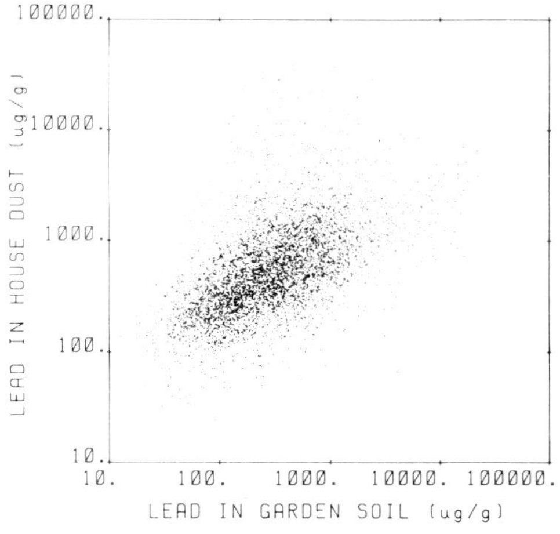

Figure 16 Relationship between concentrations of Pb in housedust and garden soil recorded in a survey of metals in urban soils and dusts in Britain

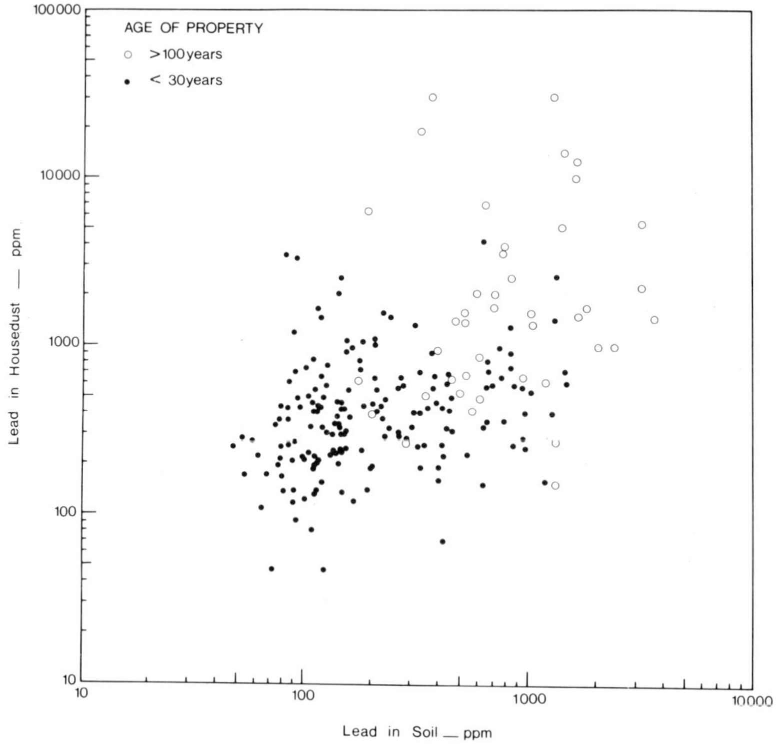

Figure 17 Concentrations of Pb in soils and housedusts in relation to age of property in 10 towns in Britain

lower lead values; this may be due to a voluntary reduction in the lead content of household paints from around that time. It should also be pointed out that these houses have a shorter history of soil contamination from bonfires and fossil fuel residues from open coal fires which were commonly used as soil ameliorants in the past. Cadmium, zinc and copper are present in significantly larger concentrations in housedusts than in garden soils, providing evidence of internal sources of these metals such as pigments in paint, domestic appliances, piping, etc.

Research carried out by AGRG has previously established the relationship between concentrations of Pb, Cd and As in garden soils of old mining areas and the metal content of dusts from the related households (Barltrop *et al.*, 1975; Thornton *et al.*, 1980; Culbard and Johnson, 1984). The percentage frequency distributions of lead and cadmium in housedusts and garden soils in a selection of the reconnaissance survey locations (referred to as the 10-town study, see Culbard *et al.*, 1983) have been compared in detail with two 'geochemical hotspots'—the old mining areas of Derbyshire and Shipham—and are presented in Figures 18 and 19, respectively. In the 10-town database the frequency of lead concentrations in housedusts and soils are generally similar, with the obvious exception of the long housedust 'tail' of

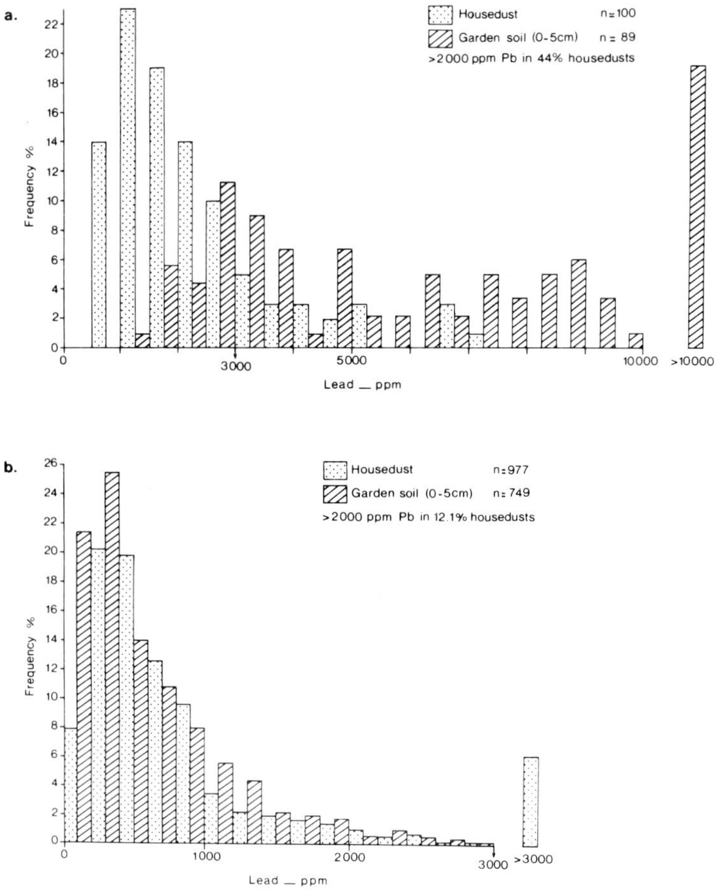

Figure 18 Frequency distribution of Pb concentrations in housedust and garden soil: (a) in Derbyshire old Pb-mining area; (b) in 10-town study

over 6% containing more than 3000 ppm Pb. A possible source of these high values in internal dusts could be paint. In the Derbyshire lead mining area both the geometric mean and range of Pb concentrations in housedusts in soils are far greater, with over 19% of soils exceeding 1% Pb and around 44% of housedusts exceeding 2000 ppm Pb.

Cadmium frequency distributions in the ten towns and Shipham show similar trends to Pb. Concentrations in housedusts from the ten towns greatly exceed those in soil (8% exceed 20 ppm Cd). In contrast, the Shipham soils exceed the housedust concentrations, even though nearly 80% of the floordust samples have cadmium concentrations above 20 ppm Cd. Examination of several housedust samples from Shipham by scanning electron microscopy and X-ray diffraction have suggested that around 20–25% by weight of this material is made up of soil particles (Johnson and Twist, 1983). This may explain the larger concentrations of cadmium in the Shipham housedusts

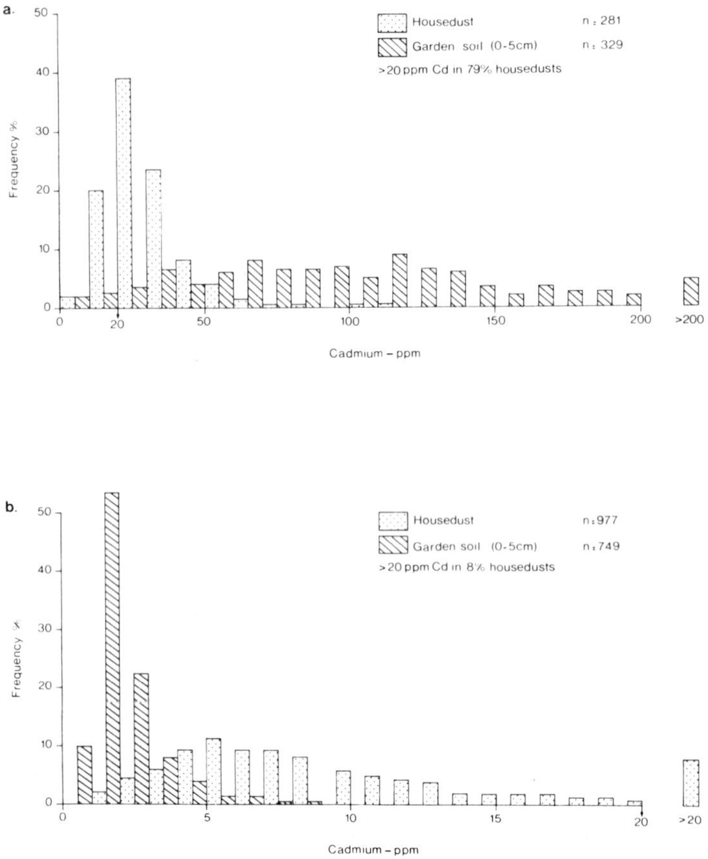

Figure 19 Frequency distribution of Cd concentrations in housedust and garden soil: (a) in Shipham old Zn-mining area; (b) in 10-town study

compared with those in the 10 towns as shown in Figure 18.

It may be concluded that in areas with geochemical enrichment, the external sources of metals can greatly enhance and also mask the possible internal components of the total indoor metal load. It should be pointed out that although these geochemical anomalies may result in increased metal exposure to the local community, no obvious health effects have as yet been shown (Barltrop *et al.*, 1975; Barltrop and Strehlow, 1982).

It has been shown that housedusts associated with properties possessing heavily contaminated garden soils appear to contain proportionately less metal than the soil, suggesting dilution due to carpet fluff and other house-derived material. Where housedusts are investigated in areas with 'normal' soil metal concentrations, they appear to have larger concentrations of metals than the soils, suggesting sizeable adventitious inputs supplementing those originating from outside the home (for example, paint flakes), or the selective transport of metal-rich material.

The elevated concentrations of metals within the home environment demonstrated in a significant number of households in the UK may possibly contribute to the body burden of young children. Judgement on the significance of these findings must await additional research.

Human health

Geochemical surveys have highlighted a number of anomalous situations, colloquially termed 'geochemical hotspots', where the metal burden of man may be augmented through foodstuffs, drinking water, dust and air. These pathways of metals to man, which may originate from natural and/or anthropogenic sources, have in part been discussed in previous sections of this chapter. The movement of metals into and through the food chain or by accidental inhalation and ingestion to man is subject to many chemical and physical processes depending on the nature and source of the elements, their mobility, solubility and availability to plants and within the human gut, chemical and biochemical interactions in the soil, plant, animal and human, and interactions with other dietary constituents.

It is frequently argued that in the modern world, health and disease in man can no longer be expected to reflect his local environment, as (a) food now comes mainly from the supermarket, derived from many parts of the world, (b) milk and water supplies are rarely of local origin, (c) movement of individuals and families from one location to another is common and people spending their lives in any one locality are the exception rather than the rule and (d) the majority of the population live in large conurbations where geochemical influences on the home environment are likely to be minimal.

However, rural communities in the UK and many people in the developing countries live much 'closer to the ground', with a greater dependence on home and locally grown vegetable and salad crops than the town dweller; in some cases water supplies may also reflect the local geochemical environment. As shown in the last section, both rural and urban communities are exposed to metals in dust and air-borne particulates; in a geochemical hotspot such as the old lead mining area of Derbyshire, lead levels in housedust reflect the extensive contamination of local garden soils (see Figure 16).

To date, there is little evidence directly linking the geochemical environment and the presence of human disease in the UK with the exception of (a) the extremely well documented inverse relationship between the fluorine content of the water supply and the incidence of dental caries (Royal College of Physicians, 1976) and (b) the much diminished prevalence of endemic goitre associated primarily with a deficiency of iodine in water supplies and food (Kelly and Sneddon, 1960). Empirical relationships are more common and include (a) a negative correlation between water hardness and the prevalence of cardiovascular disease (Crawford *et al.*, 1968; Shaper *et al.*, 1980), (b) relationships between the distribution of cancers and heavy metals in soils (Pinsent, 1968; Davies, 1973; Stocks and Davies, 1964) and (c) variations in the prevalence of dental caries and concentrations of Mo (Anderson, 1969) and Pb (Anderson and Davies, 1980). In these latter examples no causal factor linked to geochemistry has yet been established.

Two major studies have been mounted into the possible effects of increased exposure to metals in 'geochemical hotspots' related to historical mining and smelting activities in Derbyshire and Somerset.

In an area of past lead–zinc mining in Derbyshire, a collaborative study by the Paediatric Unit, St. Mary's Hospital Medical School and AGRG, showed when households were grouped according to the Pb content of their garden soils, Pb in the blood and hair of the 2–4 year old children increased with that in soil and housedust (Table 9). However, none of the Pb values in children were high enough to be considered hazardous at that time, even though the amounts in soil and dust were extremely high, peaking at 2.8 and 2.5% Pb respectively (Barltrop *et al.*, 1975). The children's Pb status does, however, reflect that of the environment, and in the absence of evidence linking the human burden with Pb in foodstuffs or water supplies, it was suggested that the major pathway of Pb into the children was through inhalation and involuntary ingestion of dust particles (Barltrop *et al.*, 1975).

A comprehensive environmental health study was conducted around the village of Shipham, Somerset where Zn was mined as the ore smithsonite ($ZnCO_3$) over the period 1700–1850. The ore, with appreciable amounts of Cd as a guest element, was washed in ponds and calcined within the village. Surface agricultural soils reclaimed from and near the old mine workings contain from 30–800 ppm Cd and range up to 7% Zn (Marples and Thornton, 1980; Matthews and Thornton, 1982). Collaborative investigations involving national and local government, Westminster Medical School and AGRG showed: (a) 60% of household garden soils with over 60 ppm Cd, greatly exceeding the levels in polluted paddy soils associated with the well documented 'itai–itai' disease in Japan; these soils also contained large

Table 9 **Lead content of garden soil, housedust, blood and hair**

Soil (ppm)	House dust (ppm)	Blood (μg/100 ml)		Hair (ppm)
		Child	Mother	Child
130–900 420 (24)	190–2450 531 (20)	15–33 21 (29)	9–40 14 (25)	2–36 8 (29)
1050–9100 3390 (40)	390–25000 1564 (36)	13–45 24 (43)	11–44 19 (41)	2–62 11 (43)
10000–28000 13969 (8)	1000–5100 2582 (8)	21–43 29 (10)	13–19 15 (8)	10–39 29 (10)

Data from Barltrop *et al.* (1975)
Range and geometric mean; number of samples in brackets

amounts of Zn with Zn : Cd ratios of around 90 : 1 (Thornton *et al.*, 1980); (b) housedust samples averaged 26 ppm Cd and 2300 ppm Zn; (c) from studies on metals in locally grown vegetables and diets, an average human intake of 200 μg Cd per week was calculated, compared with the average UK intake of 140 μg Cd per week; individual intakes rarely exceeded the World Health Organization's provisional tolerable weekly intake of 450–500 μg Cd (Sherlock *et al.*, 1983); (d) from health inventories and biochemical tests on 548 residents of Shipham and 543 controls from a nearby uncontaminated village only slight differences attributable to Cd were found (Barltrop and Strehlow, 1982). In this unique situation it is possible that the large amounts of Zn and Ca present in the environment are affording some degree of protection against possible adverse effects of cadmium.

The future

The preceeding sections have clearly shown the importance of natural geological and man-made inputs of chemical elements in the surface environment in relation to the health of plants, animals and man. In particular, the activities of our forebears through mining and smelting ore minerals have contaminated soil and water to a degree that will affect mankind far into the future.

The processes controlling the geochemical cycling of potentially toxic metals are still far from clearly understood. Research in the late 1980s and into the 1990s must address the fluxes of toxic elements in soil, water, dust and air together with their pathways to the target organism. Toxic metals rarely occur alone and their associations and interactions with one another and with other components of the environment are known to influence their availability to the plant or animal and their ultimate toxicity. At any one place and at any one time there is nearly always more than one source of a metal contributing to any one component of the environment, be it soil, air, housedust or water. Multi-sources of any one metal may give rise to different forms (speciation), which may themselves change over time. Over the past 20 years, much effort has gone into attempting to identify and quantify these different chemical species and determine their respective roles in the food chain. Chemical extractants have been and continue to be widely used for this purpose with only limited success. More recently, computer-based chemical equilibrium models have been developed for the speciation of chemical elements in waters and soil solutions. This approach is encouraging and future work on metal speciation in soil solution will hopefully lead to a greater understanding of metal 'availability' to plants.

The average gram of soil contains 10^6–10^9 bacteria and it is natural that the effect of this living mass on metal cycling and availability will be large. A greater degree of collaboration between the geochemist, soil and water chemist and microbiologist is essential if we are to further our knowledge of metal behaviour, interaction and residence time in the complex system rock : soil : plant : animal : man.

ACKNOWLEDGEMENTS

Examples are cited from AGRG's programme of research in applied environmental geochemistry funded by the Agricultural Research Council, Natural Environment Research Council, Department of the Environment, NATO and RTZ Services Ltd. The authors are grateful to many postgraduate students and colleagues who have contributed to this work.

References

Abrahams, P.W. (1983). Distribution, dispersion and agricultural significance of metal in soils of the mining regions of southwest England, *PhD Thesis, London University.*

Anderson, R.J. (1969). *Caries R.*, **3**, 75–87.

Anderson, R.J. and Davies, B.E. (1980). *J. Geol. Soc. Lond.*, **137**, 547–559.

Archer, F.C. (1971). In: *Trace Elements in Soils and Crops*, Ministry of Agriculture, Fisheries and Food Tech. Bull. 21, pp. 150–158; HMSO, London.

Associated Octel Co. Ltd. (1983). *World Wide Survey of Motor Gasoline Quality*; The Associated Octel Co. Ltd., London.

Aston, S.R. and Thornton, I. (1977). *Sci. Total Environ.*, **7**, 247–260.

Aston, S.R., Thornton, I., Webb, J.S., Purves, J.B. and Milford, B.L. (1974). *Water, Air and Salt Pollution*, **3**, 321–325.

Barltrop, D. and Strehlow, C.D. (1982). *Lancet*, **Dec.**, 18.

Barltrop, D., Strehlow, C.D., Thornton, I. and Webb, J.S. (1975). *Postgraduate Medical J.*, **51**, 801–804.

Bewley, R.J.F. (1979). *J. Gen. Microbiol.*, **110**, 247–254.

Bewley, R.J.F. (1980). *Appl. Environ. Microbiol.*, **40**, 1053–1059.

Borlase, W. (1758). *The Natural History of Cornwall*; Oxford University Press, Oxford.

Bowen, H.J.M. (1979). *Environmental Chemistry of the Elements*; Academic Press, London.

Brierley, C.L. and Thornton, I. (1979). *Miner. Environ.*, **1**, 112–119.

Cannon, H.L. and Anderson, B.M. (1971). In: *Environmental Geochemistry in Health and Disease* (Eds. H.L. Cannon and H.C. Hopps),pp. 155–177; Geol. Soc. Am. Inc. Memoir 123, Boulder.

Cannon, H.L., Connolly, G.G., Epstein, G.B., Parker, J.G., Thornton, I. and Wixson, B.G. (1978). In: *Geochemistry and the Environment, Vol. 111*, pp. 17–31; National Academy of Sciences, Washington, DC.

Carpenter, K.A. (1924). *Ann. Appl. Biol.*, **11**, 1–23.

Carpenter, K.A. (1925). *Ann. Appl. Biol.*, **12**, 1–3.

Carpenter, K.A. (1926). *Ann. Appl. Biol.*, **13**, 395–401.

CEC (1980). Directive relating to Quality of Water intended for Human Consumption 80/778/EEC, 15 July.

Chamberlain, A.C. (1983). *Atmospheric. Environ.*, **17**, 693–706.

Chopra, I. (1971). *J. Gen. Microbiol.*, **63**, 265–267.

Colbourn, P. (1976). The applications of geochemical reconnaissance data to trace metal pollution in agriculture, *PhD Thesis, London University.*

Colbourn, P. and Thornton, I. (1978). *J. Soil Sci.*, **29**, 513–526.

Colbourn, P., Alloway, B.J. and Thornton, I. (1975). *Sci. Total Environ.*, **4**, 359–363.

Crawford, M.D., Gardner, M.J. and Morris, I.N. (1968). *Lancet*, **1**, 827–831.

Culbard, E. and Johnson, L. (1984). In: *Environmental Contamination*; CEP Consultants, Edinburgh.

Culbard, E., Thornton, I., Watt, J., Moorcroft, S. and Brooks, K. (1983). In: *Heavy Metals in the Environment: Heidelberg* pp. 426–429; CEP Consultants, Edinburgh.

Davies, B.E. (1973). In: *Proc. Int. Symp. Environmental Health Aspects of Lead*, pp. 125–134; Commission of the European Communities, Luxembourg.

Davies, B.E. (1977). *Sci. Total Environ.*, **9**, 243–262.

Davies, B.E. (1980). In: *Inorganic Pollution and Agriculture*, pp. 142–156; HMSO, London.

Davies, B.E. (1983). In: *Applied Environmental Geochemistry* (Ed. I. Thornton), pp. 426–462; Academic Press, London.

Dines, H.G. (1956). *The Metalliferous Mining Region of South-west England, Vol. 1*, Memoirs Geol. Survey Great Britain; HMSO, London.

D'Itri, P.A. and D'Itri, F.M. (1977). *Mercury Contamination: A Human Tragedy*; John Wiley and Sons, New York.

Duggan, M.J. (1980). *Water Air Soil Pollut.*, **14**, 309–321.

Farey, J. (1811). *A General View of the Agriculture and Minerals of Derbyshire, Vol. 1*; Macmillan, London.

Farey, J. (1817). *A General View of the Agriculture and Minerals of Derbyshire, Vol 2*; Macmillan, London.

Gadd, G.M. and Griffiths, A.J. (1978). *Microb. Ecol.*, **4**, 303–317.

Hawkes, H.E. and Webb, J.S. (1962). *Geochemistry in Mineral Exploration*; Harper and Row, New York and Evanston.

Houba, C. and Remacle, J. (1980). *Microb. Ecol.*, **6**, 55–69.

Jarvis, S.C. (1978). *J. Sci. Food Agric.*, **29**, 12–18.

Johnson, D.L. and Twist, J.P. (1983). State Univ. New York, Syracuse, Personal communication.

Jones, L.H.P. and Clement, C.R. (1972). In: *Lead in the Environment*, pp. 29–33; Institute of Petroleum, London.

Kelly, F.C. and Sneddon, M.A. (1960). In: *Endemic Goitre* (WHO Monograph), No. 4, pp. 27–233.

Kovalsky, V.V. and Letunova, S.V. (1964). *Usp. Sovrem. Biol.*, **57**, 71.

Letunova, S.V. (1970). In: *Trace Element Metabolism in Animals* (Ed. C.F. Mills), pp. 432–437; E. & S.L. Livingstone, Edinburgh and London.

Little, P. and Wiffen, R.D. (1978). *Atmospheric. Environ.*, **12**, 1331–1343.

Marples, A.E. (1979). The occurrence and behaviour of cadmium in soils and its uptake into pasture grasses in industrially contaminated and naturally metal-rich environments, *PhD Thesis, London University*.

Marples, A.E. and Thornton, I. (1980). In: *Proc. 2nd Int. Cadmium Conf., Cannes*; Metal Bulletin Ltd., London.

Matthews, H. (1982). The distribution of cadmium and associated elements in the soil–plant system at sites in Britain contaminated by mining, smelting and metal-rich bedrock, *PhD Thesis, London University*.

Matthews, H. and Thornton, I. (1980). In: *Trace Substances in Environmental Health XIV* (Ed D. D. Hemphill), pp. 478–488; Univ. Missouri, Columbia.

Matthews, H. and Thornton, I. (1982). *Plant and Soil*, **66**, 181–193.

Mattigod, S.V. and Page, A.L. (1983). In: *Applied Environmental Geochemistry* (Ed. I. Thornton), pp. 355–394; Academic Press, London.

Metcalfe, A.P. (1983). The mobilization and dispersion of some heavy metals by surface drainage in three areas of past mining activity in England, *Ph D Thesis, London University*.

Mitchell, R.L. and Burridge, J.C. (1979). *Phil. Trans. R. Soc. Lond.*, **B288**, 15–24.

Moorcroft, S., Watt, J., Thornton, I., Wells, J., Strehlow, C.D. and Barltrop, D. (1982). In: *Trace Substances in Environmental Health XVI* (Ed. D.D. Hemphill), pp. 155–163; Univ. Missouri, Columbia.

Murozumi, M., Chow, T.J. and Patterson, C. (1969). *Geochim. et Cosmochin. Acta.*, **33**, 1247–1294.

Nelson, J.D. and Colwell, R.R. (1975). *Microb. Ecol.*, **1**, 191–218.

Nichol, I., Thornton, I., Webb, J.S., Fletcher, W.K., Horsnail, R.F., Khaleelee, J. and Taylor, D. (1970a). *Regional Geochemical Reconnaissance of the Derbyshire Area*, Rep. No. 70/2; Inst. Geol. Sci., London.

Nichol, I., Thornton, I., Webb, J.S., Fletcher, W.K., Horsnail, R.F., Khaleelee, J. and Taylor, D. (1970b). *Regional Geochemical Reconnaissance of the Denbighshire Area*, Rep. No. 70/8; Inst. Geol. Sci., London.

Nichol, I., Thornton, I., Webb, J.S., Fletcher, W.K., Horsnail, R.F. and Khaleelee, J. (1971).

Regional Geochemical Reconnaissance of Devon and Part of North Cornwall, Rep. No. 71/2; Inst. Geol. Sci., London.

Olson, B.H. (1983). In: *Applied Environmental Geochemistry* (Ed. I. Thornton), pp. 202–229; Academic Press, London.

Olson, B.H. and Thornton, I. (1982). *J. Soil. Sci.*, **33**, 271–277.

Olson, B.H., Tripp, S.C. and Thornton, I. (1983). In: *Heavy Metals in the Environment*; Heidelberg, pp. 378–381; CEP Consultants, Edinburgh.

Pinsent, R.J.F.H. (1968). Report of the Welsh Soils Discussion Group No. 9, pp. 129–129.

Purves, D. (1967). *Plant and Soil*, **26**, 380–382.

Rother, J.A., Millbank, J.W. and Thornton, I. (1982a). *J. Soil Sci.*, **33**, 101–113.

Rother, J.A., Millbank, J.W. and Thornton, I. (1982b). *Plant and Soil*, **69**, 239–258.

Rother, J.A., Millbank, J.W. and Thornton, I. (1983). *J. Soil Sci.,* **34**, 127–136

Royal College of Physicians (1976). *Fluorine, Teeth and Health*, Report of Royal College of Physicians; Pitman Medical.

Royal Commission on Environmental Pollution (1983). *Ninth Report: Lead in the Environment*; HMSO, London.

Shaper, A.G., Packam, R.F. and Pocock, S.J. (1980). *J. Enc. Path. and Toxic.*, **4**, 89–111.

Sherlock, J.C., Smart, G.A., and Walters, B. (1983). *Sci. Total Environ.*, **29**, 121–142.

Sposito, G. (1983). In: *Applied Environmental Geochemistry* (Ed. I. Thornton), pp. 123–170; Academic Press, London.

Stocks, P. and Davies, R.I. (1964). *Br. J. Cancer*, **18**, 14–24.

Summers, A.O. and Silver, S. (1972). *J. Bacteriol*, **112**, 1228–1236.

Summers, A.O. and Silver, S. (1978). *Ann. Rev. Microbiol.*, **32**, 637–672.

Thoresby, P. and Thornton, I. (1980). In: *Trace Substances in Environmental Health XIII* (Ed. D. D. Hemphill); Univ. Missouri, Columbia.

Thornton, I. (1980). In: *Inorganic Pollution and Agriculture*, pp. 105–125; HMSO, London.

Thornton, I. (1982). In: *Trace Substances in Environmental Health XVI* (Ed. D.D. Hemphill), pp. 57–67; Univ. Missouri, Columbia.

Thornton, I. (1983). In: *Applied Environmental Geochemistry* (Ed. I. Thornton), pp. 232–266; Academic Press, London.

Thornton, I. and Abrahams, P.W. (1983). *Sci. Total Environ.*, **28**, 287–294.

Thornton, I. and Abrahams, P.W. (1984). In: *Changing Metal Cycles and Human Health* (Ed. S. Bernhard), pp. 7–25; Springer–Verlag, Berlin.

Thornton, I. and Kinniburgh, D.G. (1978). In: *Trace Element Metabolism in Animals* (Ed. M. Kirchgessner), p. 499; Institut fur Ernahrungsphyiologic Technische Universitat, Munchen.

Thornton, I. and Webb, J.S. (1977). *J. Instit. Water Eng. Sci.*, **31**, 11–25.

Thornton, I., Kinniburgh, D.G., Clegg, F.G., Lewis, G., Barltrop, D. and Strehlow, C.D., *J. Agric. Sci., Camb.*, in press.

Thornton, I., Moorcroft, S., John, S., Watt, J., Strehlow, C.D., Barltrop, D. and Wells, J. (1980). In: *Trace Substances in Environmental Health XIV* (Ed. D.D. Hemphill); Univ. Missouri, Columbia.

Timoney, J.F., Port, J., Giles, J. and Spanier, J. (1978). *Appl. Environ. Microbiol.*, **36**, 465–472.

Tinker, B. (1981). *Phil. Trans. R. Soc. Lond.*, **B294**, 41–55.

Troyer, L.S., Olson, B.H., Hill, D.C., Matthews, H. and Thornton, I. (1980). In: *Trace Substances in Environmental Health XIV* (Ed. D.D. Hemphill), pp. 129–141; Univ. Missouri, Columbia.

Trudinger, P.A., Swaine, D.J. and Skyring, G.W. (1979). In: *Biogeochemical Cycling of Mineral-forming Elements* (Eds. P.A. Trudinger and D.J. Swaine), pp. 1–27; Elsevier Scientific Pub. Co., New York.

Vernadsky, V.I. (1944). In: *Problems in Geochemistry*, Vol. 11, pp. 453–517; Conn. Acad. Arts and Sci.

Webb, J.S. (1958). In: *International Geol. Cong.*, XX Session, 143–173.

Webb, J.S., Thornton, I., Howarth, R.J., Thompson, M. and Lowenstein, P.L. (1978). *The Wolfson Geochemical Atlas of England and Wales*; Oxford University Press.

West, T. (1981). *Phil. Trans. R. Soc. Lond.*, **B294**, 19–39

World Health Organization (1971). *International Drinking Water Standards*; WHO, Geneva.

Potential and Problems in Using Shellfish as Geochemical Indicators in the Marine Environment

A. DARRACOTT

Introduction

A geochemical indicator is an organism which accumulates trace metals in its soft tissues to levels above those of the environment and which can, in polluted areas, indicate contamination. Pre-eminent among such organisms are shellfish such as oysters and mussels. 'Green-sick' (Cu polluted) oysters from Cornwall have long been known (Lankester, 1886; Boyce and Herdman, 1898). Recent interest has been in the potential use of this ability to monitor metal pollution in the marine environment. The Applied Geochemistry Research Group at Imperial College, London under the former direction of Professor John Webb has played an active role in this. The Group worked initially on a trace metal (Zn) pollution problem in water used by an oyster hatchery (Elderfield *et al.*, 1971), which was shown to produce effects on larvae of the Pacific oyster, *Crassostrea gigas*, (Brereton *et al.*, 1973) and on spat (Boyden *et al.*, 1975). The source of the zinc was from old metalliferous mining activity in the catchment of the River Conwy, leading to an enhancement in zinc in the waters of the estuary used by the hatchery. The use of oysters as monitors arose during an assessment of water and sediment trace metal levels in various estuaries (Thornton *et al.*, 1975). This was further developed by Dr Charles Boyden who made a major contribution to this field with work on the effect of size on trace metal accumulation in shellfish (Boyden, 1974, 1977).

The analysis of 'indicator' or 'bioaccumulator' organisms is now a component of national and international monitoring programmes such as the Global Investigation of Pollution in the Marine Environment (GIPME). According to the GIPME programme (IOC, 1976) the purposes of analysing marine organisms are:

1. Many of the organisms are of immediate importance to man as food and hence the concentrations of pollutants in their bodies may present hazards to human health.
2. Marine organisms concentrate many pollutants and are therefore useful as indicators of the spatial and temporal distribution of pollutants.

3. The biosphere is probably an important reservoir and transport medium for pollutants, and is therefore of great importance in the attempt to determine mass balance of pollutants in the world system.

The term 'mussel watch' has been coined (Goldberg, 1975; Goldberg *et al.*, 1978) to describe such monitoring programmes.

The process of developing guidelines for the use of such organisms was initiated at international level in 1974 and a manual was produced in 1976 (Portmann, 1976). The guidelines suggested have been added to since then, particularly by the work of Phillips (1976, 1977c, 1980).

In this paper it is intended to:

1. Compare the oyster and mussel as monitors of trace metals with a discussion of why certain metals are accumulated in 'unpolluted' molluscs. Field examples from the UK and South Africa will be described.
2. Detail the advantages of using molluscs as indicators of trace metal pollution.
3. Indicate what the problems are and what guidelines have been suggested to help overcome them.
4. Outline further plans for the use of molluscs as monitors of trace metal pollution in the UK and internationally.

The indicating ability of mussels and oysters

Goldberg *et al.* (1978) considered that oysters were more effective accumulators of Ag, Zn, Cu and Ni (to this can be added Cd, see Table 1), while mussels appeared to be better concentrators of Pb. Some typical concentrations of potentially toxic trace metals in shellfish from 'polluted' and 'unpolluted' estuaries are listed in Table 1. This confirms that the mussel is a more effective accumulator of Pb (compare concentrations of mussels and oysters from the Restronguet Creek) and that the oyster *Crassostrea gigas* is better for Cd (compare Poole outer harbour levels in mussels and *C. gigas*).

Certain metals are accumulated in 'unpolluted' molluscs. In oyster soft tissue, Zn, Cu, Fe and Mn are accumulated, and in mussels, Zn and Fe, whilst in the shell, Fe and Mn (with some Zn) are accumulated in oysters, and Fe in mussels (Table 2). It seems likely that Fe in mussel soft tissue and Fe and Mn in oyster soft tissue is accumulated to allow incorporation into the shell, as has been shown for the American oyster *C. virginica* (though in this oyster greater amounts of Mn compared with Fe are incorporated) (Windom and Smith, 1972; Frazier, 1975).

Zn and Cu are accumulated to high concentrations in oysters compared to those recorded in mussels (Tables 1 and 2) and so far there has been little speculation to account for this. One possible explanation arises from the difference in the shell of the two molluscs; the shell in oysters is thicker and heavier than in mussels of comparable size. It may be that as oysters need to

Table 1 **Trace metal concentrations in the Mussel, *Mytilus edulis* and the Pacific Oyster, *Crassostrea gigas* from 'polluted' and 'unpolluted' estuaries**

Trace metal	Concentration*	Location	Reference
Mussels, *Mytilus edulis*		'polluted'	
Hg	0.41†	Firth of Forth (Site A)	Davies and Pirie (1978)
Cd	65.4	Lift Bridge, Holes Bay Poole Harbour	Boyden (1975)
Pb	6.4	Weser estuary (polluted)	Schulz-Baldes (1973)
	14	Restronguet Creek, Cornwall	Boyden (1977)
Zn	265	Restronguet Creek, Cornwall	Boyden (1977)
Cu	45	Restronguet Creek, Cornwall	Boyden (1977)
Mussels *Mytilus edulis*		'unpolluted'	
Hg	0.03†	Dornoch Firth, Scotland	Davies and Pirie (1978)
Cd	1.1–2.4	Helford, Cornwall	Boyden (unpubl.) in Phillips (1980)
	3	Port Erin, Isle of Man	Mullin and Riley (1956)
	4–6	Outer Harbour, Poole[a]	Boyden (1977)
	5	Solent	Segar *et al.* (1971)
Pb	2–2.5	Weser estuary (unpolluted)	Schulz-Baldes (1974)
Zn	97–147	Outer Harbour, Poole	Boyden (1977)
Cu	7	Outer Harbour, Poole	Boyden (1977)
Pacific Oyster *Crassostrea gigas*		'polluted'	
Hg	—	—	—
Cd	136	Holes Bay, Poole	Boyden (1975)
	27	Wareham Channel, Poole	Darracott and Watling (1975)
	26	Outer Harbour, Poole[a]	Boyden (1977)
Pb	9	Restronguet Creek, Cornwall	Boyden (1977)
Zn	7000	Restronguet Creek, Cornwall	Boyden (1977)
Cu	2000	Restronguet Creek, Cornwall	Boyden (1977)
Pacific Oyster *Crassostrea gigas*		'unpolluted'	
Hg	0.1†	West coast, Walney Is. Menai Straits	DOE (1977)
Cd	2	Helford, Cornwall	Darracott and Watling (1975)
Pb	5	Helford, Cornwall	Darracott (unpubl.)
	5	Colne, Essex	Darracott (unpubl.)
Zn	2585	Helford[b]	Thornton *et al.* (1975)
	1350	Colne	Thornton *et al.* (1975)
Cu	207	Colne	Thornton *et al.* (1975)

*$\mu g\ g^{-1}$ dry weight except where indicated; †Hg in $\mu g\ g^{-1}$ wet weight.
Note: Data from Boyden (1977) are concentrations in 1 gm dry weight.
[a]Cadmium contamination of the inner harbour at Poole also reflected in Pacific oysters from the outer harbour; no clear indication of contamination in mussels from the outer harbour.
[b]Helford Pacific oysters reflect naturally high mineralization of catchment.

Table 2 **Comparison of trace metal concentrations in the shell and soft tissue of the mussel,** *Mytilus edulis* **and the Pacific Oyster,** *Crassostrea gigas*

	Cd	Cr	Co	Cu	Fe	Pb	Mn	Ni	Zn	Mean dry weight (g)	Reference
Shell											
M. edulis	0.95	n.d.	0.15	2	290	0.4	3.6	2.1	0.04	—	Segar *et al.* (1971)
C. gigas											
Helford Oct 1971	n.d.	n.d.	n.d.	1.2	170	n.d.	12	n.d.	6.5	18	Darracott (unpubl.)*
Poole Oct 1971	n.d.	n.d.	n.d.	n.d.	178	n.d.	21	n.d.	6	21	Darracott (unpubl.)*
Soft tissue											
M. edulis	5	1.5	1.6	9.6	1700	9	3.5	3.7	91	—	Segar *et al.* (1971)
C. gigas											
Helford Oct 1971	2	1	1	643[a]	581	4	27	2	2585	0.55	Darracott and Watling (1975)
Poole Oct 1971	27[b]	1	1	396	470	4	30	2	2670	0.65	and Darracott (unpubl.)

n.d.—not detectable; concentration in $\mu g g^{-1}$ dry weight.
*Analysis: acid digestion of duplicate 1 g samples of pulverized shell. Solution made with 1 N HCL and metals extracted by sodium diethyldithiocarbamate and chloroform prior to estimation by AAS.
[a]Copper levels reflect naturally high mineralization.
[b]Cadmium contamination.

extract greater amounts of Ca from sea water they also pick up greater amounts of similar ions. Wolfe (1970) suggested that high Zn in oysters may be the result of poor discrimination at its environmental interface and efficient discrimination at the mantle/shell boundary, so that Ca is deposited in shell whilst Zn remains loosely bound to tissue proteins.

Bivalves stressed by salinity fluctuations or high concentrations of metals in their tissues are said to exhibit unusually thin shells (Frazier, 1976; Boyden, pers. comm. in Phillips, 1980). Mussels from the northern Baltic Sea had thinner shells than did those from the area to the west of Sweden; whether this was due to the lower salinity or higher metal content of mussels in the former region was unknown (Phillips, 1980). These observations indicate a close relationship between shell formation and trace metal uptake.

The role of zinc and copper in mussels and oysters

Studies of the biochemical role of zinc in the oysters *Crassostrea virginica* and *Ostrea edulis* have shown that the total amount of zinc found in soft tissue is far in excess of the amount attributed to zinc dependent enzymes (Coombs, 1972, 1974; Wolfe, 1970). Two possible reasons for the Zn accumulation have been suggested:

1. The excess may be the result of poor discrimination by the biological mechanism governing Ca uptake and shell deposition (Wolfe, 1970) (see above).
2. Zinc concentrations need to be high to overcome competition by metals such as Ca for binding sites; the presence of excess Zn would ensure constant saturation for enzyme Zn ligands and hence the maximum activity and efficiency in adverse environments such as, e.g. at shell formation sites in the mantle where Ca as well as carbonic anhydrase and alkaline phosphatase are involved (Coombs, 1974).

I would like to suggest a third:

3. That Zn and possibly Cu are involved in the process of osmoregulation in bivalves.

Most marine bivalves are osmoconformers. However, they are capable of isosmotic intracellular regulation, the mechanism of which involves changes in the concentration of intracellular amino acids and other small organic molecules which act to maintain the volume of cells constant (Bayne, 1975).

The evidence for thinking that Zn and Cu may be associated with such 'mechanisms' is as follows:

1. Forty per cent of Zn and Cu in the oyster *O. edulis* is weakly complexed to the small molecular weight compounds taurine, lysine, ATP and homarine (Coombs, 1974). Amino acids and taurine have long been considered as playing a role in osmoregulation in molluscs (Allen, 1961; Allen and Garrett, 1971; Hoyaux *et al.*, 1976; Bayne, 1975). The rest of the Zn (Cu was not detailed) is bound to tissue cells, some firmly bound, some less firmly bound and exchangeable (Coombs, 1974). A role in osmoregulation would require exchange between metals within the cell and those associated with low molecular weight compounds such as taurine.
2. Zn and Cu in 'green-sick' flat oysters (*O. edulis*) gill tissue are located separately in membrane limited vesicles where Cu is associated with S, and Zn with P (George *et al.*, 1978). It is not clear whether similar associations occur in unpolluted oysters. If there is, the association of Cu with S, a constituent of sulphonic amino acids and taurine may further hint at an involvement with osmoregulation. The association of Zn and P (and Zn and ATP in soluble form, Coombs, 1974) suggests a potential energy source for it.
3. It is well known that gill and mantle tissues contain higher concentrations of Zn and Cu than other tissues. Both of these are external, in contact with the flow of water through the mollusc, and would be the first to be exposed to low salinity and the need to osmoregulate.
4. Mussels (*M. edulis*) exposed to highly stressful salinity changes and Zn treatment showed net uptake of Zn into the soft tissue (Phillips, 1977a)

whilst mussels which had been acclimated to low stable salinities prior to exposure to the Zn treatment did not (Phillips, 1976). Possibly in the latter case Zn had already reached equilibrium with amino acids in the tissue; certainly in the time allowed for acclimation, the process of osmoregulation would have been completed prior to exposure to Zn treatment.

The hypothesis that Zn and Cu are involved in the process of osmoregulation in molluscs is speculative. Much of the evidence cited can also be interpreted in terms of accidental uptake by molluscs in areas of low salinity during the process of shell formation (e.g. mussels from the low salinity Scandinavian waters had high metal concentrations in soft tissue and thin shells (Phillips, 1980)).

Whether uptake is accidental or deliberate it has obvious implications for monitoring trace metals in low salinity areas (see Salinity, below). Hopefully, further work on the association of amino acids and metals, such as Zn, in molluscs subject to conditions of fluctuating salinity, will elucidate matters.

It should be pointed out that Zn and Cu, at least in polluted oysters, are considered to be controlled by a detoxification process (George *et al.*, 1978). A recent review discusses metal detoxification and bioaccumulation in molluscs and proposes a model for the role of ligands in metal transport into cells and their subsequent accumulation there (Simkiss *et al.*, 1982).

What are the advantages of using molluscs?

The analysis of an organism in comparison with water and sediment will provide:

1. A time integrated level of the trace metal representative of environmental conditions over a longer period than would be represented by a single water/sediment survey.
2. Faster and therefore cheaper analysis. No preconcentration is required as in water analysis.
3. It is probable that the organisms will discriminate between biologically available and non-available forms of the contaminant. Since data on contaminant levels in marine organisms are often also of relevance from a dietary intake standpoint, use of appropriately selected marine organisms may enable two purposes to be fulfilled by one sample and analysis (Portmann, 1976).

Monitoring trace metals in the marine environment using the Pacific oyster and mussels

OYSTERS IN THE UNITED KINGDOM

In 1971 a programme of geochemical investigation into five estuaries of

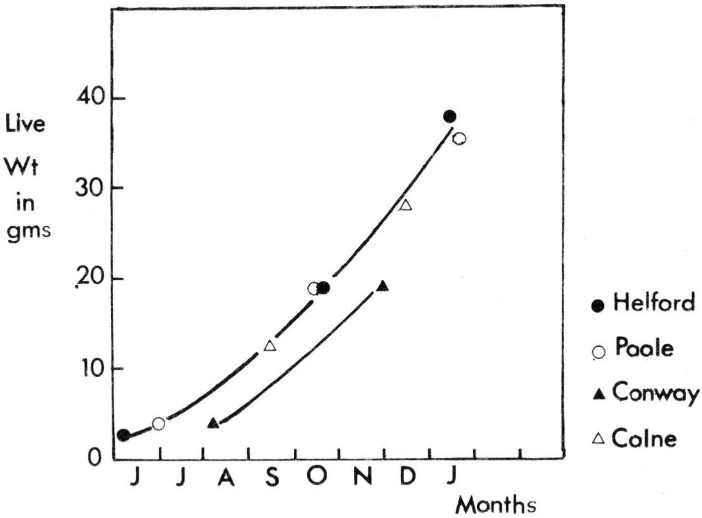

Figure 1 Growth in Live Weight of the Pacific oyster, *Crassostrea gigas*, in estuaries of differing geochemistry (live weight = shell and soft tissue)

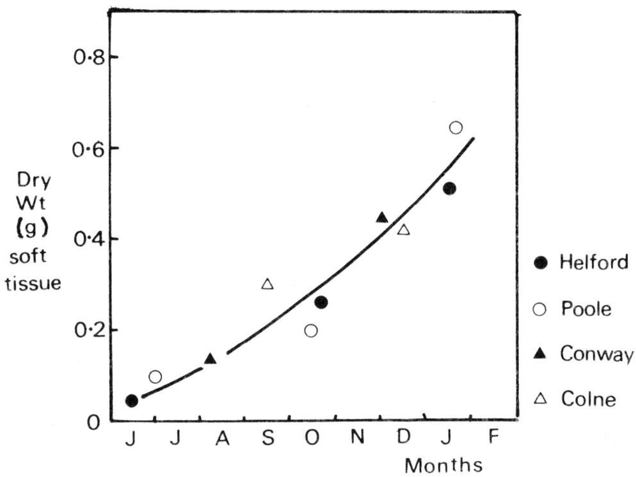

Figure 2 Growth in Dry Weight of the Pacific oyster, *Crassostrea gigas*, in estuaries of differing geochemistry

importance to oyster culture was initiated in the UK. The results of this investigation have been reported (Thornton *et al.*, 1975). In addition to water and sediment sampling, young *Crassostrea gigas*, the Pacific oyster, were placed subtidally in four of the estuaries (Helford, Cornwall; Poole Harbour, Dorset; Colne, Essex; Conwy, Gwynedd). The oysters were obtained from the White Fish Authority in Conwy. Trace metals were estimated in each group of oysters as they were placed in an estuary, and these were then sampled at about four monthly intervals.

Growth in live weight was comparable in all estuaries (Figure 1). Increase in dry weight soft tissue (Figure 2) occurred in all estuaries, though a lower figure was recorded in Poole in October. This sample also was heavily silted which may have had some influence on growth of soft tissue.

Cadmium

Oysters placed in Poole (Wareham Channel) showed a clear pattern of accumulation over eight months compared with those from other estuaries (Figure 3a). This more clearly demonstrated a pollution problem in Poole than did evidence from water and sediment surveys (Darracott and Watling, 1975).

Zinc, Copper, Iron and Manganese

These metals accumulated in oysters from all estuaries compared with initial concentrations recorded in the spat obtained from the White Fish Authority hatchery in Conwy. They are all metals normally present in oysters and their accumulation in actively growing spat might be expected, especially as they were grown subtidally in locations subject to low salinity (low tide salinities 26–28‰ with the exception of Helford which was probably nearer 30–33‰). The role of these metals in connection with shell growth has been previously discussed (see above).

Oysters accumulated Zn to similar levels in the first eight months of the study (Figure 3b). This may be due to fortuitous exposure to similar environmental levels of Zn (estuarine water samples showed similar levels for Poole, Helford, Colne, 26–29 $\mu g\,l^{-1}$ Zn, though Conwy was under 16 $\mu g\,l^{-1}$ though estuarine sediment levels were different (Thornton *et al.*, 1975). However, it is worth noting that the ability to concentrate Zn in *C. gigas* is said to be developed between 50–100 days after settlement when levels rise from *ca.* 1000 $\mu g\,g^{-1}$ to *ca.* 3000 $\mu g\,g^{-1}$ (George and Coombs, 1975), the latter level being regarded as 'normal' for adult oysters. Later oyster samples taken from the Helford and the Colne did show a difference in Zn level (Helford, June, 1972 2153 $\mu g\,g^{-1}$ dry weight, compared to Colne, May, 1972 1350 $\mu g\,g^{-1}$ dry weight).

Copper accumulation was greatest in Helford oysters (Figure 3d) in line with higher environmental levels in the estuary (Thornton *et al.*, 1975).

The peak in Fe recorded for Poole oysters in October (Figure 3c) was probably due to these oysters being contaminated with silt. Iron levels in sediment are generally much higher than those in oyster tissue so contamination by sediment will increase apparent tissue levels. Iron and Mn are accumulated in the oyster shell (see Table 2).

The Cd contamination problem discovered in oysters from the Wareham Channel in Poole Harbour was followed up to discover the source of the pollution. This was found to be in Holes Bay and two species of oyster *O. edulis* and *C. gigas* placed near to the entrance of the bay both accumulated Cd (Boyden, 1975).

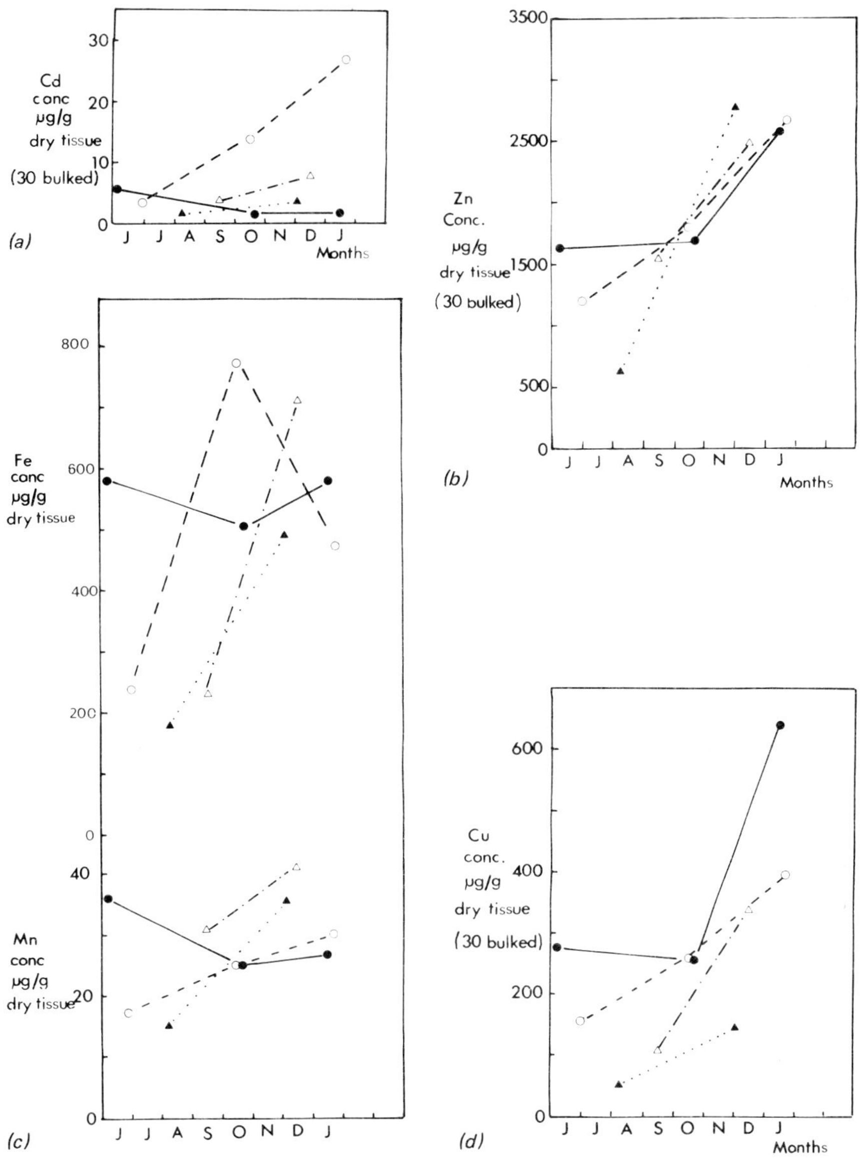

Figure 3 Trace metal concentrations in the Pacific oyster, *Crassostrea gigas* from estuaries of differing geochemistry: ● Helford; ○ Poole; ▲ Conway; △ Colne. (a) cadmium; (b) zinc; (c) iron and manganese; (d) copper

OYSTERS IN SOUTH AFRICA

Trace metal concentrations in the Pacific oyster grown in two UK estuaries (the Helford, Cornwall and Wareham Channel, Poole Harbour) are

Table 3 **Trace metal concentrations in the Pacific oyster, *Crassostrea gigas* grown in two estuaries in the UK and a lagoon in South Africa**

Location and sampling date	Mean wet weight (g)	Mean concentration*									Reference
		Zn	Cu	Fe	Mn	Cd	Ni	Pb	Co	Cr	
Helford (H) Oct 1971	5.4	247	63	55	2.6	0.2	0.3	0.5	<0.1	0.1	Darracott and Watling (1975)
Poole (P) Oct 1971	4.6	380	56	67	4.2	3.8†	0.3	0.6	0.1	0.2	and Darracott (unpubl.)
Langebaan Lagoon (L) Sept 1974	6.4	81	8	42	2.4	2.2	0.3	0.6	0.25	0.1	Watling and Watling (1974)
Ratio H(P): L		3(5)	8(7)	1(2)	1(2)	<1 (2*)	1	1	1	1	

* $\mu g\,g^{-1}$ wet weight.
†Poole oysters contaminated.
Note: All samples were subtidal. The UK oysters grown in Jonex bags raised slightly off the bottom are compared with oysters grown on the plastic tray nearest to the bottom (G5) in South Africa.
Sample size: UK bulk analysis of 30 oysters; South Africa individual analysis of 9 oysters.

compared with concentrations in the same oyster grown in Langebaan, South Africa. Langebaan Lagoon connects to the sea via Saldanha Bay in the West Cape. An ore terminal has been constructed in the bay and was to be used for the export of iron ore from Sishen and copper/lead/zinc ore from Aggeneis, Namaqualand. In this section trace metals will be expressed in $\mu g\,g^{-1}$ wet weight to facilitate comparison with data on 'unpolluted' molluscs from the UK (DOE, 1977).

Concentrations of Zn and Cu are greatly different, the UK oysters containing 3–5 fold more Zn than the South African oysters and Cu 7–8 fold (Table 3). Cadmium concentrations were, however, slightly higher in the South African oysters in comparison with those from the Helford, though not as high as the Cd polluted Poole oysters. Iron and Mn were comparable or slightly elevated (1–2 fold) in the UK oysters as might be expected from the fact that growth in live weight (the major constituent of which is shell growth) is comparable (Figure 4). (South African data from A. du Plessis, Fisheries Development Corporation, pers. comm.) other metals (Ni, Pb, Co, Cr) were comparable.

As the Pacific oysters used in the UK and South Africa were obtained from the same UK hatchery (White Fish Authority Hatchery in Conwy) genetic differences would not apply. It was considered that the South African lagoon was unpolluted with respect to heavy metals (Watling and Watling, 1974). Low metal concentrations were recorded for the sediments in comparison with those from the UK estuaries (Table 4) as might be expected for a lagoon environment with no terrigenous metal input in river water. Low metal

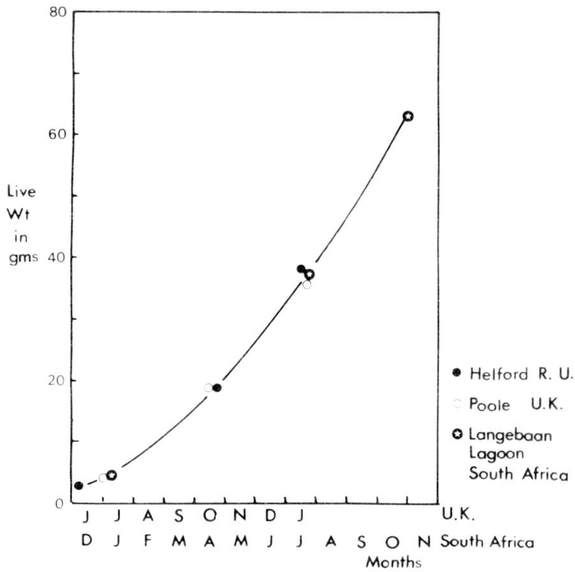

Figure 4 Growth in Live Weight (shell and soft tissue) of the Pacific oyster *Crassostrea gigas* from the UK and South Africa

concentrations in oysters and sediment levels were also recorded for an estuary in South Africa (Watling and Watling, 1976b) where it was noted that there was an absence of silt in the river water (Day, 1967) which would imply little terrigenous input.

The slight elevation of Cd was not commented on in Watling and Watling (1974). It may be that environmental conditions in the lagoon (summer water temperatures of >24°C and salinities of >37‰ have been recorded (Shannon and Stander, 1977)) could influence uptake. Experiments with the American oyster *C. virginica* found that a combination of high salinity (35‰) and high temperature (35°C) severely stressed the mollusc and resulted in increased Cd concentrations whilst other metals such as Zn and Cu decreased (Breedeveld, 1971). Although this duplicates to some extent the situation found in the lagoon, it must be said that there is no evidence that these oysters were under stress. The oysters grew well in the lagoon (A. du Plessis, pers. comm.) as might be expected for a subtropical mollusc.

The Pacific oysters from the UK estuaries might be expected to show higher trace metal levels as they had been deliberately placed in areas subject to low salinities where there would be exposure to terrigenous sources of metals. However, a further comparison of the Pacific oysters from the UK, considered to be unpolluted (DOE, 1977), with the same species grown subtidally and intertidally in South Africa, shows the same pattern of lower levels of Zn, Cu (and Hg) in the latter environment (Table 5). The UK Marine Pollution Monitoring Management Group considers that Zn in oysters occurs in naturally high concentrations and is not a reflection of pollution of the marine environment (DOE, 1982).

Table 4 **Trace metal concentrations in sediment (-200 μm fraction) from estuaries in the UK and a lagoon in South Africa**

Location		Range and mean ($\bar{x}$) in $\mu g\ g^{-1}$ (Fe in %)								Ref.
		Zn	Cd	Cu	Pb	Fe %	Mn	Ni	Co	Cr	
Helford River, UK		180–620	<1–3	37–278	50–500	2.3–3.7	266–1250	50–106	20–40	50–200	1
	$\bar{x}$	353	2	179	177	3	599	81	32	147	
Poole Harbour, UK		41–85	<1–2	6–20	33–200	0.8–1.8	71–850	40–63	8–20	25–130	1
	$\bar{x}$	60	1	12	104	1.1	340	47	16	65	
Langebaan Lagoon, South Africa		3.5–14	0.4–1.4	2–7	9–32	—	4–31	4–10	1–7	—	2
	$\bar{x}$	9	0.7	3	17	—	12	6	3	—	

Refs: 1—Thornton *et al.* (1975); 2—Watling and Watling (1976a)

Table 5 **Trace metal concentrations in unpolluted Pacific oysters,** ***Crassostrea gigas*** **and Mussels,** ***Mytilus edulis*** **and** ***Choromytilus meridionalis*** **from the UK and South Africa**

Location	Mean concentration*					No. analyses	Ref.
	Hg	Cd	Pb	Zn	Cu		
Pacific oysters *Crassostrea gigas*							
UK							
East Coast	0.09	0.6	0.2	160	37	10 (500 bulked)	DOE (1977)
West Coast	0.1	0.6	0.4	250	30	4 (225 bulked)	DOE (1977)
South Africa Langebaan Lagoon							
Intertidal (G1–2)	0.02	1.5	0.5	77	6	18	Watling and Watling (1974)
Subtidal (G3–5)	0.02	1.7	0.6	80	6	27	
Mussels							
UK *Mytilus edulis*							
East Coast	0.04	0.2	0.7	13	1.2	4 (200 bulked)	DOE (1977)
West Coast	0.02	0.7	0.4	14	1.2	2 (45 bulked)	DOE (1977)
South Africa *Choromytilus meridionalis*							
Saldanha Bay (near Langebaan Lagoon entrance)	0.01	0.2	0.3	14	1.4	24	Watling and Watling (1974)

*$\mu g\ g^{-1}$ wet weight.

It would seem that the oyster is a very much more sensitive indicator than hitherto considered, but its use on a global scale may be complicated by this as it could reflect 'natural' variations in geochemical run-off into estuaries and the distinguishing of 'natural' from actual contamination could be difficult. This seems more likely than assuming that oysters from many coasts exhibiting higher metal levels than in the South African oysters are 'polluted' as suggested by Watling and Watling (1976c). It is also possible that different environmental conditions of temperature and salinity could influence uptake of metals by, for example, effects on metal-metal interaction.

MUSSELS

Unfortunately, as far as I am aware there are no published data on trace metal concentrations in mussels (*Choromytilus meridionalis*) from the Langebaan Lagoon to compare with UK mussels (*Mytilus edulis*). Mussel samples from Saldanha Bay near to the entrance to Langebaan Lagoon, although of a different species, showed similar Zn, Cu and Hg concentrations to 'unpolluted' mussels from the UK (Table 5), confirming the conclusion of Watling and Watling (1974) that metal concentrations in other mussels are comparable to *C. meridionalis*. They did not shown the slightly elevated Cd concentrations as might be expected, as environmental conditions in the bay would not be directly comparable to those in the lagoon. Oysters are anyway regarded to be better accumulators of Cd compared to mussels (see above).

Some problems in using shellfish as geochemical indicators, and possible solutions

There are certain difficulties in using biological accumulators in monitoring programmes such as variations due to size, age, sex etc. of the organisms used. The condition of the animal may also be affected by the general quality of its environment and this in turn might affect either its uptake rate or saturation capacity (Portmann, 1976). The problems involved in using organisms for monitoring were investigated by Phillips (1976, 1977c, 1980) who concluded that trace metal levels in an organism could be affected by size, season, sampling position, environmental variables such as salinity and temperature and metal-metal interaction. That some of these variables are interlinked should become obvious from the following.

SIZE

Boyden (1974, 1977) investigated the influence of body size upon whole tissue metal content in a variety of marine and estuarine molluscs. He found that in some cases metal concentrations were independent of size whilst in others concentrations were greatest in smaller individuals (Table 6, Figure 5). The same results have not invariably been found by others. Boyden (1977) attributed some of the different results to the smaller size range covered by the other workers, e.g. in the case of *C. commercialis* and *Saccostrea cucullata* where Zn as well as Cu and Cd was found to decrease with increase in size (Mackay *et al.*, 1975). The same reason would account for the different results of Cossa *et al.* (1979) concerning Cd in the mussel, *Mytilus edulis*. Contrary to the results of Boyden (1977) they found that Cd decreased with increase in

Table 6 **Effect of size on the trace metals, Zn, Cu and Cd, in three species (from Boyden, 1977)**

Species	Zn	Cu	Cd
Pacific Oyster, *Crassostrea gigas*	I	D	D
Flat Oyster, *Ostrea edulis*	I	I	I
Mussel, *Mytilus edulis*	D	D	I

I—independent of size; D—decreases with increase in size.

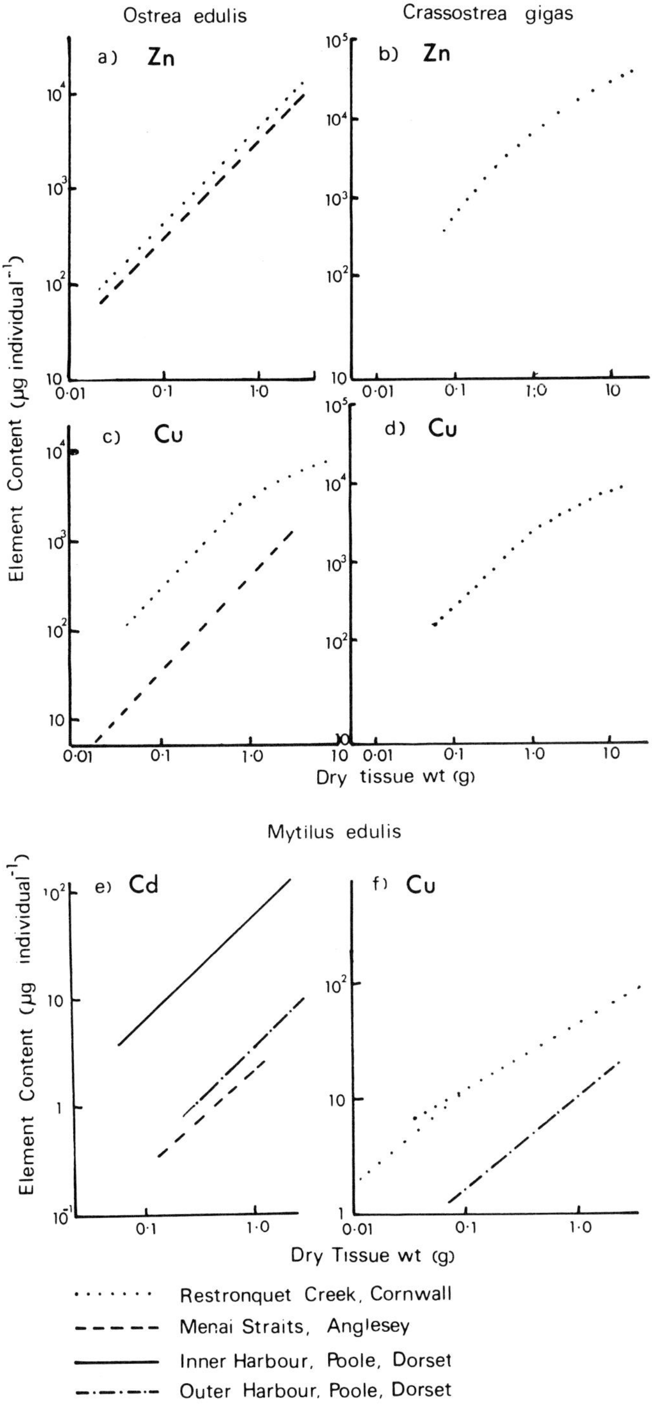

Figure 5 Relationships between element content and body weight in the flat oyster, *Ostrea edulis*, the Pacific oyster, *Crassostrea gigas* and the Mussel, *Mytilus edulis*. *Note*: in 5e and 5f the regression line for Menai Straits data has been omitted as it does not agree with tabulated data in Boyden (1977)

size. However, in two sites Cd concentrations remained steady after a certain size (i.e. 0.07 g dry weight), a size associated with the onset of maturity. Such 'steady' levels are comparable to Boyden (1977) finding that Cd in mussels is independent of size. Boyden (1977) plotted only limited data on Cd levels in mussels of 0.07 g dry weight or less (Figure 5e) and may not have dealt with Cd in immature mussels. Copper levels in mussels were plotted (Figure 5f) and show an interesting similarity with the pattern of metal concentration changing at around 0.08 g dry weight, the difference being that below that weight, mussels showed an increased concentration of Cu. It may be that at this size the mussel switches its energies from extracting Ca from the water for shell building to accumulation of food reserves and formations of gonads with consequent effects on trace metal accumulation. Cadmium was also found to decrease with increase in size by Boalch *et al.*, (1981). No information on maturity was given but some were probably immature (wet weight 0.5–5 g).

Curved graphs were obtained relating metal content to body size in the oysters *C. gigas* and *O. edulis* which had been introduced into the contaminated environment of the Restronguet Creek in Cornwall (Figure 5). It was considered that the curves were due to large individuals requiring longer than 5 months to equilibrate to the exceptionally high environmental concentrations (Boyden, 1977). It is thought that larger bivalves may have a decelerated mean rate of transport over the gill compared with smaller ones (Jorgensen, 1976) and such a phenomenon could account for the lack of equilibration in larger molluscs in comparison with smaller ones. The relatively modest elevation in Zn from polluted (Restronguet) compared to unpolluted (Menai Straits) in the oysters *O. edulis* (Figure 5a) will be discussed further (see Minor–minor, below).

Boyden (1977) considered that by acknowledging the influence of size, examination of a standard sized individual (e.g. 1 g dry weight) allows comparison between samples and species. He thought that seasonal variations are likely to influence slopes relating trace element content to body weight in view of the differences over a season in contribution of various tissues to a changing total weight. However, working on seasonal samples of the limpet *Patella* sp and mussels *Mytilus edulis*, he found that the slopes remained fairly constant, though the intercepts (represented by 1 g dry weight) varied with season. He concluded that seasonal variations may not always influence slopes relating element content to body weight and that further investigation is clearly required as intuitively one would suggest that the slopes should change throughout the year.

Three alternative ways have been suggested by Phillips (1980) to cope with the problem of size in monitoring schemes:

1. Animals of the same size, or similar to be taken from each location studied (also recommended in IOC, 1976).
2. Animals of a large range of sizes are selected and analysed in bulk.
3. Animals of a large size range are selected and analysed individually to generate a size : concentration regression for each metal.

Phillips (1980) considered the best alternative to be the selection of a restricted size range of individuals for study while pointing out that a complete understanding of the overall picture of metal contamination cannot be attained using this method. This alternative has been selected for use in the UK monitoring programme (Topping, 1983) and in the international marine monitoring programme (GIPME) (IOC, 1976). The UK guidelines also note that data will be normalized using the relationships developed by Boyden (1977).

SEASON

As Phillips (1980) has pointed out, seasonal fluctuations in trace metal levels in aquatic biota can arise from weight changes in the organisms and direct effects of salinity, temperature and other water quality parameters which alter seasonally, as may pollutant delivery to the environment.

Seasonal weight changes in shellfish can be due to seasonal variations in stored food reserves (glycogen) which 'dilute' trace metal concentrations, and to weight loss after the release of gametes. With regard to the latter, it has been suggested that metal levels in whole soft parts of bivalves increase when tissue weight drops because of loss of gametes which have relatively low metal contents (Phillips, 1980).

However well defined, seasonal variation in trace metal content is not always seen (Goldberg *et al.*, 1978; Boalch *et al.*, 1981). Fowler and Oregioni (1976) could find no evidence of the involvement of the reproductive cycle in seasonal variation of metals in the Mediterranean mussel *M. galloprovincialis*, and suggested that peak metal concentrations were the result of high precipitation and increased run-off bringing particulate metals to the sea. Recently, seasonal measurements of Zn in *M. edulis* from the Tyne estuary demonstrated a small Spring peak and an annual low in September (Lobel and Wright, 1982). Zinc concentrations were found to be significantly lower in gonadal tissue than in non-gonadal tissue. Although gonadal dry weight was strongly negatively correlated with gonadal Zn concentration there was no significant relationship between gonadal dry weight and whole body Zn levels, indicating that gonadal state had little effect on Zn variability in these animals. Construction of a gonadal index showed that the seasonal peak Zn concentration in May occurred before the major spawning had taken place.

It may be that mussels take up metals more actively during reproduction, especially a metal such as Zn which is normally present to quite high levels. This would account for the seasonal peak Zn occurring before major spawning had taken place. Mussel gill appears to show increased Cu uptake during reproduction (Delhaye and Cornet, 1975). It would be interesting to know whether filtration rates increase during reproduction, perhaps in order to disseminate gametes; such an increase could conceivably have an effect on metal uptake.

Assuming the problem of when and how seasonal variation of trace metals in molluscs can be resolved for the temperate regions where most of the work so far mentioned was done, there remains the problem of what happens in warmer regions. The American oyster, *C. virginica* from Florida (Lat. 30°N) is known to feed and grow for longer during the year than the same oyster from Long Island Sound (Lat. 41–42°N) and has a longer reproductive period. Copper concentrations in this oyster from the Gulf and South Atlantic States were lower than from the North Atlantic States (Galtsoff, 1964). Are these two facts connected or is it due to the greater industrialization of the north (as suggested by Galtsoff). Such problems need to be resolved to ensure comparability in international monitoring programmes.

Phillips (1980) has suggested that different populations of the same species can exhibit different degrees of metal fluctuation due to differences in seasonal fluctuation of ambient trace metals and differences in weight lost at spawning.

Sex has also been found to influence trace-metal concentrations. Female mussels *Choromytilus meridionalis* have been found to have higher trace metal levels than males (Watling and Watling, 1976a; Orren *et al.*, 1980), a difference particularly marked in June (southern winter) (Orren *et al.*, 1980). In November, levels of trace metals had increased in both sexes and the animals were considered to be in the process of spawning (Orren *et al.*, 1980).

Suggested ways of resolving problems caused by seasonal variation in trace metal levels in molluscs are:

1. To enable inter-site comparison, obtain molluscs (*sic* mussels) in the same condition (Phillips, 1980). Sampling in January to March (in UK) is suggested when all would be in a pre-spawning state. It is also pointed out that this is the period of greatest tissue weight and therefore lowest seasonal concentration of metals.
2. Use of immature molluscs which could be transplanted into the area of study (Darracott and Watling, 1975; Cossa *et al.*, 1979; Orren *et al.*, 1980).
3. Use of single tissue analysis (suggested by Simkiss *et al.*, 1982). Gill and mantle tissues are, in bivalve molluscs, in direct contact with sea water and are well known to accumulate metals to higher levels than other tissues in the mollusc body (e.g. Ikuta, 1968; Romeril, 1971; George *et al.*, 1978).

   A comparison of gill Cd levels with those in whole soft tissue in *C.gigas* from Poole (Figure 6) demonstrates a pattern of elevated levels in the gill compared with the whole tissue. It is felt that if parallel observations of gill and whole soft tissue are made the former would provide information on trace metal levels independent of any weight changes due to spawning or accumulation of food reserves.

   Gills are to be preferred to mantle tissue as the latter are known to function as a glycogen store (Galtsoff, 1964) and in mussels are also the site of gonad development (as noted in Cossa *et al.*, 1979). However, even with the use of single tissue problems may still occur. Information in Delhaye and Cornet (1975) indicates that increased Cu uptake occurs in mussel gill during reproduction. It is also very likely that gill tissue will pick up more metals, e.g. Zn from the environment when exposed to low salinities (see above).

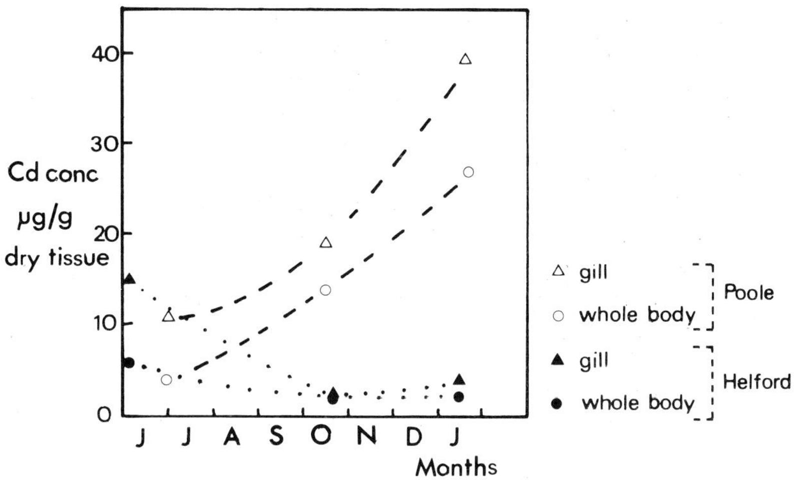

Figure 6 Comparison of cadmium concentrations in gill and whole soft tissue in the Pacific oyster, *Crassostrea gigas* from the Wareham Channel, Poole Harbour and the Helford River, Cornwall (from Darracott and Watling, 1975)

Guidelines on the use of organisms for marine monitoring in the UK (Topping, 1983) propose the use of alternative (1) above, sampling once a year in late Winter. They do also mention the possibility of using transplanted animals, though maturity is not specified.

SAMPLING POSITION

Phillips (1980) considers that the results of monitoring surveys may be affected by the vertical position of sampling of organisms, either intertidally or subtidally at the shore line, or in the water column away from the shore. He points out that most examples of variation in vertical sampling known to date concern water bodies with known vertical stratification; the differences may be explained on the basis of such stratification (i.e. depths affected most by low salinity are more likely to contain molluscs with higher metal levels).

Evidence from Pacific oysters *C. gigas* grown on racks in a fully saline lagoon in South Africa supports the above. There was no distinct relationship between metal concentration and depth of tray (Table 5) though oysters on the lowest subtidal level (G5) showed the greater metal concentration (Table 3) and the lowest weight gain. This was considered to be due to increased sediment load and solution of metals from the bottom sediment (Watling and Watling, 1974).

It is known that the growth efficiency of *M. edulis* varies with the tidal level they occupy. The longer they are exposed during low tides the slower they grow and the smaller size they maximally attain (Seed (1968) in Jorgensen, 1976). Growth approaches zero when mussels are exposed about half the time (Baird (1966) in Jorgensen, 1976). However, slower growing mussels have a longer life expectancy (Jorgensen, 1976).

It has been suggested that to avoid differences in pollutant concentration due to the vertical position of sampling on the shore line, all samples should be taken from the same vertical position (relative to tide levels), so preventing interference from this effect with conclusions concerning inter-site differences in pollutant availability (Phillips, 1980).

Guidelines on the use of organisms for marine monitoring in the UK (Topping, 1983) have recommended sampling from shore positions where animals are exposed for 6 hours each tidal cycle. This means they will be exposed for half the time (average tidal cycle is 12 hours) and will thus, at least in mussels, show zero growth rate (see Baird, 1966) and will have a long life expectancy (Jorgensen, 1976).

ENVIRONMENTAL EFFECTS

Salinity

Phillips (1980) considers that the effect of salinity on the seasonality of trace metals in aquatic biota may be divided into two components:

1. The influence of run-off on delivery of trace metals to the marine environment, there being an inverse relationship between metal levels in the water and salinity.
2. Direct effects on the physiology of the aquatic organisms, not dependent on the presence of higher ambient metal levels occurring concomitantly with lowered salinities; effects which could include, for example, changes in filtration rate.

Fieldwork in Scandinavia on Zn (and Cd) concentrations in *M. edulis* resulted in the observation that mussel tissues in areas of low stable salinity contained higher metal concentrations than those from areas of high stable salinity, whilst mussels from an area of fluctuating salinity separating these two had intermediate concentrations (Phillips, 1977b). Similarly, a study of the areal distribution of metals (Cu, Cd, Zn) in various river systems in Virginia showed that as fresh water was approached, the concentration of these metals in soft tissue of *C. virginica* increased (Huggett *et al.*, 1973). Reasons to account for this are discussed elsewhere. Salinity is likely to have an influence on trace metal levels in shellfish not only in the individual estuary but on a larger scale. In the warmer parts of the world well-defined wet and dry seasons exist and biota will be exposed to seasonal variation in mean salinity leading to marked changes in metal abundance in biota (Phillips, 1980). As he points out, delivery rates of metal to estuaries during high run-off would compound this effect (in the tropics run-off during the rainy season is associated with a massive loss of terrigenous material into estuaries).

Two suggested ways of resolving the problem of salinity effects are:

1. Sampling molluscs from areas of similar salinity or taking both

subhalocline samples as well as those in areas of freshwater inflow–to find out whether results of surface samples are due to spurious temperature/salinity effects or to actual contamination of the freshwater layer (Phillips, 1976)—though it is not clear how the latter would be achieved.

2. The use of concentration ratios of two metals as a means of indicating whether metal concentrations in a mollusc are from natural or man-made sources (Huggett *et al.*, 1973; Simkiss and Taylor, 1981).

No special precautions are mentioned in Topping (1983).

Temperature

In general, increases in temperature cause an increased net uptake of trace metals (Phillips, 1980). In laboratory experiments with mussels exposed to two salinities (15‰ and 35‰) and two temperatures (10°C and 18°C) and a four-metal solution (Zn, Cu, Cd, Pb), it was found that the temperature effect was insignificant for Zn and Pb, but Cd uptake at lower salinities was increased by higher temperatures; the behaviour of Cu was erratic (Phillips, 1976). Increase in temperature (10°C and 20°C) also resulted in increased Cd uptake from solution in the bivalves *Mya arenaria* and *Mulinia lateralis* though little effect was noted for *M. edulis* (Jackim *et al.*, 1977). However, in another experiment combining different salinities and temperatures it was found that Cd was picked up more in the lower salinity (20‰) and that Cd levels in mussels *M. edulis* subject to the higher temperatures were greater in both salinities (Jackim *et al.*, 1977). A possible field example of a temperature effect on Cd uptake in oysters is given under Oysters in South Africa, above.

There is no obvious means of resolving temperature effects on trace metal uptake. It seems likely that Cd uptake is influenced by temperature and this could have an influence in lagoon situations. Whether localized heat, e.g. from a power station outflow, could have an effect is not known. It may be that in such situations, any clear effect due to temperature can only be estimated by measuring trace metal levels in molluscs inside and outside the affected area.

METAL-METAL INTERACTION

Major-minor

There is ample evidence of interaction between the major ion Ca and minor ions, e.g. Zn, in bivalve molluscs. Uptake of Cd by the mussel *M. edulis* exposed to experimental low salinities was considered due to changes in the ratio of major ions to added trace metal and that the uptake of Cd might be responsive to ambient Ca availability (Phillips, 1976, 1980). This is similar to the results of Duke *et al.* (1967) where uptake of ^{65}Zn by the American oyster *C. virginica* (oysters acclimated for 24 hours) showed an inverse relationship between salinity and uptake. It was suggested that at the higher salinity (35‰) the water contained many more ions to compete with ^{65}Zn ions for sorption sites, thus suppressing concentration of ^{65}Zn.

It may be significant that Ca, Zn, Cd and Hg are all Group 2 elements in the periodic table and all can be accumulated by molluscs (see Table 1). Simkiss (1981) suggested that similarities in ionic radii and electronegativity between Ca and other metals such as Zn could account for the latter's accumulation. Phillips (1980) considers that further research on the relationship between major and minor ions in aquatic biota would be worthwhile.

Minor-minor

Competition between minor ions also occurs. In experiments with *M. edulis* and *Mulinia lateralis* it was found that uptake of ^{109}Cd was inhibited by increase in ambient Zn especially in the former mollusc (Jackim *et al.*, 1977). Experiments exposing *M. edulis* to a mixture of 4 metals (400 $\mu g\,l^{-1}$ Zn; 20 $\mu g\,l^{-1}$ Cu; 40 $\mu g\,l^{-1}$ Cd; 20 $\mu g\,l^{-1}$ Pb) over a period of 35 days indicated that uptake of Cd, Pb and Zn was unaffected by the presence of other metals but that Cu uptake was inhibited by the high concentrations of Zn and Cd (Phillips, 1976). However, results from the Restronguet Creek, Cornwall where Zn concentrations in the water at low tide were comparable to those used by Phillips (though Cu was higher), i.e. 570 $\mu g\,l^{-1}$ Zn; 65 $\mu g\,l^{-1}$ Cu; 13 $\mu g\,l^{-1}$ Cd (Thornton *et al.*, 1975), did not show any inhibition of Cu uptake by molluscs, in fact rather the reverse. When Zn and Cu concentrations in unpolluted molluscs (*C. gigas, O. edulis* and *M. edulis*) are compared with those recorded from the Restronguet Creek and a ratio obtained it can be seen that in each case the ratio for Cu is higher than for Zn (Table 7). This phenomenon was pointed out by George *et al.* (1978) for Zn and Cu in *O. edulis*.

It may be that Zn is regulated more than Cu as, compared to the latter, Zn is accumulated less in polluted environments and in metal impoverished environments, it does not decrease as much as Cu (see Table 3). Zinc regulation in oysters has been suggested by Chipman *et al.* (1958) and Preston (1966).

Experimental results that high ambient Zn suppress Cd uptake by molluscs (Jackim *et al.*, 1977) finds support in the same field example. Cadmium uptake by *M. edulis* and possibly *O. edulis* from the Restronguet Creek (Table 7) in the presence of high ambient Zn concentration did not reflect the fact that the water also contained high Cd concentrations.

There is no easy means of resolving the effect of interactions on trace metal uptake in relation to monitoring. Metal-metal interaction is regarded as a major problem associated with the use of bivalve molluscs as indicators of trace metals (Phillips, 1980). The only way to be aware of a possible effect would be to have available trace metal data for the water in which the molluscs were living, and in the case of the major-minor effect, data on salinity.

Wet weight or dry weight?

Both ways of reporting trace metals concentrations have been used in this paper. The basic problem is that for public health purposes it is necessary to

Table 7 **Comparison of the trace metals Zn, Cu and Cd in shellfish from the metal polluted Restronguet Creek, Cornwall with those from 'clean' estuaries and a derivation of ratios**

Trace metal	Restronguet Creek (R)	Outer Harbour Poole (P)	Helford (H)	*Ratio R:P(H)
Pacific Oyster, *Crassostrea gigas*				
Zn	7000	2570	2585[a]	3(3)
Cu	2000	261	—	8
Cd	12	26	2[a]	<1(6)
Flat Oyster, *Ostrea edulis*				
Zn	11900[b]	1966[c]		5–6 using
	10621[d]			maximum
	4358			values
	Range 1816–11200			
Cu	3800[b]	86[c]		19–44
	1629[c]			
	3000			
Cd	8[b]	6[c]		1
	6			
Mussel, *Mytilus edulis*				
Zn	265	97–147		2–3
Cu	45	7		6
Cd	1	4–6		<1

*$\mu g\ g^{-1}$ dry weight.
Data from Boyden (1977) (concentration in 1 g dry wt tissue) except for:
[a]Thornton *et al.* (1975) included to provide additional cadmium data as oysters from the Outer Harbour, Poole show evidence of cadmium contamination.
[b]Darracott (unpubl.) (mean of 8 oysters taken from the bed of the creek) included to provide data on indigenous oysters to complement data in Boyden (1977) on shellfish transplanted for 5 months into the Restronguet Creek.
[c]Boyden (1975).
[d]George *et al.* (1978) maximum values.

quote data in terms of wet weight, but most of the scientific work has been reported in terms of dry weight. The reason seems to be to reduce variability of data. Boyden (1977) reported that almost twice as much variability was recorded with wet weight data compared to dry weight. However, other workers (Mackay *et al.*, 1975) obtained a high correlation between wet and dry weight and saw no advantage in using the latter. The reason for the difference may lie in the way the molluscs were handled. Boyden (1977) analysed individual molluscs and these were frozen until required for analysis, then subsequently thawed. The soft parts were then dissected from the shell, washed and surface dried with paper tissue and then were placed in a flask for drying and analysis. Mackay *et al.* (1975) shucked twenty live oysters into a container including all liquids from inside the shell. This bulk sample was then homogenized and prepared for analysis. Two factors would account for the good correlation between wet and dry weight observed in the latter case: bulk sampling (which would damp down the effects of individual variation) and the use of live oysters. It is known that oysters lose water very rapidly in the first half hour after removal from the shell (Galtsoff, 1964). It seems likely that the variation in water loss would be greater the more the soft tissue was handled, depending on the degree to which an individual mollusc was punctured. It could also be that when molluscs are frozen prior to analysis metals could be lost into the liquids which are then discarded in sample preparation. The use of live oysters removes any effect due to preservation prior to analysis.

The UK Marine Pollution Monitoring Management Group reports data on trace metals in terms of wet weight (DOE, 1977; Topping, 1983). The international monitoring programme GIPME recommended reporting results in dry weight but including the wet/dry ratio (IOC, 1976). This allows both figures to be calculated (Portmann, 1976). Whatever reporting method is used it would seem that to obtain representative results minimum handling of mollusc soft tissue is necessary.

Future plans involving the use of shellfish as monitoring organisms

UK MONITORING PROGRAMME

Mytilus edulis, the mussel, is included as one of the recommended organisms for assessing contamination (Topping, 1983). The mussel is more freely available around UK shores than the oyster; this facilitates sampling. However, the possibility of using transplated molluscs in the UK programme has been mentioned (Topping, 1983) a procedure which has previously yielded good results (Darracott and Watling, 1975), and which would allow the use of oysters as well as mussels.

Comprehensive fish and shellfish monitoring programmes have been undertaken in the past (1975 and 1976). Current practice is for a simpler surveillance of one or two commercial species plus a more detailed investigation of 'hot spot' areas. Trace metals monitored are Hg, Pb, Cd, Zn, Cu and Ni, Cr and As on particular occasions. The information obtained is of national use and enables the UK to fulfill international obligations to, for example, the International Council for the Exploration of the Sea (ICES), and the Joint Monitoring Group of the Convention on the Prevention of Marine Pollution by Dumping. It will also be required by a number of European Community Directives (DOE, 1982). This report notes that a baseline study of contaminant levels in fish and shellfish from the North Sea and North Atlantic has been proposed for 1985 (under the auspices of ICES) in which the UK will be participating.

INTERNATIONAL MONITORING PROGRAMMES

The aims of international monitoring programmes have been discussed in the Introduction above. Arising from the Global Investigation of Pollution in the Marine Environment (GIPME) is the development of a Marine Pollution Monitoring System (MARPOLMON). A component of MARPOLMON is monitoring contaminants (trace metals, organochlorines, hydrocarbons) in organisms for public health and water quality purposes and the involvement of developing countries is being sought. At a meeting of the GIPME Group of Experts on Methods, Standards and Intercalibration (GEMSI) in 1982 the following recommendations were made (IOC, 1982):

1. Because of the low level of analytical capability available in developing countries at present for the measurement of contaminants in biological tissue this type of monitoring should not be immediately implemented in MARPOLMON.
2. IOC should set up both an intercomparison programme and a training exercise for labs in developing countries interested in participating in this aspect of MARPOLMON.
3. Suitable manuals should be prepared for use in sampling and analysis (in relation to training exercise) bearing in mind those produced by FAO (Portmann, 1976).

These manuals should help ensure comparability of data.

Conclusion

The use of shellfish, particularly bivalve filter feeding molluscs, such as oysters and mussels has become an integral part of monitoring programmes for trace metals. A great deal of effort has been devoted to developing guidelines for their use (IOC, 1976; Portmann, 1976; Phillips, 1976, 1977c, 1980; Topping, 1983). In this paper an attempt has been made to outline the potential and the problems in using shellfish as monitors of trace metal pollution with an indication of the guidelines suggested for solving the latter. The interest in shellfish as monitors of pollution is still strong and work by the scientific community in testing the findings on which guidelines are based should enable them to be updated.

References

Allen, R.S. (1961). Amino acids in the Molluscs, *Am. Zoologist*, **1**, 253–261.
Allen, J.A. and Garrett, M.R. (1971). Taurine in marine invertebrates, *Adv. mar Biol.*, **9**, 205–253.
Baird, L.H. (1966). Factors affecting the growth and reproduction of mussels (*Mytilus edulis* L.), *Fish. Invest. Ser. II*. **25**, 1–33.
Bayne, B.L. (1975). Aspects of physiological condition in *Mytilus edulis* L. with special reference

to the effects of oxygen tension and salinity, *Proc. 9th Europ. mar biol. Symp.* (Ed. H. Barnes), 213–238.

Boalch, R., Chan, S., and Taylor, D. (1981). Seasonal variation in the trace metal content of *Mytilus edulis, Mar. Pollut. Bull.*, **12**, 276–280.

Boyce, R. and Herdman, W.A. (1898). On a green leucocytosis in oysters, *Proc. R. Soc. Lond.*, **62**, 30–38.

Boyden, C.R. (1974). Trace metal content and body size in molluscs, *Nature*, **251**, 311–314.

Boyden, C.R. (1975). Distribution of some trace metals in Poole Harbour, Dorset, *Mar. Pollut. Bull.*, **6**, 180–187.

Boyden, C.R. (1977). Effect of size upon metal content of shellfish, *J. Mar. Biol. Ass. UK*, **57**, 675–714.

Boyden, C.R., Watling, H. and Thornton, I. (1975). Effect of zinc on settlement of the oyster *Crassostrea gigas*, *Marine Biology*, **31**, 227–234.

Breedeveld, J.F.V. (1971). Changes in solids, fats, ash and metal content in *Crassostrea virginica* (Gmelin) at different temperatures and salinities. In: *Symposium 'A Preliminary Investigation: The Effect of Elevated Temperature on the American Oyster* Crassostrea virginica (*Gmelin*)' (Ed. J.A. Quick), pp. 36–97; Florida Dept. of Natural Resources, Professional Paper Series, 15.

Brereton, A., Lord, H., Thornton, I. and Webb, J.S. (1973). Effect of zinc on growth and development of larvae of the Pacific oyster, *Crassostrea gigas*, *Marine Biology*, **19**, 96–101.

Chipman, W.A., Price, T.R. and Price, T.J. (1958). Uptake and accumulation of radioactive zinc by marine plankton, fish and shellfish, *Fish. Bull. US Fish. Wildl. Serv.*, **58**, 279–292.

Coombs, T.L. (1972). The distribution of zinc in the oyster *Ostrea edulis* and its relation to enzymic activity and to other metals, *Marine Biology*, **12**, 170–178.

Coombs, T.L. (1974). The nature of zinc and copper complexes in the oyster *Ostrea edulis*, *Marine Biology*, **28**, 1–10.

Cossa, D., Bourget, E. and Piuze, J. (1979). Sexual maturation as a source of variation in the relationship between cadmium concentration and body weight of *Mytilus edulis* L, *Mar. Pollut. Bull.*, **10**, 174–176.

Darracott, A. and Watling, H. (1975). The use of molluscs to monitor cadmium levels in estuaries and coastal marine environments, *Trans. R. Soc. S. Afr.*, **41**, 325–335.

Davies, I.M. and Pirie, J.M. (1978). The mussel *Mytilus edulis* as a bioassay organism for mercury in sea water, *Mar. Pollut.*, **9**, 128–132.

Day, J.H. (1967). The biology of Knysna estuary, South Africa. In: *Estuaries* (Ed. G.H. Lauff), pp. 397–407; Publs Am. Ass. Advmt. Sci., 83.

Delhaye, W. and Cornet, D. (1975). Contribution to the study of the effect of copper on *Mytilus edulis* during reproductive period, *Comp. Biochem. Physiol.*, **50A**, 511–518.

DOE. (1977). Monitoring the marine environment of the United Kingdom. First report of the marine pollution monitoring management group, 1975–76, *Pollution Report*, **2**.

DOE. (1982). Monitoring the marine environment—into the eighties. *Pollution Report*, **14**.

Duke, T., Willis, J., Price, T. and Fischler, K. (1967). Influence of environmental factors on the concentrations of ^{65}Zn by an experimental community. *Proc. 2nd Nat. Symp. Radioecology* (Eds. D.J. Nelson and F.C. Evans), Ann Arbour, Michigan, 355–362.

Elderfield, H., Thornton, I. and Webb, J.S. (1971). Heavy metals and oyster culture, *Mar. Pollut. Bull.*, **2**, 44–47.

Fowler, S.W. and Oregioni, B. (1976). Trace metals in mussels from the NW Mediterranean, *Mar. Pollut. Bull.*, **7**, 26–29.

Frazier, J.M. (1975). The dynamics of metals in the American oyster, *Crassostrea virginica* I. Seasonal effects, *Chesapeake Science*, **16**, 162–171.

Frazier, J.M. (1976). The dynamics of metals in the American oyster *Crassostrea virginica* II. Environmental effects, *Chesapeake Science*, **17**, 188–197.

Galtsoff, P.S. (1964). The American oyster, *Crassostrea virginica*, Gmelin, *Fish. Bull. US Fish. Wildl. Serv.*, **64**.

George, S.G. and Coombs, T.L. (1975). A comparison of trace metal and metalloenzyme profiles in different molluscs and during development of the oyster, *Proc. 9th Europ. Mar. Biol. Symp.* (Ed. H. Barnes), 433–449.

George, S.G., Pirie, B.J.S., Cheyne, A.R., Coombs, T.L. and Grant, P.T. (1978). Detoxication of metals by marine bivalves: an ultrastructural study of the compartmentation of copper and zinc in the oyster *Ostrea edulis*, *Marine Biology*, **45**, 147–156.

Goldberg, E.D. (1975). The mussel watch—a first step in global marine monitoring, *Mar. Pollut. Bull.*, **6**, 111.

Goldberg, E.D., Bowen, V.T., Farrington, J.W., Harvey, G., Martin, J.H., Parker, P.L., Risebrough, R.W., Robertson, W., Schneider, E. and Gamble, E. (1978). The mussel watch, *Environmental Conservation*, **5**, 101–125.

Hoyaux, J., Gilles, R. and Jeuniaux, C. (1976). Osmoregulation in molluscs of the intertidal zone, *Comp. Biochem. Physiol.*, **53A**, 361–365.

Huggett, R.T., Bender, M.E. and Slone, H.D. (1973). Utilizing metal concentration relationships in the eastern oyster (*Crassostrea virginica*) to detect heavy metal pollution, *Water Research*, **7**, 451–460.

Ikuta, K. (1968). Studies on accumulation of heavy metals in aquatic organisms III. On accumulation of copper and zinc in the parts of oysters, *Bull. Jap. Soc. Sci. Fish.*, **34**, 117.

IOC (1976). A comprehensive plan for the global investigation of pollution in the marine environment and baseline study guidelines, *IOC Technical Series*, 14.

IOC (1982). Global Investigation of Pollution in the Marine Environment (GIPME)/Group of Experts on Methods, Standards and Intercalibration (GEMSI), *Fourth Session Curacao, March 1982*. Summary Report, UNESCO.

Jackim, E., Morrison, G. and Steele, R. (1977). Effects of environmental factors on radiocadmium uptake by four species of marine bivalves, *Marine Biology*, **40**, 303–308.

Jorgensen, C.B. (1976). Growth efficiencies and factors controlling size in some mytilid bivalves, especially *Mytilus edulis* L. Review and interpretation, *Ophelia*, **15**, 175–192.

Lankester, E.R. (1886). On green oysters, *Q. J. Microsc. Sci.*, **26**, 71–94.

Lobel, P.B., and Wright, D.A. (1982). Gonadal and nongonadal zinc concentrations in mussels, *Mar. Pollut. Bull.*, **13**, 320–323.

Mackay, N.J., Williams, R.J., Kacprzac, J.L., Kazacos, M.N., Collins, A.J. and Auty, E.H. (1975). Heavy metals in cultivate oysters. (*Crassostrea commercialis Saccostrea cucullata*) from the estuaries of New South Wales, *Aust. J. Mar. Freshwat. Res.*, **26**, 31–46.

Mullin, J.B. and Riley, J.P. (1956). The occurrence of cadmium in sea water and in marine organisms and sediments, *J. Mar. Res.*, **15**, 103–122.

Orren, M.J., Eagle, G.A., Hennig, F-K.O., and Green, A. (1980). Variations in trace metal content of the mussel *Choromytilus meridionalis* (Kr.) with season and sex, *Mar. Pollut. Bull.*, **11**, 253–257.

Phillips, D.J.H. (1976). The common mussel *Mytilus edulis* as an indicator of pollution by zinc, cadmium, lead and copper. 1. Effects of environmental variables on uptake of metals, *Marine Biology*, **38**, 59–69.

Phillips, D.J.H. (1977a). Effects of salinity on the net uptake of zinc by the common mussel *Mytilus edulis*, *Marine Biology*, **41**, 79–88.

Phillips, D.J.H. (1977b). The common mussel *Mytilus edulis* as an indicator of trace metals in Scandinavian waters. 1. Zinc and cadmium, *Marine Biology*, **43**, 283–291.

Phillips, D.J.H. (1977c). The use of biological indicator organisms to monitor trace metal pollution in marine and estuarine environments—A review, *Environ. Pollut.*, **13**, 281–317.

Phillips, D.J.H. (1980). *Quantitative Aquatic Biological Indicators*; Pollution Monitoring Series, Applied Science.

Portmann, J. (Ed.) (1976). Guidelines for the use of biological accumulators in marine pollution monitoring, *FAO Fish. Tech. Pap.*, **150**.

Preston, A. (1966). Concentration of ^{65}Zn in the flesh of oysters related to the discharge of cooling pond effluent from the CEGB Nuclear Power Station at Bradwell-on-Sea, Essex. In: *Radio-ecological Concentration Processes. Proceedings of the International Symposium, Stockholm 1966* (Eds. B. Aberg and F.P. Hungate), pp. 995–1004; Pergamon Press.

Romeril, M.G. (1971). The uptake and distribution of ^{65}Zn in oysters, *Marine Biology*, **9**, 347–354.

Schulz-Baldes, M. (1973). Die miesmuschel *Mytilus edulis* als indikator fur die bleikonzentration in Weserastuar und in der Deutschen Bucht, *Marine Biology*, **21**, 98–102.

Schulz-Baldes, M. (1974). Lead uptake from sea water and food, and lead loss in the common mussel *Mytilus edulis*, *Marine Biology*, **25**, 177–193.

Seed, R. (1968). Factors influencing shell shape in the mussel *Mytilus edulis*, *J. Mar. Biol. Ass. UK*, **48**, 561–584.

Segar, D.A., Collins, J.D. and Riley, J.P. (1971). The distributions of the major and some minor elements in marine animals. II. Molluscs, *J. Mar. Biol. Ass. UK*, **51**, 131–136.

Shannon, L.V. and Stander, G.H. (1977). Physical and chemical characteristics of water in Saldanha Bay and Langebaan Lagoon, *Trans. R. Soc. S. Afr.*, **42**, 441–459.

Simkiss, K. (1981). Cellular discrimination processes in metal accumulating cells, *J. Exp. Biol.*, **94**, 317–327.

Simkiss, K. and Taylor, M. (1981). Cellular mechanisms of metal ion detoxification and some new indices of pollution, *Aquatic Toxicology*, **1**, 279–290.

Simkiss, K., Taylor, M. and Mason, A.Z. (1982). Metal detoxification and bioaccumulation in molluscs, *Mar. Biol. Lett.*, **3**, 187–201.

Thornton, I., Watling, H. and Darracott, A. (1975). Geochemical studies in several rivers and estuaries used for oyster rearing, *Science of the Total Environment*, **4**, 325–345.

Topping, G. (1983). Guidelines for the use of biological material in first order pollution assessment and trend monitoring, *DAFS Fish. Res. Rep.*, **28**.

Watling, J. and Watling, H.R. (1974). Environmental studies in Saldanha Bay and Langebaan Lagoon, No. 1. Trace metal concentrations in selected molluscs and algae, *CSIR Report FIS*,**70**.

Watling, H.R. and Watling, J. (1976a). Trace metals in *Choromytilus meridionalis*, *Mar. Pollut. Bull.*, **7**, 91–94.

Watling, H.R. and Watling, J. (1976b). Trace metals in oysters from Knysna estuary, *Mar. Pollut. Bull,* **7**, 45–48.

Watling, H.R. and Watling, J. (1976c). Trace metal studies in Knysna estuary. *First Interdisciplinary Conference Marine and Freshwater Research in Southern Africa*, Port Elizabeth, 5–10 July 1976, S122 Fiche 9.

Windom, H.L. and Smith, R.G. (1972). Distribution of iron, magnesium, copper, zinc and silver in oysters along the Georgia coast, *J. Fish. Res. Bd. Canada*, **29**, 450–452.

Wolfe, D.A. (1970). Zinc enzymes in *Crassostrea virginica*, *J. Fish. Res. Bd. Canada*, **27**, 59–69.

Geochemistry and Human Health in the 1980s

R. G. CROUNSE

It may be more than coincidence that the only physician to address the 1978 Royal Society Discussion Meeting in London on Environmental Geochemistry and Health (Bowie and Webb, 1979) was Dr E.J. Moynahan, a fellow dermatologist from Guy's Hospital, London. In the same year, Dr Moynahan addressed a symposium on 'Hair, Trace Elements and Human Illness' which I organized in the United States (Brown and Crounse, 1980). It happens that the major nutritional syndromes in man are frequently signalled by changes in the hair, skin and nails, and are thus especially likely to be noted first by dermatologists. This is particularly true for trace element deficiencies and excesses, both in man and animals.

This contribution is concerned with only a limited segment of the broad topic of geochemistry and human health. A comprehensive overview covering all aspects is contained in the recent text, *Applied Environmental Geochemistry* (Thornton, 1984). The most exciting developments which relate geochemical information and events to human health and illness in recent years are concerned with the trace element selenium.

The selenium story begins with the expeditions of Marco Polo, who observed that his own pack animals imported by the expedition to the Eurasian plains suffered symptoms such as crumbling hoofs, loss of hair, gait disturbances and ultimately death. Animals indiginous to the area had no such difficulties. This was deciphered to indicate that the local animals had learned to avoid eating poisonous plants, while imported animals did not. These plants are known (now) to be very high in Se by virtue of accumulating high levels from the local soil. The modern anology is the syndrome known in the upper midwest USA as 'alkali disease', or 'blind staggers', occurring in animals grazing on the accumulator plant species in these high soil Se regions.

Despite relatively few reports on Se and animal or human health during the 600 or so years since Marco Polo's observations, it is possible from within the last decade or two to assemble a bibliography on Se and health which numbers in the thousands of references. The occasion for this profusion of literature is simply that Se has now been discovered to be *essential* to the health of both animals and man. If the last decade (in relation to essential trace elements) could be called the 'Zn decade', we are now in the 'Se decade'. The danger is that the decade's promise might go unfulfilled; the current evidence for believing otherwise is presented in this chapter.

Zinc deficiency was announced dramatically just a few years ago to be the basis for the dermatological medical entity known as 'acrodermatitis enteropathica', a fact discovered by Moynahan (1974), and copper deficiency (or cellular malutilization) (LaBadie, 1981) to be the basis for 'Menkes' syndrome', a complex neuroectodermal disorder which includes 'twisted' hair (pili torti) (Collie *et al.*, 1980). Both these milestones in the uncovering of the etiology of two inherited disorders of human trace element metabolism (neither related directly to geochemistry) have been virtually eclipsed by a more recent finding. This is the discovery by Chinese biomedical and epidemiological investigators (Chen *et al.*, 1980) that an obscure childhood cardiomyopathy (abnormalities of the heart and its functioning) with frequently fatal outcome could be totally prevented by dietary supplementation with small doses of Se. As human population disease/intervention studies go, the magnitude of this discovery ranks as one of the most significant medical detective stories of the times. It also happens to represent an example of geochemistry and health 'par excellance'. Through careful long-term investigation and clinical trials it was established that individuals affected by this heart disorder resided in geographic clusters; that nearby populations were not necessarily affected; that the common factor seemed to involve consumption of locally grown food; that this factor linked to local water and soil conditions; and finally that the local factors included very low levels of Se in water, soil, produce (Table 1) and finally in blood and hair from affected individuals compared to those not affected, or from other areas. Proof of the etiologic role of Se deficiency was provided by dietary supplementation with small doses of oral Se, 0.5–1.0 mg per week depending upon age. This supplementation first diminished in frequency, then virtually abolished the occurrence of 'Keshan disease', the local name for the cardiomyopathy. Thousands of cases and deaths have since been prevented. Much attention is now being given to the possibility of soil seeding of an appropriate form of Se as an alternative to the need for oral feeding.

The scientists involved have been careful not to attribute the syndrome entirely to simple Se deficiency, but to allow the possibility that the cardiomyopathy, characterized by apparent degeneration of subcellular energy-producing particles (mitochondria), could be the result of other factors. An example might be a virus infection conditioned by Se deficit. Since Se deficiency has been demonstrated in many animals to result in impaired

Table 1 **Se content in staple foods grown in geographic areas with and without Keshan Disease**

	With	Without
Corn	*0.005(178)	0.036(82)
Rice	0.007(49)	0.024(49)
Wheat	0.008(116)	0.032(30)
Soybean	0.010(134)	0.069(31)

*ppm(N) (from Chen *et al.*, 1980)

immunological defence mechanisms (Shamberger, 1983), this caution is admirable.

Since initial publication in the English language of this commanding summary of human Se deficiency, there are isolated case reports of probable Se deficiency occurring, especially during total intravenous replacement of all essential dietary ingredients (Kien and Ganther, 1983). This medical technique, called 'total parenteral nutrition' (TPN) has proved to be a significant source of uncovering new dietary requirements. When all the dietary substances which may enter the body for months or even years must be pre-mixed in bottles for intravenous delivery, it has happened more than once that an eventually critical substance has been omitted, either because of uncertain requirements (as with several trace elements, e.g. Zn, Cu, Mo, Se) or potential mixing incompatabilities or safety considerations (e.g. essential fatty acids). Not unexpectedly from the viewpoint of a dermatologist predominant signs of such deficiencies are reflected in abnormalities of hair, skin and nails. These occasions reflect simply the imperfect knowledge of every aspect of human nutrition; TPN is a life-saving technique in many circumstances Enlightened observation of its consequences is providing a unique opportunity for advancement of knowledge of human nutritional requirements.

Chinese researchers have provided us not only with the first detailed information on human Se deficiency, but also with the first good documentation of human Se excess with a direct geochemical link (Yang *et al.*, 1983). Once again, predominant clinical signs have included abnormalities of hair (easily broken), nails (distortions and weakness) and skin (a peculiar 'rash'). The geochemically localized concentration of Se has led to human consumption of large amounts of Se. Selenium rich coal in highly limited geographic areas of China (Enshi County, Hubein Province) apparently released excessive Se by weathering. The Se was especially available for crop uptake by virtue of the use of lime as a soil ameliorant in the region, and amplified by a drought causing failure of the rice crop. The local population ate more high Se corn and other vegetables, some of which such as turnips contained extraordinary amounts of Se (457 mg/g!) as compared to rice (4 mg/g). Even the latter amount is 4–100 times the 'normal' concentration in rice. Concentrations of Se in hair and blood (Table 2) approximated ten times higher than those in unaffected individuals from the same area, one hundred times higher than individuals from non-seleniferous areas, and one thousand times higher than individuals with clinical Se deficiency (Keshan disease) (Chen *et al.*, 1980). Correlations were generally quite good with concentrations of Se in soil and water and food.

Thus in just a few years, both human Se deficiency and toxic excess have been clearly linked to geochemical origins. Yet these discoveries may be only the beginning of the 'Se story', because clinical deficits and excesses represent only the opposite and extreme poles of the interaction of Se and man. Currently under investigation is the whole 'middle ground', that large range of Se levels (up to thousand fold) between the two extremes. It is in this area that

Table 2 **Concentrations of Se in hair and blood and human intake in seleniferous and selenium deficient regions of China** *

	Clinical state	Se(ppm) Hair	Se(ppm) Blood	Average intake (mg/day)
Seleniferous	Sick	32.20	3.18	4.99
Region	Well	3.69	0.44	0.75
Se-deficient	Well	0.16	0.09	0.12
Regions	Sick	0.07	0.02	0.01

*adapted from Yang *et al.* (1983)

the most exciting discoveries regarding Se and man may yet be found. There is much provocative evidence that (a) low environmental Se levels without evidence of human deficiency may be related to high incidence of many cancers and at least some kinds of cardiovascular diseases; (b) low Se intake in animal studies can enhance tumorogenesis, and high intake may retard or prevent development of many kinds of cancer.

A detailed summary of numerous animal tumor/Se experiments has been published recently (Shamberger, 1983). In brief, many kinds of animal cancers can be reduced in incidence or inhibited by increased dietary Se, while Se deficiency often has the opposite effect. Interactions with Vitamin E and with fatty acids of differing degrees of unsaturation are many and complex. Both stages of carcinogenesis (the 'initiating' event and later 'promoting' influences) can be affected. The precise mechanisms involved are not clear; most speculation centres around protection from peroxidation, immunologic alteration of both humoral (antibody production) and cellular (T cell functions) types, and prostaglandin metabolism. All three mechanisms may be true, and others remain to be discovered. In animal studies at the Department of Surgery, East Carolina University (Howells *et al.*, 1982) for example, Se added in drinking water was able to prolong significantly survival time of mice with a transplanted myeloma type of tumor (in conjunction with immunostimulation), and the effect was completely abolished by addition of Indomethacin, a known inhibitor of prostaglandin synthesis. Furthermore, in mice supplemented with oral Se, the *in vitro* responsiveness of their lymphocytes to standard agents causing T cell blastogenic transformation (an *in vitro* correlate of delayed non-humoral hypersensitivity) was greatly enhanced. The Se effects might best be considered as 'modulating' effects, i.e. working via other systems. There is no real evidence for direct interaction of Se with either carcinogen or cancer. What makes the animal cancer data so intriguing is that there is increasing information regarding a role for Se in modulation of human cancer, with at least a tentative suggestion of a geochemical link.

There are basically three kinds of evidence that low Se status is related to high cancer incidence in humans. The first (and earliest) is epidemiologic correlation between low environmental Se levels (and estimated dietary intake derived indirectly) and increased cancer based upon mortality rates from high and low Se areas (Shamberger, 1983). The evidence is intriguing, but not compelling (New Zealand is a notable exception to the inverse correlation), and in any case cannot prove a causal relationship.

The second is case control data showing low blood Se levels in patients with known cancer (Shamberger, 1983). However, it is entirely likely that low tissue or blood Se levels in people already affected by cancer may be due to sequestration of Se by the tumor tissue, resulting in secondary decreases in other tissues or fluids (Robinson *et al.*, 1979). One study however, in which the Department of Surgery, East Carolina University participated (Clarke *et al.*, 1982b) avoided the problems of tumor sequestration of Se by choosing a type of cancer which could not effect such an alteration. We investigated Se plasma levels in patients with skin cancer only (Table 3). The cancer had been in most cases completely removed, but even if a recurrence or new cancer had developed, its cellular mass would be incapable of altering Se levels. In this study, patients with the lowest (decile of) Se levels were four times more likely to have had skin cancer (basal cell epithelioma or squamous cell carcinoma) than patients in the highest decile of Se levels, confirming the relationship between low Se and high cancer reported from the world-wide epidemiologic studies. The earlier studies, incidentally, did not contain information on skin cancer, since skin cancer is not normally reflected in cancer mortality rates (i.e. it does not normally cause death).

Although skin cancer is rarely a fatal cancer, the number of cases per year in the United States alone is nearly 500 000, and the total cost of treatment is therefore in the millions of dollars per year. The high incidence makes prospective prevention studies feasible, and skin cancer may possibly serve as a model for the other epithelial cancers, e.g. lung, bladder etc. It is even possible that trace elements might play a direct role in the pathogenesis of lung cancer. Cadmium levels are known to be higher in smokers than in non-

Table 3 **Wilson, NC skin cancer study**

Age	Case	Plasma Se concentration*†	
		Low	High
30–54	Cancer	54(72%)	21(28%)
	Control	31(47%)	34(53%)
55–75	Cancer	93(81%)	21(19%)
	Control	62(70%)	27(30%)

*dichotomized at mean (0.145 mg/ml plasma)
†adapted from Clark *et al.* (1982a,b)

smokers, and cigarette tobacco from countries with high cancer incidence contains significantly less Se than tobacco from countries with low lung cancer incidence (Bodgen *et al.*, 1981). A potential interaction of Cd and Se is mentioned also in connection with high blood pressure (see below).

The third (and most recent) evidence is that serum Se determinations in blood obtained *before* development of internal cancer (in the course of a long term prospective study of patients being followed for high pressure) showed lower Se levels in patients who ultimately developed cancer than in patients who did not develop cancer (Willett *et al.*, 1983). The increased risk of cancer was twice that for patients in the lowest quintile of Se levels compared to those in the highest quintile of Se levels; indeed the increased risk was nearly limited to the lowest quintile. This suggests strongly that the subpopulation with lowest serum Se might benefit from additional Se, assuming that there is a causal relationship. Many other factors however, could influence the outcome, some alluded to in the study (such as Vitamins A and E), others not accounted for (such as type and dose of carcinogen, co-carcinogen or promoter).

As the story of Se and cancer continues to unfold, the evidence is sufficiently strong to warrant an attempt to decrease skin cancer in patients at high risk (those with two or more previous skin cancers) by adding a small amount of supplementary Se to their daily intake, and it is intended to begin such at East Carolina University.

Trace element researchers are more and more concerned that measurement of a single element may be experimentally inadequate in many situations. Particularly interesting are increasing numbers of examples of antagonistic, or inversely interactive, elements. Hyperzincemia (increased blood zinc) seems often correlated with hypocupremia (decreased blood copper). If both were not estimated, one might inadvertantly attribute signs or symptoms to elevated zinc which in fact might be due to decreased copper. More pertinent in the context of this chapter, however is the apparently 'antagonistic' relationship of Se and As. Indeed, it has been suggested that an excess of either one might be treated by administering the other! In the case of our skin cancer studies, we are interested in the possible role of As as a potential skin carcinogen. Environmentally-derived epidemics of human As excess have been well documented in the Taiwan studies (Yeh, 1973) and others. Cutaneous (and probably some internal) cancers are clearly related, and the patients in our studies represent a population exposed to considerable As, widely used as a tobacco plant spray only two or three decades ago, and in a region known to be one of the lowest Se regions in the USA (Southeastern Coastal Plain). Many of the study patients had small hard circular skin lesions (Keratoses) on their palms, associated in our area with previous As exposure (Clarke *et al.*, 1982a). Perhaps the ratio of As/Se plays a role in the development of cancer, at least in this specific geochemical area.

Similar considerations are being given to possible Se/Cd interactions in both cancer and vascular disease. The potential relationship of Se to heart and blood vessel disorders is based upon evidence somewhat analagous to the cancer data. Epidemiological studies (Shamberger, 1983) show a high prevalence of several kinds of cardiovascular disease in geographic areas with forage crops relatively low in Se, and the inverse in areas rather higher in Se. The deficiency syndrome described from China is a primarily cardiovascular syndrome, and very recent studies from Finland (Salonen *et al.*, 1982), a country low in Se and high in cardiovascular disease, purport to show an increased risk of death from cardiovascular disease in patients with lower serum Se levels. Young adults with high blood pressure living in the southeastern USA show higher Cd levels than 'normal', all have relatively low Se levels (Table 4) (Andrews *et al.*, 1980). Years ago Cd was shown to induce high blood pressure in certain rat strains (Shamberger, 1983). Though not a universally accepted model, other animal experiments with different types of high blood pressure have also shown ameliorative effects of Se; Vitamin E, unsaturated fatty acids and prostaglandins are also under study. This whole subject has also been recently summarized (Shamberger, 1983).

One of the interesting aspects of the relationship between geochemistry and human health is that it is always changing. Modern transportation and refrigeration can change food habits and food sources dramatically. Weather changes can alter availability of geochemical elements. Changing patterns of energy production can change both land utilization and food production relationships. In another aspect of our own research, Se release into the cooling water for a coal-powered electric station in a lake in the state of North Carolina is the apparent cause of rapidly increased mortality of young fish in the waters. Current research is examining Se and other protein-binding elements from these fish as a potential indicator of industrial pollution (Dr John Bray and Mr Frank Reilly, pers. comm.). It is uncertain whether this excess Se will find its way into human foods grown downstream.

In this account, I have really only highlighted the Se 'story', and cannot give reference or due credit to the many investigators throughout the world, from chemists and biochemists to physicians and epidemiologists who have contributed to this subject area. The dynamics of intellectual advances in biomedical sciences have always been a human spectacle. 'Fortune-telling gurus' (those sometimes brilliant, sometimes erroneous predictors of advances in human biomedical knowledge) often have been predictably

Table 4 **Concentrations of Se and Cd in red blood cells (ng/g DM) high school seniors, coastal Georgia***

	Hypertensive	Normotensive
Se	558±112	571±109
Cd	9.3±4.7	4.4±3.0
	$p < 0.01$	

*adapted from Andrews *et al.* (1980)

maligned by rank-and-file physicians and scientists. In the fields of trace elements, vitamins, nutrition and even immunology, insightful predictors have often been accused of inadequate science, over-enthusiastic prediction, or both. It is too early to tell if predictions of essentiality of unlikely absurd substances, for example As, Pb, Si, Ni will survive the test of time, doubt and continuing critical investigation. I suspect they may; who would have believed that the apparently toxic substance Se would be proven essential.

Whatever the outcome, some advanced thinkers will probably be awarded for their insights posthumously. It is therefore altogether fitting that some true leaders in the field receive their recognition in person. Such a leader is the man we are here to honour, John Webb. His contributions are unquestioned, his wisdom already acknowledged. We students can only emulate his performance.

References

Andrews, J.W., Hames, C.G., Metts, J.C. and Davis, J.M. (1980). Selenium, cadmium and glutathione peroxidase in blood of cardiovascular diseased and normal subjects from the cardiovascular belt of southeastern USA. In: *Trace Substances in Environmental Health-XIV* (proceedings) (Ed. D.D. Hemphill), pp. 38–42; University of Missouri, Columbia, Missouri.

Bodgen, J.D, Kemp, F.W., Buse, M., Thind, I.S., Louria, D.B., Forgacs, J., Llanos, G. and Terrones, I.M. (1981). Composition of tobaccos from countries with high and low incidences of lung cancer. I. Selenium, polonium-210, alternaria, tar and nicotine, *J. Nat. Cancer Inst.*, **66**, 27–31.

Bowie, S.H.U. and Webb, J.S. (1979). *Environmental Geochemistry and Health*; Phil. Trans. R. Soc. Lond. **288**.

Brown, A.C. and Crounse, R.G. (1980). *Hair, Trace Elements and Human Illness*, pp. 1–360; Praeger Publishers, New York.

Chen, X., Yang, G., Chen, J. Chen, X., Wen, Z. and Ge, K. (1980). Studies on the relations of selenium and Keshan disease, *Biol. Trace Element Res.*, **2**, 91–107.

Clark, L.C., Graham, G. and Crounse, R.G. (1982a). Selenium arsenical keratoses and skin cancer in Eastern North Carolina. Presented at the Society for Environmental Geochemistry and Health, Greenville, NC, USA.

Clark, L.C., Graham, G., Crounse, R.G., Grimson, R. and Shy, C.M. (1982b). A case control study of skin cancer and selenium in Eastern North Carolina. Presented at the annual meeting, American Public Health Association, Montreal, Canada.

Collie, W.R., Goka, T.J., Moore, C.M. and Howell, R.R. (1980). Hair in Menkes Disease: a Comprehensive Review. In: *Hair, Trace Elements and Human Illness* (Eds. A.C. Brown and R. G. Crounse), pp. 197–210; Praeger Publishers, New York.

Crounse, R.G., Pories, W.J., Bray, J.T. and Mauger, R.L. (1984). Geochemistry and Man: Health and Disease. I. Essential Elements. II. Elements possibly essential, those toxic and others. In: *Applied Environmental Geochemistry* (Ed. I. Thornton), pp. 262–328; Academic Press, London and New York.

Howells, J.M., Crounse, R.G., Thomas, J.M., Whitley, T.K., Finn, M., Bray, J.T., Smith, A.M. and Jones, C. (1982). Prolonged survival of Balb/c mice with a transplanted tumor (MOPC 467) induced by oral selenium plus immunoactivation with C. parvum (poster session). Presented at The Society for Environmental Geochemistry and Health, Greenville, NC, USA.

Kien, C.L. and Ganther, H.E. (1983). Manifestations of chronic selenium deficiency in a child receiving Total Parenteral Nutrition, *Amer. J. Clin. Nutrition*, **37**,319–328.

LaBadie, G.U., Beratis, N.G., Price, P.M. and Hirschhorn, K. (1981). Studies of the copper-binding proteins in Menkes and normal cultured skin fibroblast lysates, *J. Cell. Physiol.*, **106**, 173–178.

Moynahan, E.J. (1974). Acrodermatitis Enteropathica. A Lethal Inherited Human Zinc Deficiency Disorder, *Lancet*, **ii**, 399–400.

Robinson, M.F., Godfrey, P.J., Thomson, C.D., Rea, H.M. and van Rij, A.M. (1979). Blood selenium and glutathione peroxidase activity in normal subjects and in surgical patients with and without cancer in New Zealand, *Amer. J. Clin. Nutrition*, **32**, 1477–1485.

Salonen, J.T., Alfthan, G., Huttunen, J.K., Pikkarainen, J. and Puska, P. (1982). Association between cardiovascular death and myocardial infarction and serum selenium in a matched-pair longitudinal study, *Lancet*, **ii**, 175–179.

Shamberger, R.J. (1983). Selenium and cancer. In: *Biochemistry of Selenium*, pp. 207–271; Plenum Press, New York.

Thornton, I. (Ed.) (1984). *Applied Environmental Geochemistry*; Academic press, London and New York.

Willett, W.C., Polk, B.F., Morris, J.S., Stampfer, M.J., Pressel, S., Rosner, B., Taylor, J.O., Schneider, K. and Hames, C.G. (1983). Prediagnostic serum selenium and risk of cancer, *Lancet*, **ii**, 130–134.

Yang, G., Wang, S., Zhou, R. and Sun, S. (1983). Endemic selenium intoxication of humans in China, *Amer. J. Clin. Nutrition*, **37**, 872–881.

Yeh, S. (1973). Skin cancers in chronic arsenicism, *Human Pathology*, **4**, 469–485.

Concluding Address

R. DAVIS

This 'Applied Geochemistry in the 1980s' meeting has been a happy and successful occasion. Over the past two days we have heard and discussed excellent papers which display the range and growth of applied geochemistry. There are always new horizons beckoning to the resourceful, and I feel sure that John Webb will leave this meeting of friends secure in the knowledge that others will continue with vigour along the way which he has pointed.

When Iain Thornton invited me to wind up this meeting, my thoughts went back thirty years to my meeting with John Webb in Uganda in April 1954. I should explain that before joining Imperial College in 1966, I had spent half of my working life in the mining industry. As a company geologist in 1954, I needed the new geochemical prospecting techniques to help find extensions to the Kilembe copper–cobalt deposit in the densely vegetated Ruwenzoi mountains. John Webb needed research fields and funds. Our communication suitably lubricated, led rapidly to mutually beneficial arrangements. This sort of interface with the practical industrial scientist, combined with his enthusiasm and powers of persuasion, have over many years been a hallmark of John's success in combining scientific vision with the ability to raise money to support applied research projects. In my own case, the university/industry association which was established in 1954 continued fruitfully into Zimbabwe and Zambia. I remember in the late fifties donating to the Geochemical Prospecting Research Center (as it then was) a large set of drainage sediment samples from the Livingstone–Namwala concession in Zambia. This company collection was put to good use by Ian Nichol in pioneering regional geochemical mapping under John's direction.

I now hold up before you a familiar draft of geochemical sample packet. (This particular one is a souvenir of mine, being inscribed by John Webb and John Butler, who jointly appoint me 'Prospector in Simplo Electo—but that is another story.) To the perfection of the design of this small but important item so familiar in the world of applied geochemistry, John Webb devoted characteristically thorough attention. It symbolizes for me some of the evergreen guidelines to which John has adhered, and of which we are again reminded by the papers presented over these past two days, viz:

1. Attention to detail in design.
2. The importance of orientation studies.
3. Effective communication between industry and research groups.
4. Awareness of advances in related technologies, e.g. analytical procedures, data processing.
5. The power of overlap between the disciplines of pure and applied science.
6. The fact that the patterns arising from all forms of geochemical survey should make geological sense.

Unlike many present here today, I would not describe myself as a geochemist, fully occupied with one or more of the lines along which the subject has evolved. But like many others in the business of finding ore and making mines, I have found applied geochemistry an invaluable tool, to be coordinated with other search techniques in the management of mineral exploration. As a tool-user—a favourite phrase of John's—I salute all those tool-makers whose high standards have brought applied geochemistry to the present admirable state of the art.

John Webb's praises have been loudly sung over the past two days, and I will sing no more. I am sure that John would prefer us to recognize, instead, the excellent work done over the years by all the members, past and present, staff and students, of the tightly knit group which he created. The fact that so many distinguished members of the group have made time and travelled long distances to be here is evidence enough of their sense of loyalty and achievement. Other distinguished colleagues and associates, also, have come; it is not my intention to pick on individual names, but I feel sure that I will have your unanimous approval in not allowing this opportunity to pass without inviting at least Herb Hawkes to rise and take a bow.

In coming to the close of Geochemistry in the Eighties as a tribute to John Webb, three cheers are not perhaps entirely fitting, but three more words I must say on behalf of all of us here today for whom this has been a memorable meeting. First, a word of deep appreciation and fondness to John Webb, coupled with good wishes for future health and happiness to John and his wife Jean who continues her strong support as always. My second word is of confidence in the future development of applied geochemistry in all its forms in the service of man. And last but not least, a word of thanks to Dr Iain Thornton and all his helpers for conceiving and organizing this meeting which we have all enjoyed so much.